Remanufacturing Engineering

Remanufacturing Engineering

Edited by **Jude Stinton**

SYRAWOOD
PUBLISHING HOUSE
New York

Published by Syrawood Publishing House,
750 Third Avenue, 9th Floor,
New York, NY 10017, USA
www.syrawoodpublishinghouse.com

Remanufacturing Engineering
Edited by Jude Stinton

International Standard Book Number: 978-1-68286-363-3 (Hardback)

Printed in China.

Contents

	Preface	**VII**
Chapter 1	**A manufacturing framework for capability-based product-service systems design** Gokula Vijaykumar Annamalai Vasantha, Hitoshi Komoto, Romana Hussain, Rajkumar Roy, Tetsuo Tomiyama, Steve Evans, Ashutosh Tiwari and Stewart Williams	**1**
Chapter 2	**A remanufacturing process library for environmental impact simulations** Nurul Hanna Ismail, Guillaume Mandil and Peggy Zwolinski	**33**
Chapter 3	**Developing performance management systems for the green supply chain** Michael Mutingi, Herbert Mapfaira and Robert Monageng	**47**
Chapter 4	**Conditions of emergence of OEM's reverse supply chains** Akram El Korchi and Dominique Millet	**67**
Chapter 5	**An integrated fuzzy approach for evaluating remanufacturing alternatives of a product design** Xiaojun Wang and Hing Kai Chan	**84**
Chapter 6	**Remanufacturing process for used automotive electronic control components in China** Junjun Wang and Ming Chen	**103**
Chapter 7	**Offering branded remanufactured/recycled products: at what price?** Leila Hamzaoui-Essoussi and Jonathan D Linton	**120**
Chapter 8	**Develop a cost model to evaluate the economic benefit of remanufacturing based on specific technique** Yuchun Xu and Wei Feng	**135**
Chapter 9	**Exploring inter-organizational relationships in automotive component remanufacturing** Sebastian Lind, David Olsson and Erik Sundin	**147**
Chapter 10	**Examination of demand forecasting by time series analysis for auto parts remanufacturing** Mitsutaka Matsumoto and Akira Ikeda	**161**

Chapter 11 **The manufacturer's value chain as a service – the case of remanufacturing** **181**
David Opresnik and Marco Taisch

Chapter 12 **Sustainable manufacturing for Indonesian small- and medium-sized enterprises (SMEs): the case of remanufactured alternators** **204**
Yun Arifatul Fatimah, Wahidul Biswas, Ilyas Mazhar and Mohammad Nazrul Islam

Chapter 13 **Reverse logistics network design: a holistic life cycle approach** **215**
Joanna Daaboul, Julien Le Duigou, Diana Penciuc and Benoît Eynard

Chapter 14 **Forecasting product returns for remanufacturing systems** **230**
Xavier Liang, Xiaoning Jin and Jun Ni

Chapter 15 **A business process model of inspection in remanufacturing** **248**
Mark Errington and Stephen J Childe

Permissions

List of Contributors

Preface

Every book is a source of knowledge and this one is no exception. The idea that led to the conceptualization of this book was the fact that the world is advancing rapidly; which makes it crucial to document the progress in every field. I am aware that a lot of data is already available, yet, there is a lot more to learn. Hence, I accepted the responsibility of editing this book and contributing my knowledge to the community.

In the past ten years, the remanufacturing engineering has gained pace in the industrial area as it is one of the most effective and potential strategies for the "end-of-life" product management. This book presents an overview of the remanufacturing engineering. It details topics like remanufacturing processes, product designs and environmental analysis. This book aims to compile a thorough survey and provide a foundation for further researches. It covers theoretical as well as practical aspects to understand the applications of this field. Students will find this text beneficiary.

While editing this book, I had multiple visions for it. Then I finally narrowed down to make every chapter a sole standing text explaining a particular topic, so that they can be used independently. However, the umbrella subject sinews them into a common theme. This makes the book a unique platform of knowledge.

I would like to give the major credit of this book to the experts from every corner of the world, who took the time to share their expertise with us. Also, I owe the completion of this book to the never-ending support of my family, who supported me throughout the project.

Editor

A manufacturing framework for capability-based product-service systems design

Gokula Vijaykumar Annamalai Vasantha[1], Hitoshi Komoto[2], Romana Hussain[3], Rajkumar Roy[4*], Tetsuo Tomiyama[4], Steve Evans[5], Ashutosh Tiwari[4] and Stewart Williams[6]

* Correspondence: r.roy@cranfield.ac.uk
[4]Manufacturing and Materials Department, Cranfield University, Cranfield, Bedfordshire MK43 0AL, UK
Full list of author information is available at the end of the article

Abstract

Manufacturers aim to design product-service systems (PSS) which integrate services with products to attain sustained competitive advantage from a life cycle perspective. PSS design should be customised solutions which are aligned to integrated stakeholders' capabilities, a subject which the extant literature has not sufficiently addressed. This paper proposes a systematic framework for the PSS solution provider to address this aim and operationalizes this through software developed for PSS design which models stakeholders' individual activities and simulates their occurrences depending on their relations. The framework stresses that integrated stakeholders' capabilities define continuing ability to generate a desired operational outcome for the customers. The paper reports a PSS design case for a laser system manufacturer and then applies the framework to it. The industrial experts' views on this framework reveal that it helps to develop PSS design from a holistic systems approach which facilitates a change in the designer's mindset from a product-centric to a systems-centric. The level of trust and transparency required for this framework is argued to be absent in most industrial sectors, being one of the foremost limitations for implementation of PSS.

Keywords: Product-service system; Design; Framework; Simulation; Representation

Background

Particularly in the light of recent economic downturns, manufacturers require alternative strategies to cope with globalization and reduced profit margins, and for retaining and attracting customers. One promising approach in helping manufacturers to achieve these objectives is product-service systems (PSS). This approach facilitates manufacturers in bundling products and services together to create a sustained competitive advantage. It aims at providing more value by fostering the optimal use of resources which can be sustained for both consumption and production. Major advantages for the PSS solution provider include prolonged and strategic relationships with the customer and product/service improvements based on the improved understanding of customer needs. However, Neely's [1] findings from the analyses of a large industrial database were that designing, implementing and managing PSS is a huge challenge to the provider as there are distinct possibilities of economic downturns.

A review of current PSS literature [2] reveals that the theories and methodologies to aid the design of PSS are still in their initial stages of development, and substantial research is required to develop a practical PSS design methodology along with supporting tools. Also, from the interviews which were conducted with 15 industrial maintenance experts of large technical systems such as aerospace engines and related systems, naval ships, land vehicle systems, trains and trucks, our understanding is that, currently, PSS conceptual design in practice is *ad hoc* (fairly intuitive) and lacks a systematic approach in allowing heterogeneous (tangible and intangible characteristics) aspects to be reflected within the PSS design process. The interaction of the stakeholders in the design process and the unique characteristics of products, services, networks of players as well as the supporting infrastructure [3] which are involved in the design of PSS all demand new theories, methodologies, tools and techniques.

The aim of our research is the formal development of a PSS design framework for the PSS solution provider to create customised PSS designs aligned to integrated stakeholders' capabilities. The proposed PSS framework has a view to developing PSS designs to support long-term business solutions from a capability viewpoint whilst stressing the development of additional value to customers by fostering the optimal use of integrated stakeholders' resources. A framework has been developed to address gaps identified in the literature. The framework encompasses a systems thinking perspective, which aims to improve overall PSS design on a system level and avoid sub-optimized solution towards any single activity across the whole life cycle-including remanufacturing. Therefore, this work did not focus specifically on the remanufacturing phase. To help PSS designers' model and simulate PSS designs, a framework has been implemented in a software environment. Service CAD integrated with a life cycle simulator (ISCL) [4] has been chosen because it provides both modelling and simulation facilities. A simulation facility is used to quantitatively evaluate the performance of the PSS designs regarding the requirements. This paper is structured into seven sections detailing: the gaps identified in the literature, understanding industrial challenges in PSS development, defining the constituents of PSS design, a step-by-step illustration of the proposed framework, application of the proposed framework to an industrial laser system case study, academic and industrial experts' views on this framework, and conclusions with future directions of research. Figure 1 illustrates the structure of the paper with the research methodology description.

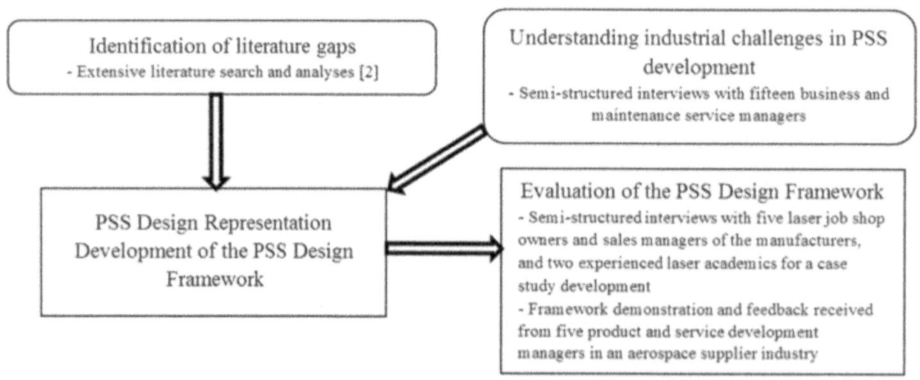

Figure 1 Structure of the paper with the research methodology description.

Related PSS design research

In this paper, four state-of-the-art methodologies which have been proposed for PSS design are reviewed: Komoto and Tomiyama [4,5] proposed a method to design and analyze business models of manufacturers by focusing on the variations of services in product life cycles (e.g. rental, sharing, maintenance and upgrade services, and pay-per-function sales). The business models are modelled using Service CAD and are quantitatively analyzed using a life cycle simulation technique [6]. Sakao and Shimomura [7] and Shimomura et al. [8,9] developed Service Explorer for service engineering to design products with a higher added value from enhanced services. Service Explorer is also a Service CAD software which employs discrete event simulation methods for the evaluation of PSS design. The scope of evaluation is different from the sequence of activities of customers in a specific service environment (e.g. restaurants) [9] to those of the activities of manufacturers and users of products during their entire life cycle [4]. Maussang et al. [10] presented a PSS design model to assist engineers in the joint development of physical products and interacting services to generate more added value. Alonso-Rasgado et al. [11] and Alonso-Rasgado and Thompson [12] proposed a total care design process to develop innovative offerings consisting of hardware and services integrated to provide complete functional performance. Compared to other methodologies in the literature, these four methodologies are detailed, demonstrated through industrial examples, published in refereed journals and widely discussed in the literature. Table 1 compares four state-of-the-art PSS design methodologies with reference to six characteristics. The chosen six characteristics are important for discussion based on which the proposed framework is developed and nurtured. Table 1 illustrates the differences between the approaches proposed in the literature and aids to identify the literature gaps.

This table shows that all of the authors define PSS in terms of increasing the value of hardware (functional entities that carry out the elementary functions of the system) by focusing on services (entities that will ensure the smooth functioning of the whole system). Some of the aforementioned methods can deal with capabilities of stakeholders as parameters of PSS models employed in their methods. However, they do not provide specific methods or guidelines for PSS design based on the measurement, control and increase of capabilities. Furthermore, a common problem in PSS definition is the usage of different terms to define constituents of PSS. Various terms such as environment, activity, provider, receiver, channel, content, receiver state parameter (RSP), agent, and relationship among RSPs, function, entity and attribute parameters can create confusion and misunderstanding with regards to defining and communicating PSS design amongst research and industrial practitioners - a simple and unified PSS definition illustrating its constituents is required. Also, a PSS design representation technique should be commonly accepted to implement and develop a computer-supported PSS design platform for the effective evaluation of PSS performance and the capture and reuse of PSS design knowledge. The next section summarizes the industrial challenges in the PSS development.

Industrial challenges in designing PSS

This section summarizes the challenges which have been observed by the experts in business and maintenance of several providers of large, technical, capital-intensive and sensored product-service systems. Specifically, these are challenges observed in

Table 1 Comparison of state-of-the-art PSS design methodologies

Characteristics of the proposed methods	Komoto and Tomiyama [4,5]	Shimomura et al. [7-9]	Maussang et al. [10]	Alonso-Rasgado et al. [11,12]
PSS definition	A set of services in the life cycle of products, whose characteristics are customized with respect to the services	Service/product engineering as a discipline seeking to increase the value of artefacts by focusing on service	PSS are composed of physical objects and service units that relate to each other	Total care products as integrated systems comprising hardware and support services
Aim	Support the design and analysis of integration of services with a product life cycle and the identification of the characteristics of products	Focuses on service engineering to design products with a higher added value from enhanced services	Assists engineers in the joint development of physical products and interacting services to generate more added value	Develop innovative offerings consisting of hardware and services integrated to provide complete functional performance
The first step	Define goal(s) and quality as specified by product users	Define the state change of the receiver	Customer expectations, needs and specifications involved in the whole life cycle	Business ambitions of the client
PSS variables	Stakeholders in a product life cycle and the activities (e.g. production, use and services)	RSP, sequential chain of agents, relationships among RSPs, function, entity and attribute parameters	External functional representation, specifications of the physical elements	Customer's business needs, business solutions, clearer view of the hardware and/or services
PSS design representation technique	A graph description based on service formulation	Business process markup language, service blueprint	Scenarios and FBD	No representation technique is mentioned
Evaluation of PSS designs	Life cycle simulation considering multi-objectives (e.g. economic and environmental)	AHP, Dematel and Petri nets (discrete event simulation)	No evaluation approach proposed	Business case validation and evaluation of alternatives

RSP, receiver state parameter; FBD, functional block diagram; AHP, analytic hierarchy process.

maintenance planning. The maintenance experts were interviewed to understand engineering services issues because maintenance service occupy nearly three quarters of acquisition and support costs as compared to other aftermarket services [13]. Semi-structured interviews were held with five companies: aerospace engines and related systems, naval ships, land vehicle systems as well as, trains and trucks. Overall, manufacturers generally lack the competence to address the challenges of PSS-type contracts. The reasons for this situation are listed below:

- Customers now have higher expectations from manufacturers.
- Product-orientated manufacturers tend to be product-centric and so do not have the mindset to develop and deliver PSS.
- There tends to be very few technical employees of manufacturers interacting with customers.
- Customer issues could take months or years to be resolved if they are to be addressed by services.
- Current PSS design methodologies tend to be *ad hoc* and tend not to start with the business case.
- The design of the product and service is not completely performed simultaneously and maintenance is mostly an afterthought; only slight modifications to the product are considered following a decision to create a PSS.
- Manufacturers tend not to perform enough modelling to fully understand maintenance activities sufficiently to undertake PSS-type contracts.
- There is a lack of high-level strategic decisions to, for example, trade-off between design, maintenance and supply network solutions for the efficiency of the overall solution.
- A common understanding of PSS-type contracts is lacking across teams.
- The framing of competitive maintenance offerings is a challenge. Most importantly, the consideration of value-added benefits to customers and a suitable operative model between stakeholders to ensure the throughput of inputs as well as reasonable profits tends to be lacking.

These challenges present obstacles to the design of PSS solutions offered by these companies. As a result of an investigation of the state-of-the-art methodologies and challenges in designing PSS in the literature, we have framed the following research questions to be answered in this paper:

- What constitutes PSS design? (This question intends to define characteristics and properties of the system).
- How can customised PSS solutions be designed to be aligned to integrated stakeholders' capabilities?

The next section discusses our definition and constituents of PSS design.

What constitutes PSS design?

PSS design aligned to integrated stakeholders' capabilities is mandatory to achieve a viable and sustained solution for an intended duration. A capability can be defined as the continuing ability to generate a desired operational outcome [14]. The definition of capability exemplifies how the joint capability of all the stakeholders could achieve the desired outcome required by the PSS customer and for the PSS provider to design economically sustainable PSS. Considering capabilities as a core element, we defined PSS design as a process to synthesize and create sustained functional behaviour through tangible products and intangible services. Sustained functional behaviour represents the degree to which a system can continuously achieve its purpose by adapting its capabilities. To represent sustained functional behaviour, an activity-based modelling approach is proposed (Figure 2). An activity could be defined as an action incorporated or influenced in the customer's system. The activity-based modelling approach is in-line with the definition of the PSS design processes provided by Komoto and Tomiyama [4], in which designers define the activity to meet a specified goal and quality, and also define environment as being the circumstance within which that activity is realized. Furthermore, Matzen [15] and Tan [16] emphasized activity systems in modelling PSS development. Tan focused on customer activity cycles whereas Matzen viewed activity systems with a broader and general view encompassing both customer and company activities. Both Matzen and Tan conceptualized PSS solutions by considering artefact-, activity- and actor-based domains with slight differences. In this work, capability is mapped through resources, competences, responsibilities undertaken and outcomes. Inputs to a particular activity are mapped through customer needs and precedence activities' outcomes. Other influential parameters on an activity are enforced through an environmental variable. In Figure 2, the coloured boxes represent modification incorporated in the existing system.

We propose in this work that a network of seven parameters, namely *activities, customer needs, environmental influences, resources, competences, responsibilities, inputs (precedence activities' outcomes) and outcomes*, are sufficient to model PSS design. We have chosen these familiar and simpler terminologies for easy acceptance within industrial practitioners. These parameters map the required information for mapping products, services, processes and interactions of all the elements involved within the system.

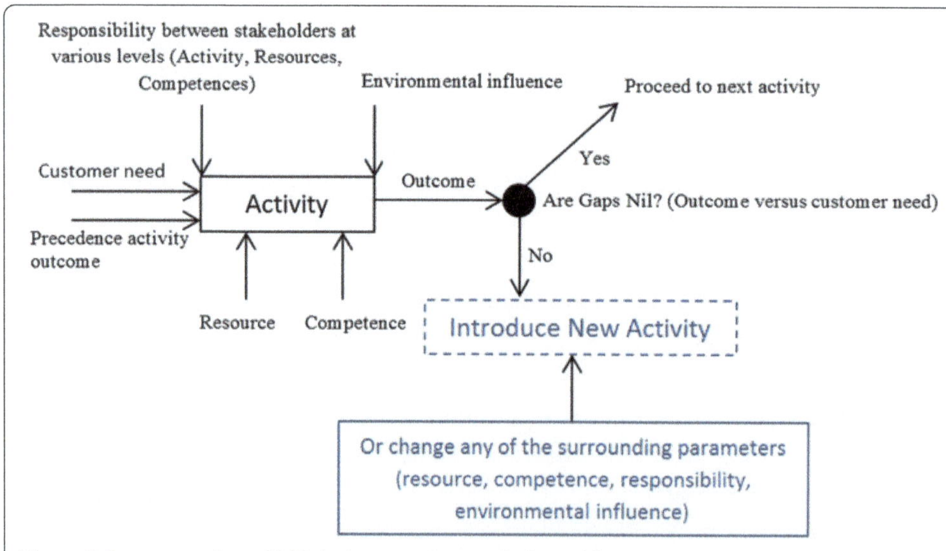

Figure 2 Representation of PSS design constituents (coloured boxes represent modification to the existing system).

Mapping of these parameters helps to understand the gaps within the current customer's system (PSS user) and aids the development of innovative PSS designs to satisfy customer business needs. The derivation of these parameters in PSS design is presented in the proposed framework. The framework is detailed step by step in the next section.

A capability-based PSS design framework

The proposed framework aims to support manufacturers in designing customer-adjusted PSS designs which are aligned to stakeholders' capabilities. The core principles supporting this framework to realize this aim are as follows:

- Gap analysis: Identification of value addition required in the customer's system (PSS user) by understanding their needs through assessing customer's business activities and constructing relative key performance indicators [17]. This initial step lays a foundation to develop customised solutions.
- Generation of new and/or re-designs of integrated product and service solutions along with conditions and consequences of each design. These designs take into consideration the partial substitution of product and service shares over the life cycle.
- Responsibility assignment which considers the capabilities of all the stakeholders involved at various levels: activity, object (resources), and parameter (competences). This assignment enables the derivation of innovative function-, availability- or result-oriented business models.
- Synthesis-generated solutions in each gap to improve the overall PSS design on a system level and avoid a sub-optimized solution towards any of single activity, stressing the importance of resource effectiveness.
- Standard representation of PSS designs is required for effective communication across all the stakeholders.

These core principles are structured into 10 steps in the proposed framework (Figure 3). The steps mentioned in this framework are highly inter-dependent, and feedback loops exist between every step. The following sub-sections describe each step individually and detail the course of action involved.

Step 1: understand customer needs

The first step in deriving customer needs from their complete business activities and existing systems is not taken into consideration in most of the PSS methodologies; commonly, customer needs are deemed to be the requirements of products and services. Note that the term 'customer' represents the PSS user throughout this framework. Figure 4 illustrates a typical product-centric life cycle. The product life cycle within the customer's business process only partly covers operation. Although the consideration of the product life cycle improves the understanding of what is required of products and services, extra value could be offered by considering the customer's goals as revealed by their business processes (Figure 4). Moreover, understanding the customer's business processes reveals the 'need behind the need' [18] of the customer that has to be fulfilled. Tan [16] has emphasized that PSS solutions may be conceptualized by considering the product life phase, customer activities and actor network.

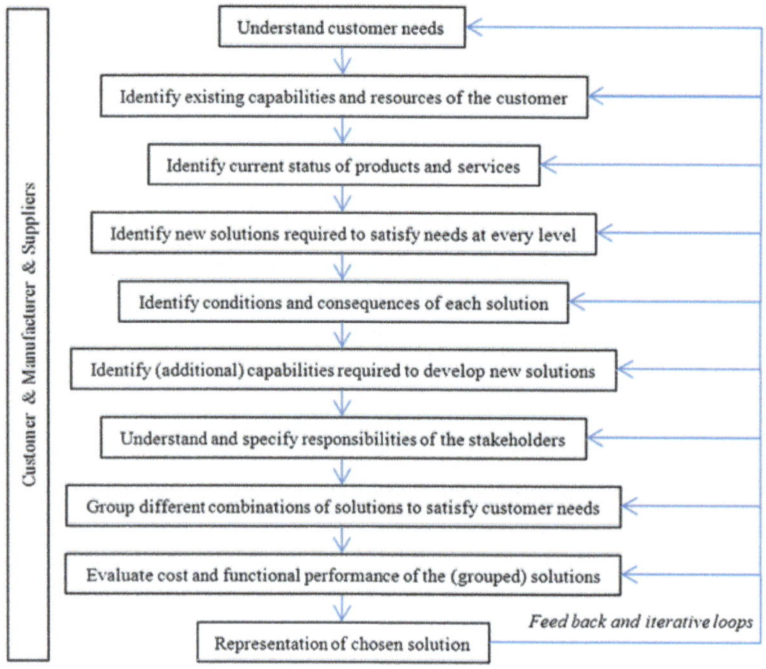

Figure 3 The proposed capability-based PSS design framework.

Figure 4 illustrates the point that it is the consideration of the capabilities that are required to use a product that presents opportunities to add value. Identifying and understanding the customer's overall needs should therefore be the foremost step in the design process. Once the overall customer needs have been identified by focusing on their business processes, the next step would be to identify the current capabilities of the customer.

Step 2: identify existing capabilities and resources of the customer

Identifying the customer's needs is followed by understanding of the customer's existing capabilities. Such an understanding helps to develop designs which are more aligned to their capabilities. A capability can be defined as the continuing ability to generate a desired operational outcome [14]. Capabilities can be realized through people, processes, tools and technology [19]. It should be noted that these elements are highly coupled and should be visualized together. Integration of these elements will be facilitated if the customer's activities are identified and the efficiency of each activity is measured. For this reason, capabilities should be mapped depicting their outcomes and the reasons for deficiency. Some of the parameters to assess each activity could be performance, time taken, reliability, responsiveness, expense and quality [20,21]. Such analyses will highlight the gaps within the customer capabilities that need to be filled by a PSS design. In the next step, in-depth analyses of existing products and services are performed to ascertain the degree to which the needs can be addressed.

Step 3: identify current status of products and services

The identification of existing products and services (whether they are on the market or just being developed) that could help to address the identified customer needs is

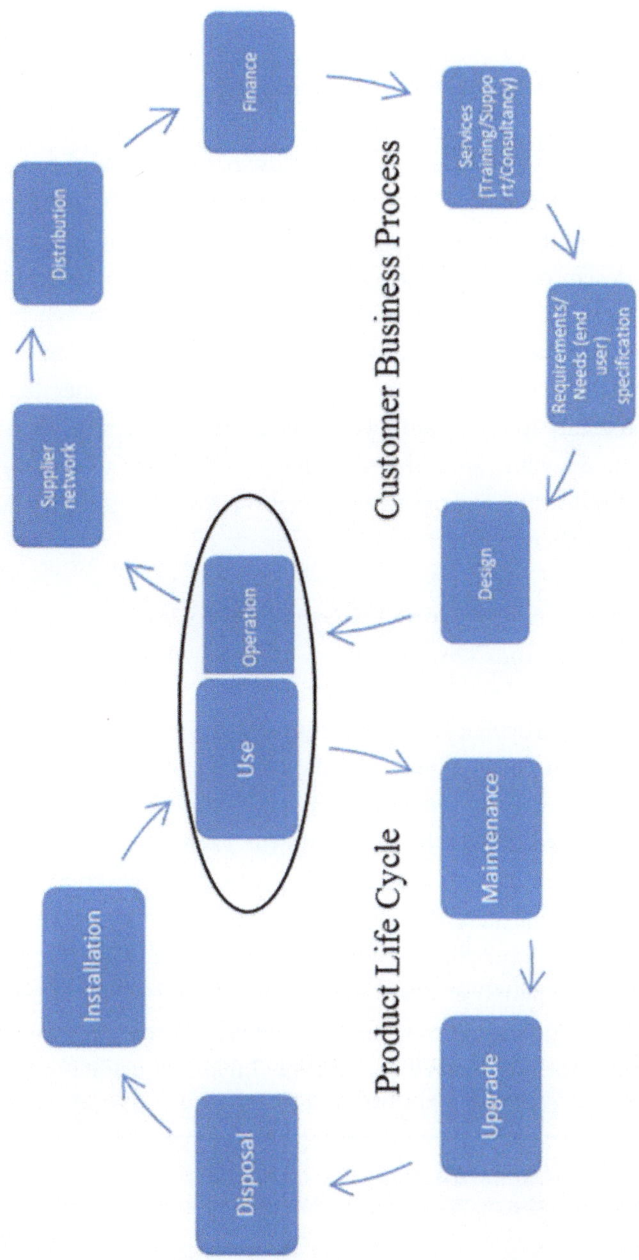

Figure 4 Integration of product life cycle in customer business process.

mandatory. The next step is to identify and highlight how these offerings could be changed or added to meet customer needs even more closely; this would involve identifying sub-systems within those products and services that need to be considered to achieve this change. This outcome could be achieved by:

- Comparing the key performance indicators for these products and services against those of the solution to meet customer needs.
- Performing root cause analyses to find out which sub-system within the products or services is responsible for failing to fully meet customer needs.

The data required for this step is available in most organizations through condition based monitoring and effective data management systems. The outcomes from these points should help to inform design as to how these offerings could be improved. It should help in the design of the right product and service mix to satisfy the needs of customers. For example, Figure 5 illustrates the capability shifts between machine capability and maintenance service. In scenario 1, the customer finds the amount of maintenance unacceptable as there is too much disturbance to business operations. Scenario 2 shows how the capability for a certain level of availability has shifted from the maintenance service to the machine: here, the machine is re-designed to require less maintenance. The next step details on approaches to develop PSS designs based on the gaps identified in the last two steps.

Step 4: identify new design required to satisfy needs at every level

The outcomes from steps 1 to 3: customer needs and capabilities and the properties of existing apposite products and services, will highlight the current gaps which could be filled by PSS designs. In particular, the capability map of the customer's processes will pinpoint the focus areas for development. Based on the conceptualization of PSS design through the network of seven parameters described in the 'What constitutes PSS design?' section, a new design should be generated for each gap by either introducing and/or modifying the current system in terms of activity, resource, competence and environmental influence. In this step, PSS designers have to identify possible designs for each gap identified. The designs should aim to eliminate or reduce the gap. For each gap, designs need to be explored at all of these levels. Design generation is a creative activity. To structure this process at the conceptual stage, we mapped resources for objects and competences for parameters and values. Table 2 illustrates sample designs

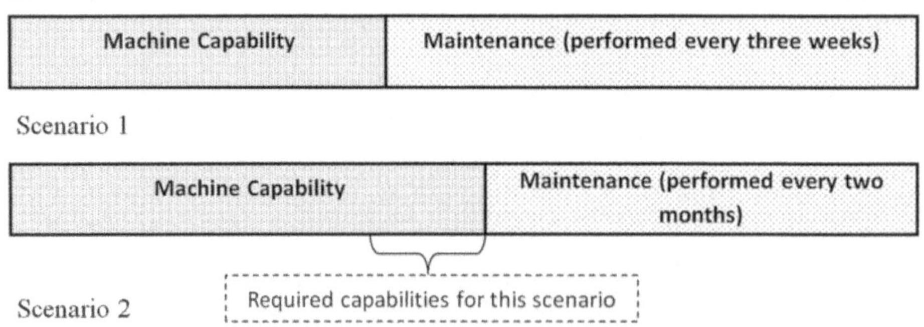

Figure 5 Scenarios 1 and 2: shifted availability division between product and service capability.

Table 2 Illustration of usefulness of mapping elements in design generation for a gap

Gap	Mapping elements	Designs
Difficulty in cutting parts with varying complexity and different shapes	Objects (resources)	1. New system with inbuilt knowledge
		2. Modify feature extraction mechanism
	Activities	1. To develop a technique to group parts
		2. New technical assistance team
	Parameters and values (competences)	1. The customer should specify the cutting parameters and respective values
	Environment	1. Availability of using off-the-shelf standard parts

generated for a gap in these elements. The subsequent sections illustrate steps in the framework to understand, evaluate, chose and represent the design.

Step 5: identify conditions and consequences of each design

In this step, the conditions and consequences of each generated design from the last step need to be identified. This is an important step in understanding the ramifications of every generated design. The conditions should detail the circumstances which are required by the particular design. The consequences express the changes which occur in the system or which could affect another system if the particular design is executed. Since the overall system impacts upon both the conditions and consequences, these could be expressed in terms of environmental factors, product and service attributes, as well as economic and socio-cultural factors. At the conceptual stage, it can be difficult to identify the conditions and consequences. For this reason, the heuristic judgement of experts can play a vital role in predicting the conditions and consequences of each design; organizing a group exercise between experts would enrich the specification of conditions and consequences. From the conditions and consequences, the additional capabilities required for each design are derived in the next step.

Step 6: identify (additional) capabilities required to develop new designs

The identified designs that have been selected (given the understanding of the customer's capabilities) will help the provider to develop their own capabilities along with those of the supply network. The commonalities and differences between the stakeholders' capabilities need to be explicitly shared and understood by the stakeholders. The following processes are involved in explicating the capabilities required:

- For each design, a detailed list of activities to be carried out to implement the design should be documented.
- The evaluation of the required resources and their efficiency in each activity and which stakeholders are best placed to offer them.
- Identification of the parameters, which can be monitored and controlled by stakeholders, within each design considering past, present and future scenarios.

Shifts in the capabilities between the stakeholders which may then require additional resources should be carefully aligned and integrated. Any difficulties the stakeholders may have in meeting increased capability demands could be compensated for with variations in resources and time constraints. In the next step, the additional capabilities required are assigned between stakeholders.

Step 7: understand and specify the responsibilities of the stakeholders

Based on the required capabilities identified from the previous step and the preferences and views of stakeholders, responsibilities have to be assigned. Responsibilities could be taken at various levels: activity, object and parameter levels. These assignments form a core part of the PSS design process. This step develops various business models based on the responsibilities undertaken by the stakeholders. This alignment of responsibilities will precisely define network relationships. Various soft elements play vital roles in such relationship development such as trust, confidence, commitment and culture. The development of an open network would be valuable as the responsibility map should ideally be visible to all in the network, since, ultimately, all of the responsibilities are the concern of all in the network. In the next step, different PSS designs and responsibilities of the stakeholders across identified gaps are merged to meet customer needs.

Step 8: group different combinations of designs to satisfy customer needs

While analyzing product life cycle activities along with the customer business process, it is important to improve PSS at an overall system level and avoid sub-optimizing any individual activity. To avoid sub-optimizing any single activity and to satisfy the identified customer needs, different combinations of designs addressing different activities could be grouped based on the responsibilities assigned. Although synthesizing different designs could produce a variety of possible designs, this could expand exponentially and, therefore, could be difficult to manage. To systematically explore this process, Taguchi's factorial method [22] could be used to synthesize the different designs generated. Taguchi's method intends to ensure sufficient performance at the design stage of products and services. It is a systematic procedure in which all controllable factors (except one) are held constant as a variable factor is altered discretely. The controllable factor is considered to influence the simulation response, and its level can be controlled by designers. The control parameters identified in each design in step 6 could be used to vary the values to generate multiple options within each design. This synthesizing process also helps in understanding the sensitivity of important variables in the performance of the proposed PSS designs. At the conceptual stage, the detailed application by the development of Taguchi's orthogonal matrices may not be required. Each synthesized design should be evaluated in detail in the next step.

Step 9: evaluation of the cost and functional performance of the (grouped) designs

Evaluation should be part of every step in the proposed framework. To emphasize this evaluation process, it is dealt with separately in the framework. The evaluation should focus on three primary dimensions: economic, social and environmental [20]. From a business perspective, the major evaluation criteria will be profit, customer satisfaction, quality of products and services, value-in-use and risk reduction [23]. Both the tangible and intangible merits and demerits should be evaluated. To evaluate and choose the generated designs for an intended period, the sequence of activities involved in the grouped design has to be specified. By varying the conditions and consequences of designs, the responsibility assignments, and the combinations of designs for each gap, a variety of designs could be evaluated. To understand variable changes at important distinct points for an intended period, discrete event simulation is recommended at this stage to identify the performance of each design. In the final step, the evaluated and chosen design should be represented clearly to ensure a common interpretation across stakeholders.

Step 10: representation of the chosen design

The final step is to represent the chosen design in a format which could be easily generated and commonly interpreted by the stakeholders involved. The primary motivation in PSS modelling is to co-produce conceptual models that can be systematically shared by stakeholders. By analyzing various elements discussed from the above steps, a common representation map through extended IDEF0 modelling has been developed (Figure 2). The representation is based on the mapping of each activity in PSS design into inputs, outcomes, resources, competences, responsibilities, environmental variables and customer needs. The sequence of activities is linked as each activity is based on the satisfaction of preceding activity's outcomes. This representation emphasizes important PSS parameters and the interactions amongst them. This representation helps to highlight the current and modified system states, and assesses the performance of each activity involved in the grouped design. This representation appears to be simple, flexible and easy to maintain. Advantages of this framework and representation are illustrated using an industrial case study in the following section.

Framework corroboration with a laser system case study and implemented in ISCL

To corroborate the proposed framework, a step-by-step application of the framework to laser systems for cutting operations which are used in manufacturing is used. A number of interviews were conducted with laser job shop owners (users of laser cutters), sales managers of manufacturers of laser cutters and also with experienced academics within this field to determine the current level of servitization within laser job shops. Table 3 provides details about the approaches used to build this laser system case study and for its evaluation. A step-by-step illustration to implement a higher level PSS is offered.

To assist PSS designers computationally, the framework has been illustrated through ISCL [4]. ISCL has been chosen over other software proposed in the literature for the following reasons:

- ISCL is well aligned with our framework especially with regard to the elements used: entities in a service environment, attributes, specifications and activities changing attribute values and realizing specifications.
- ISCL supports life cycle simulation through a quantitative and probabilistic approach, which is important for assessing PSS designs. Life cycle simulation evaluates product life cycles from an integrated view of economic profitability and environmental awareness and optimizes the life cycles [6]. In ISCL, life cycle simulation is implemented through a discrete event simulation technique applied to life cycle design, such as the selection of end-of-life options (e.g. reuse, recycle and remanufacturing), the design of product modularity considering the options, and the timing and contents of service during the contract period.

Table 3 Details about the approaches used to collect necessary data

Purpose	Roles and responsibilities	Number of interviews	Approaches	Duration
To build a laser system case study to apply the developed framework	Laser job shop owner (laser system user)	3	Semi-structured interviews - questions covered broad concerns of job shops: business needs, laser system specification, available resources, types of customers, applications, usage, current business solutions, types of services, laser system life cycle and laser cutting process parameters	Approximately 1 h each
	Laser system manufacturer	2	Semi-structured interviews - questions covered the broad topics of company background, types of laser system manufactured, types of customers, types of services, types of business solutions, supplier network and costing process	Approximately 1 h each
Evaluation of the developed laser case study along with the proposed framework	Academic researchers on laser system and processes	2	Discussion and feedback received through a PowerPoint presentation	Approximately 1 h each
Evaluation of the framework with industrial practitioners	Product development team	3	Discussion and feedback received through a PowerPoint presentation and also filled an assessment sheet to rate the proposed framework in a 5-point scale for potential usefulness, completeness, usability and clarity	Approximately 30 min to 1 h
	Business programme operation team	1		
	Technology development team	1		

- Standard process modelling and simulation approaches are not encouraged in this work because the PSS domain requires a specialized software environment defining its own terminologies and incorporating specific methods to support development. It helps to develop and integrate PSS research knowledge generation and understanding into a specific platform for wider uses and support tools evaluation.
- Although there is significant scope for improvements, from the authors' opinion, ISCL is a mature, reliable and accessible PSS software which well integrates modelling and simulation modules.
- Finally, the software is currently in the public domain and is actively supported by the developers, which is an incentive for industries to use it in practice.

Before describing the case study, the PSS modelling method employed in ISCL is briefly explained. A PSS-based business model in ISCL consists of activities, scenes and entities which are described along with their attributes (what the entity owns) and specifications (what the entity aims are) (Figure 6). Scenes represent partial states of the service environment as defined by the attribute values of entities. Activities treat scenes as the execution condition and change the value of attributes and realize specifications. As shown, these elements have several ports that are connected with lines. These lines have 11 different relations between elements instantiated during life cycle simulation. Figure 7 illustrates these links, and Table 4 shows the meanings (for details of the modelling method and grammar of the simulation codes, please refer to [24] and the user manual on the supporting website [25], respectively). Designers can create these elements on the canvas of ISCL and also move, inspect and delete these elements on the canvas.

Step 1: understanding customer needs

The laser systems under consideration in this case study are mature products as are the laser processes which are structured and mostly inbuilt to the system. The customers are laser job shop owners who procure laser systems from the original equipment manufacturer and supply semi-finished goods to the end product manufacturer. Figure 8 illustrates stakeholders' map of the laser system case study. As the laser job shops have many years of experience in this field, they are able to precisely specify their requirements of laser-cutting systems. The transaction type between the laser system manufacturer and the laser job shop customer is business-to-business. The semi-structured interviews with the laser job shops revealed the importance of the need behind the need, experiences, required state change and business ambitions of the laser

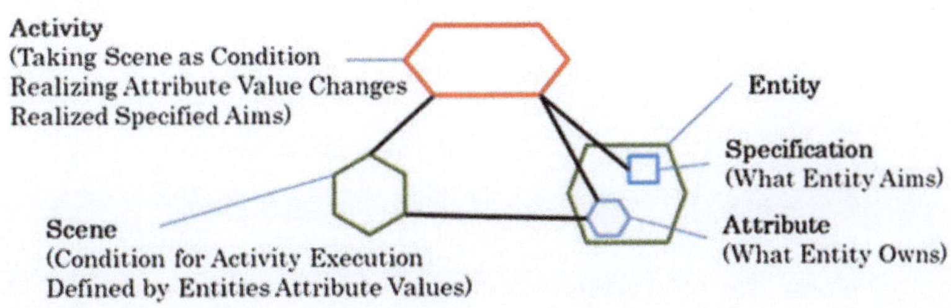

Figure 6 Elements in a business model on ISCL.

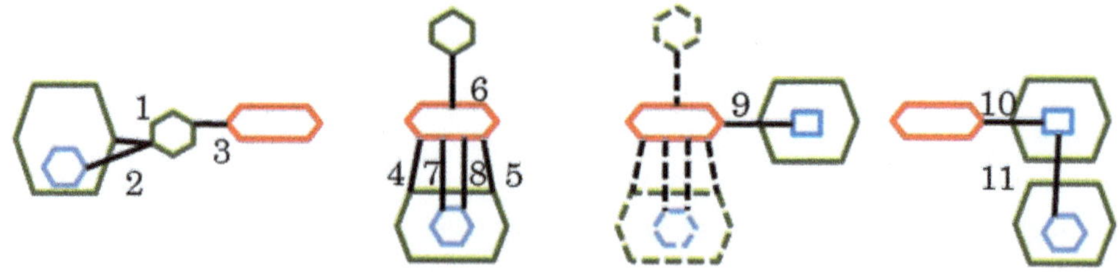

Figure 7 Relations in a business model on ISCL (please read Table 4 for link description).

job shop. We developed a PSS customer needs specifications of the laser system by using overall equipment effectiveness which is a multiplication of availability, performance and quality. The required values are mapped to be:

- Overall equipment effectiveness ($A \times P \times Q$), 50% to 60%
- Availability (Mean time between failure (MTBF) / (MTBF + Mean time to repair (MTTR)), 85% to 95%
- Performance (Working speed / Designed speed), 55% to 65%
- Quality (Good units within tolerance / Total units produced), 95% to 99%
- Laser system usage period, 5 years

At this point, ISCL supports the designer to define the specifications of the users of laser systems as well as the attributes of the laser systems, which are identified at this step. Figure 9 details the specifications and the attributes of the laser systems with examples. In Figure 9a, three entities: 'Manufacturer', 'LaserSystem' and 'User', are shown. The model includes an activity 'Use' to deliver 'Function' as a specification targeted by User. LaserSystem already includes the overall equipment effectiveness specification and the relevant attributes such as *MTBF* and *MTTR*. The value of these attributes can be individually calculated with respect to each instance of LaserSystem during the life cycle simulation. Figure 9b shows the objectives of the model, which are statistical values obtained as a result of life cycle simulation. For instance, 'AvrOEE' is the average overall equipment effectiveness of all laser systems in the market with respect to

Table 4 Definition of relations in a business model on ISCL in relation to Figure 7

Figure 7 link number	Element 1	Element 2	Meaning
1	Scene	Entity	Scene includes all instances of Entity
2	Scene	Attribute	Entity instances included in Scene are specified by the value of Attribute
3	Activity	Scene	Scene is regarded as the execution condition of Activity
4	Activity	Entity	Activity creates instances of Entity
5	Activity	Entity	Activity deletes instances of Entity
6	Activity	Scene	Activity refers to (calls) an instance of Entity specified by Scene
7	Activity	Attribute	Activity gets the value of Attribute
8	Activity	Attribute	Activity assigns the value of Attribute
9	Activity	Specification	Specification is realized as a result of execution of Activity
10	Activity	Specification	Specification is evaluated during the execution of Activity
11	Specification	Attribute	The value of Specification is related with Attribute

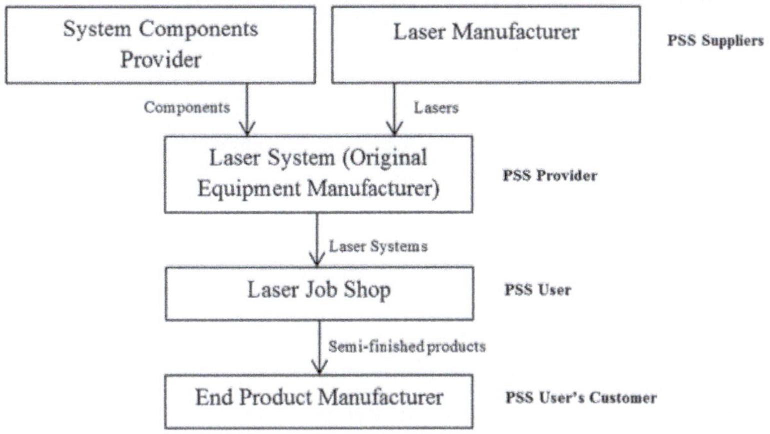

Figure 8 **Stakeholders' map of the laser system case study.**

simulation time, which is defined with the window in Figure 9c. At this moment, the dynamic behaviour of laser systems such as physical deterioration has not been defined yet. In the next step, to understand whether required needs levels are achievable through existing customer's capabilities, they are noted.

Step 2: identification of the existing capabilities and resources of the laser job shop

To understand the laser job shop's capabilities, currently performed activities are mapped with resources, outcomes and the reasons for deficiency. The important tasks which are mapped include machine calibration, CAD file preparation, identification of process parameters, material preparation, work piece (un)loading, work piece alignment, machining operation, daily and planned maintenance, repair work, and material

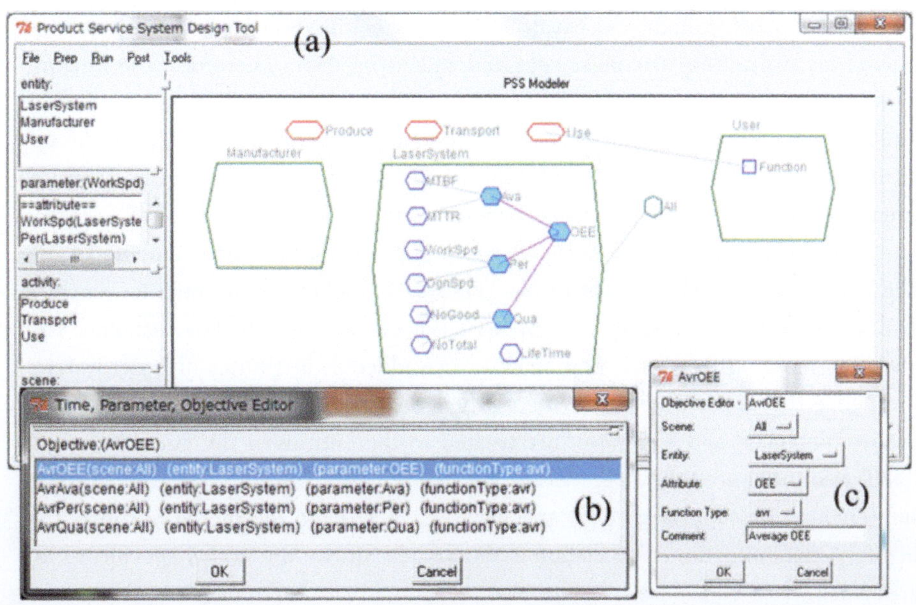

Figure 9 **Modelled attributes of laser systems and their objectives in ISCL.** **(a)** Entities of the model. **(b)** Objectives of the model. **(c)** Average overall equipment effectiveness.

Table 5 Mapping existing capabilities of laser job shops

Activities	Resources	Outcomes	Reasons
Machine calibration	Limited skills. One employee	Trial and error	Not having enough understanding of the machine. The system is partially protected by the manufacturer
Work piece loading/unloading	Manual process. Two unskilled employees	Time-consuming	Automated machine unavailable

and gas procurement. Some of the activities to map the laser job shop's capabilities are detailed in Table 5. The complete list of activities and the respective status of each helps us to understand the capability gaps of laser job shops. Along with these gaps, current products and services are assessed in the next step.

Step 3: identify current status of laser systems and services

A laser system is an assemblage of a laser generator unit, beam delivery system, beam manipulation system, motion system, process monitoring system and a control system (Figure 10). The key performance indicators (KPIs) for these systems are failure rate, repair time, degradation rate, redundancy and reliability. Mapping these KPIs to the sub-systems shows that beam delivery and beam manipulation systems have to be improved. Identifying root causes through discussions with the experts revealed that mirror misalignment, laser instability, variation within suppliers, operator's error and a mismatch in cooling needs could be possible problems to be addressed. Mapping the KPIs of services (frequency, number of technicians available, time consumed, spare parts and tools availability, and location) with provided services (training, planned maintenance, technical assistance and repair activity) revealed several potential improvement areas like the operator's knowledge of the machine, complexity regarding the different shapes to be machined, constraints in space requirements, the probability of making mistakes being high and escalated expense of module replacement. In this study, there was restriction to collect required industrial data. Therefore, the current scenario was simulated through data collected from the interviews. If data could have been available, ISCL supports to import these data through .CSV format. From the interviews, it was revealed that the overall equipment effectiveness at the required level is not maintained and, in particular, that performance should be improved.

At this point, ISCL supports the designer in adding other necessary elements to describe the current design. The refined model (Figure 11) includes new activities, entities and attributes so that the model can represent the current design. For instance, 'Fail(ure)Rate', 'Deg(radation)Rate' and 'Rep(air)Rate' are added as attributes of LaserSystem, which influence the degradation in terms of the availability, performance and quality. Furthermore, 'Engineer' and 'Operator' are treated as the entities of the current design. Their 'Skill' also influences the degradation and its recovery through activities 'Repair' and 'Maintenance'. By supplying codes specific to each activity, the model becomes the input of life cycle simulation. For instance, Figure 12a shows the codes specific to an activity Repair. These codes are used to automatically create links between the model elements as shown in Figure 12b, which is helpful to the designer in debugging these codes. An explanation of coding as the input to life cycle simulation is beyond the

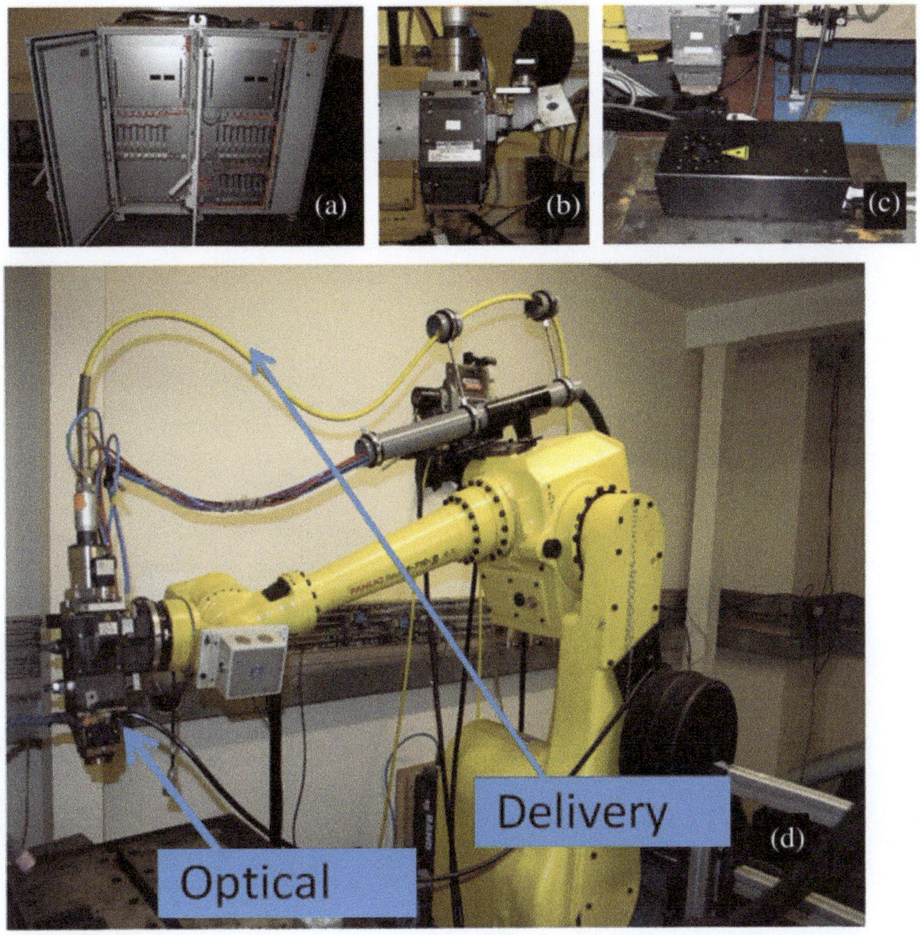

Figure 10 Laser system. (a) Laser unit (IPG YLR-8000). **(b)** Optical head. **(c)** Monitoring system. **(d)** Robot motion system.

scope of this paper. Figure 13 shows the simulation result of the current design, the average overall equipment effectiveness with respect to simulation time. At this stage, the simulation results are partly based on fictive parameter values, which should be specified in the design process. The gaps identified in these first three steps facilitate generation of PSS designs in steps as described in the following sections.

Step 4: identification of new designs required to satisfy the needs at every level

From steps 2 to 3, 10 activities have been identified for improvement. Within the 10 activities, 18 problems are observed. Through group brainstorming with the researchers, 54 designs have been generated. From the perspectives of adding and modifying activities, enriching the laser systems through support systems and focusing on specific parameters helped to generate many designs. These designs have been checked and rated by the researchers in preference with the feasibility scope. The highlighted circles in Figure 14 are some of the preferred designs. In order to add these designs in ISCL, corresponding model elements should be added to the model to represent the current design. Alternatively, corresponding model elements can be separately modelled and stored in the knowledge base of ISCL. The stored model elements are instantiated

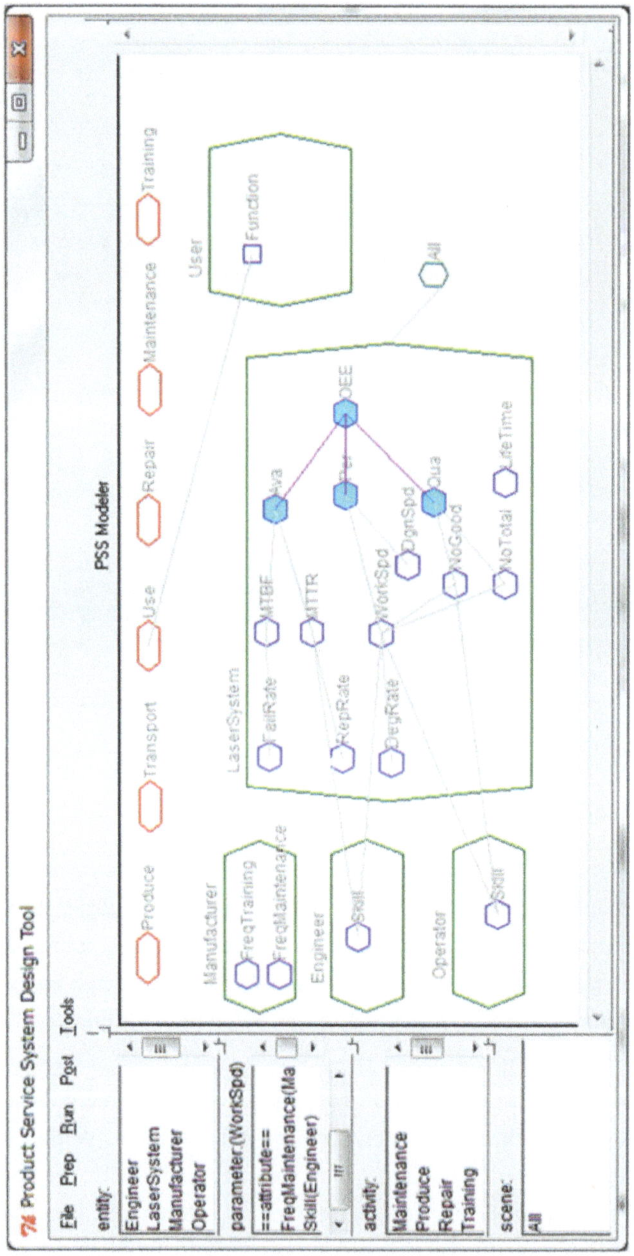

Figure 11 Modelling the current design in ISCL.

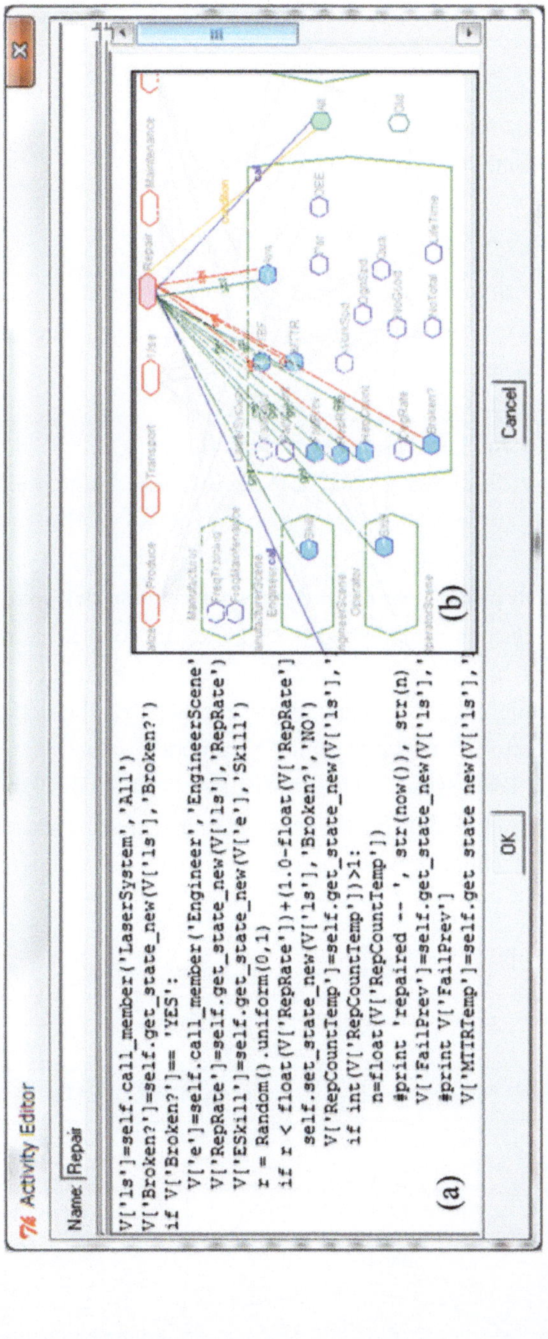

Figure 12 The detail of an activity repair in the current design. (a) Codes specific to an activity Repair. **(b)** Links between the model elements.

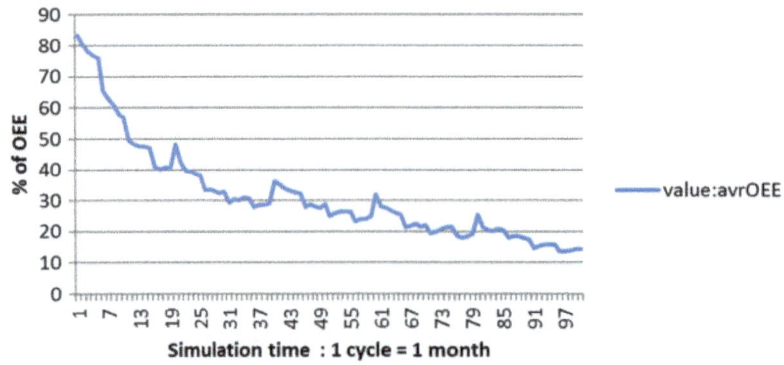

Figure 13 The simulation results of the current situation.

when necessary. Furthermore, the formalization of this step can be useful for ISCL to automatically synthesize new designs using the knowledge base. The subsequent steps help to develop and assess these designs.

Step 5: identify conditions and consequences of each design

Steps 5 and 6 are jointly discussed in the next section due to greater continuity between them.

Step 6: identify (additional) capabilities required to develop new designs

Even though preferred designs are highlighted in the previous step, the potential of every design is identified by specifying conditions and consequences. The conditions are mainly specified through the frequency and complexity of the activities, laser system usage period and patterns, required skills, parameters values, man-days required and the condition of the laser system. The consequences of each of the proposed designs are expressed in terms of the impact on availability, performance and quality to set the overall equipment effectiveness and incurred costs. The generated conditions

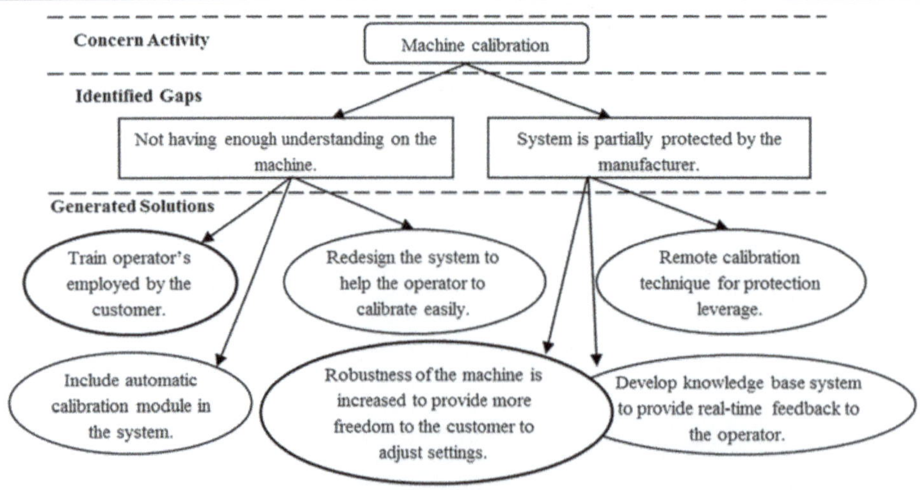

Figure 14 Representation of generated designs for two problems in the concern activity.

and consequences have been evaluated by the heuristic judgements of two experts. From the conditions and consequences, the capabilities required by each design are derived. Table 6 illustrates the conditions, consequences, capabilities, resources and control variables derived for new designs which have been generated. Conditions and consequences are inherent processes in ISCL. Support should be developed to highlight the new capabilities achieved, and the additional resources should be incorporated in the PSS laser system modelling. Developing new integrated business models between the laser job shop, the manufacturer and suppliers through the alignment of capabilities for each design will be the next step.

Step 7: understand and specify responsibilities of the stakeholders

Responsibilities have to be assigned precisely between the laser job shop, the manufacturer and suppliers (based on the identified required resources) for the capability shift to occur to satisfy the proposed design and eventually to satisfy the required level of overall equipment effectiveness. These assignments are precisely defined either at the level of activities, objects or parameters. Based on inputs of preferences and views from the laser job shops and manufacturers, various possibilities of responsibility assignment are generated. The outcome of such a PSS design process has stark impact upon business model generation. Table 7 illustrates responsibilities aligned within a new design generated. Using ISCL, such responsibilities are defined by selecting appropriate entities (such as Manufacturer) as the service provider of specific activities. This assignment results in modification of financial flow. In some cases, the expected service may not be realized in life cycle simulation, because some entities may not satisfy the conditions to provide the expected service (such as the level of Skill of Engineer). Synthesizing designs generated for 18 problems through responsibility alignment is the next step.

Step 8: group different combinations of designs to satisfy customer needs

To satisfy the required overall equipment effectiveness levels throughout the usage period, a wide variety of designs are generated by synthesizing the designs identified for each gap. The synthesis process is systematically carried out through Taguchi's factorial method. The control variables identified in each design in step 6 are used to vary values to generate multiple options within each design. Table 8 details the combinations possible across four activities. In Table 8, the four columns explain the chosen four activities (in which gaps exist in the current system) to be addressed in PSS designs. The six cells in each column explain two solutions with three improvement levels each (based on Taguchi's factorial method) to fill the gap in each activity.

Similar to step 4, the automatic synthesis of the new design by combining model elements stored in the knowledge base including these activities can be implemented in ISCL. Currently, ISCL supports the automatic synthesis process, when the conditions and consequences of added activities are defined and these activities solely influence the state of a single entity with a hierarchical structure. For instance, [4] shows a procedure implemented in ISCL to generate possible functional upgrading services combined with repair services for medical equipment systems. The next step evaluates combinations to understand the satisfaction of required overall equipment effectiveness levels.

Table 6 Outcomes from steps 5 and 6 for the derived new design

Solutions	Conditions	Consequences at each time	Capability shift	Resources	Control variables
Technical assistance to set efficient process parameters	Laser parts (new shape and size parts)	Man-days, +0.2 (24-h day); performance, +0.5%; quality, +1%; set (overall equipment effectiveness (laser system)); total cost, +£50	Efficient process parameters irrespective of varying shapes and sizes	To employ an additional three technical employees to support process query	Frequency of new shapes and sizes, technical employees, number of queries

Table 7 Responsibility assignment at various levels for new designs

Responsibilities	Laser job shop	Manufacturer	Supplier
Activities	Identifying machined part variety	Measurement of operator skills	To develop learning content
Objects (resources)	Experienced operators	Web-based support system	Control system for error identification
Parameters (competences)	Skills of operators	Frequency of training	Queries redirection

Step 9: evaluate cost and functional performance of the (grouped) designs

The primary questions to be answered in the evaluation process for the laser system case study are as follows:

- Which designs and combination of designs satisfy the required level of overall equipment effectiveness for a specific period?
- How much does each successful implemented design cost for the specific period?
- How should cost be shared between the laser job shop, manufacturer and suppliers for each design based on responsibility assignment?

Answering these questions provides predictable costs, cost transparency and maximal security that are important factors to be considered in laser business model selection. ISCL provides an exceptional environment to carry out life cycle simulation. Discrete event simulation is carried out by sequencing activities involved in the grouped designs. By varying possible conditions and consequences of designs, responsibility assignments and combination of designs for each gap, a variety of designs are evaluated. Figure 15 points out the increase in overall equipment effectiveness levels for each design identified to improve machine calibration activity. All of the identified 54 designs are evaluated, and the best designs for each gap are synthesized to achieve the aforementioned overall equipment effectiveness intervals. The final synthesized chosen design should prove to be an improvement of operator skills, module repair, as well as daily, preventative and repair maintenance by advanced and frequent training, an efficient diagnostic and prognostic system and enhanced support systems. Figure 16 shows that the overall equipment effectiveness depreciation of this combined design would satisfy the required value throughout 5 years. The cost sharing between the laser job shop,

Table 8 Combination of designs possible for four activities of focus and variable change in brackets

	Chosen four activities in which gaps exist in the current system			
	Machine calibration	**Process path/ parameters**	**Daily maintenance**	**Repair work**
Two solutions with three improvement levels each (control variable in bracket)	S1-CalibrationTraining-1 (*operator's skill level*)	S3-GroupTech-1 (*system's reliability*)	S5-OperTraining-1 (*operator's skill level*)	S7-DiagnosSys-1 (*system's reliability*)
	S1-CalibrationTraining-2	S3-GroupTech-2	S5-OperTraining-2	S7-DiagnosSys-2
	S1-CalibrationTraining-3	S3-GroupTech-3	S5-OperTraining-3	S7-DiagnosSys-3
	S2-Calibrateassist-1 (*service level*)	S4-TechAsst-1 (*service level*)	S6-SysRedesign-1 (*system's reliability*)	S8-MTTRIncres-1 (*time to repair*)
	S2-Calibrateassist-2	S4-TechAsst-2	S6-SysRedesign-2	S8-MTTRIncres-2
	S2-Calibrateassist-3	S4-TechAsst-3	S6-SysRedesign-3	S8-MTTRIncres-3

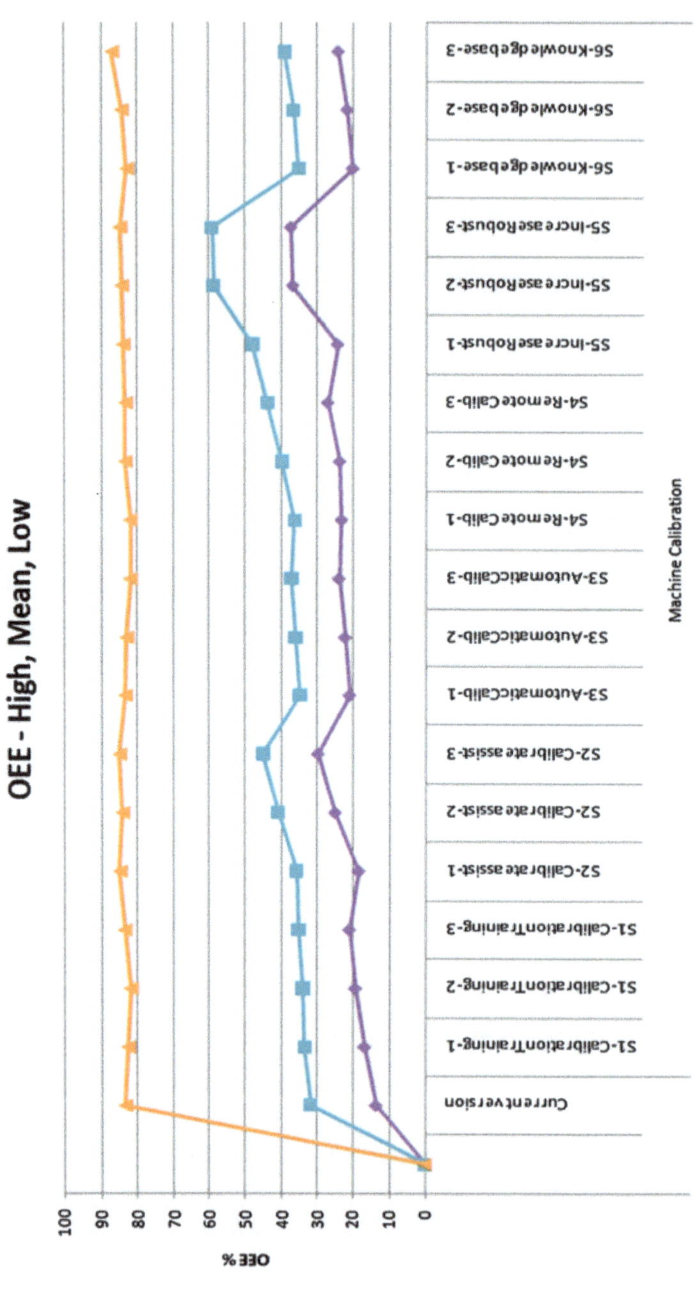

Figure 15 Overall equipment effectiveness increases due to the implementation of the new design (for three improvement levels each).

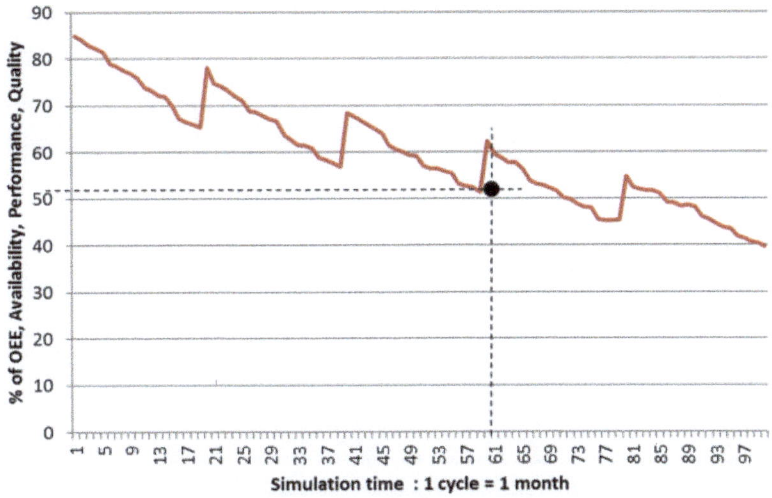

Figure 16 Overall equipment effectiveness depreciation for the proposed finalized design. The black dot represents the required level of overall equipment effectiveness to be maintained.

manufacturer and suppliers is shown in Figure 17. The representation of this chosen design is detailed in the last step.

Step 10: representation of chosen design

The chosen design is represented by an extended IDEF0 format for common interpretation amongst the stakeholders involved (Figure 18). It highlights the activities of repair, maintenance, replacement module, daily maintenance, operator training and preventive maintenance along with customer needs, environmental variables, responsibilities, resources, the current and reached states, outcomes of a particular activity and interactions amongst them. The modifications to the current system are highlighted through dotted boxes (reduced daily maintenance) and colour changes. It represents alternatives between activities (repair maintenance and module replacement) and achievable outcomes. A module should be developed in ISCL to implement the extended IDEF0

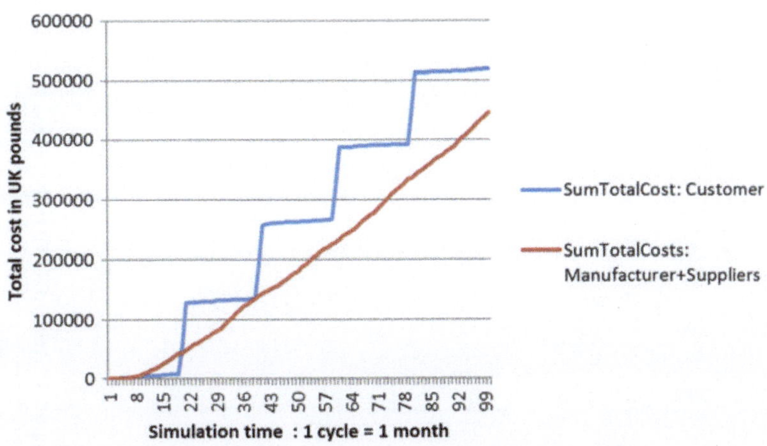

Figure 17 Cost sharing between the laser job shop, manufacturer and suppliers.

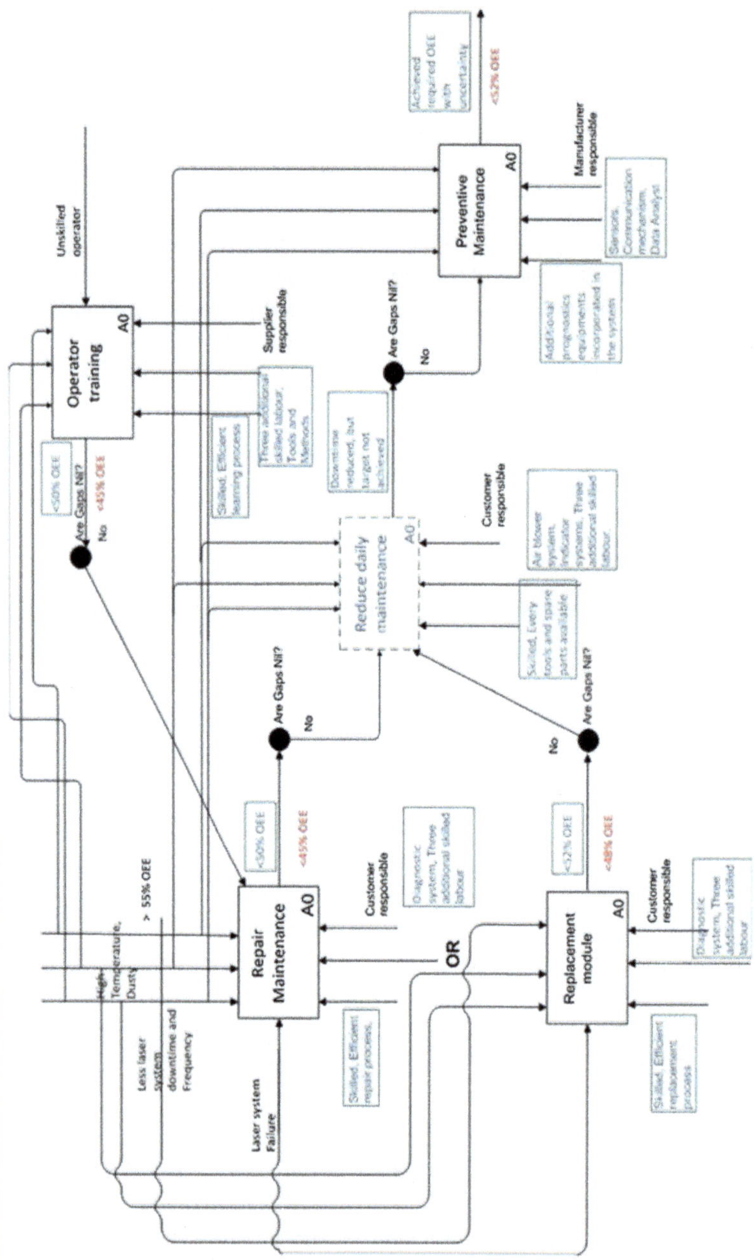

Figure 18 Extended IDEF0 model of the synthesized chosen PSS design.

representation. Also, this representation helps to frame the final contract using terms and conditions that are relevant to all of the stakeholders involved. The academic and industrial views of these steps and the results produced are detailed in the next step.

Academic and industrial views

To understand merits and limitations of the proposed framework, it has been corroborated by a step-by-step application with two laser system experienced academics and five industrial experts who are knowledgeable in PSS. The overall feedback is encouraging and potential applications of this framework have been stressed; notable benefits are that the structure and flow of the framework are clear and robust, the framework helps to develop PSS design from a holistic systems approach which facilitates a change in the designer's mindset from product-centric to systems-centric, the framework could be used as a general problem solving approach which incorporates the co-production process between relevant stakeholders, and the IDEF0 representation is perceived to be clear and informative. Potential usefulness, completeness, usability and clarity are highly graded by the experts (Figure 19). The limitations that have been expressed are that the level of trust and transparency required for this framework is absent in most industrial sectors, there exists a need to measure intangible benefits and to perform more quantitative data analysis to assess gaps, and the assessment of the skill set of industries and the development of tutorials to train employees to apply this framework along with obtaining the necessary information required for each step could all be problematic.

All of the discussed methodologies have novel steps in designing PSS. However, compared to other methodologies, the merits of the proposed framework are that its starting point is an existing system and that the customer needs are elicited as deficiencies within that system, it develops PSS designs which are aligned to the stakeholders'

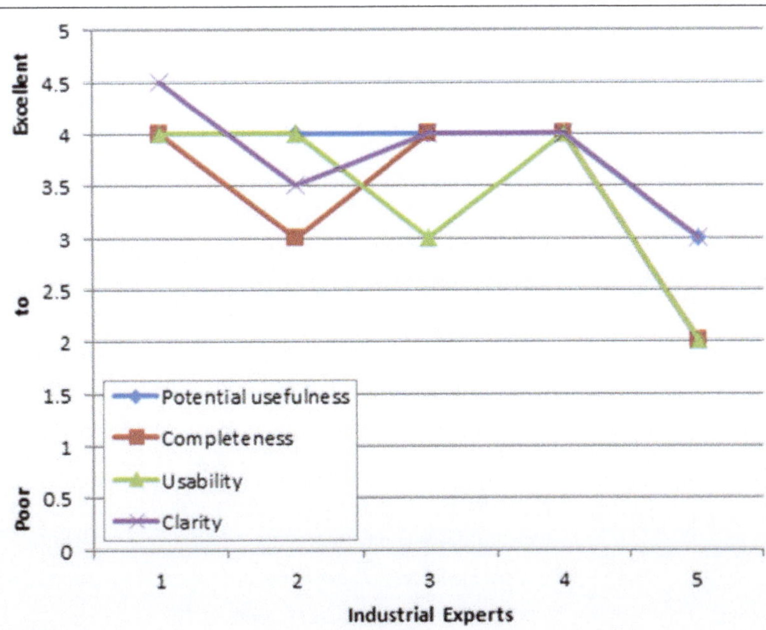

Figure 19 Five industrial experts' ratings for the proposed framework.

capabilities, and it synthesizes solutions to produce optimized designs which can encompass the whole of the customer business process. We believe that the proposed framework could be further developed by applying it to industrial case studies to develop functional and sustained PSS designs. As most of the proposed methodologies in the literature have not been formally applied to industrial case studies and demonstrated, a comparison using performance metrics is not possible. Further work in validation would involve a methodological effectiveness comparison to measure performance metrics. A list of performance metrics that could be potentially applied in this comparative exercise are tabulated in Table 9.

Discussion and conclusions

Based on identified gaps from the literature, a capability-based PSS design framework is proposed. This framework highlights the important features required in designing

Table 9 PSS design methodology performance metrics parameters

	Performance metrics parameters
Methodological applicability	Potential usefulness
	Completeness
	Usability
	Clarity
	Systematic approach
	Reliability
	Effectiveness of feedback loops
	Change in designer's perspective from product-centric to PSS-centric
	Avoidance of re-work
	Time consumed in designing and delivering PSS
Methodological outcomes	PSS design functional performance for a specified duration
	Customer satisfaction (requirements and solutions)
	Enhanced value-in-use
	Agility and responsiveness of designed PSS
	Reliability of PSS
	PSS revenue and profit generation
	Whole life cycle PSS design
	Optimization of costs and resources
	Effectiveness of integrated system and co-production
	Common interpretation of PSS
	Innovative and usable PSS designs
	Minimum number of loopholes in PSS
	Robust stakeholders network and responsibility adherence
	Sustainable PSS solutions by stakeholders' capabilities
	Quality of PSS
	Prolonged relationship with customers
	Product/service improvements
	Understanding and control of the existing and developed system
	Validity of simulation results in real time
	Diffusion of product and service integration

PSS to help meet customer goals. A capability-based approach generates a wide variety of PSS designs which are intended to produce sustained functional behaviour in the proposed system to achieve its purpose continuously and create an innovative value addition for the customer. It encourages a broader customer business perspective in designing PSS as compared to the traditional approach where the functional availability of products is considered given their deterioration during the use stage. Details of a capability map of the customer's processes help to pinpoint the focus areas for PSS development. It stresses the capability shift and mix between products and services and also between stakeholders in order to develop feasible and enduring PSS designs.

The framework emphasizes the importance of taking into account the preferences and views of the stakeholders and the apposite alignment of responsibilities. Organized feedback loops between the various steps aid PSS design modelling and performance evaluation. The framework facilitates the exploration of a wide variety of designs by synthesizing designs that have been identified for various gaps. The generation and evaluation of designs are well structured in the framework. The results of the simulation using ISCL provide good insights into PSS designs. We believe that this framework would facilitate and structure the interactions between the customer, manufacturer and supplier. It also helps in understanding the capabilities of the stakeholders and aids an understanding of the value of PSS designs as appreciated by the customer. The representation of PSS designs through extended IDEF0 was found to be useful in providing a common understanding among the stakeholders.

The application of this framework to a real industrial laser system case study helps to demonstrate the benefits of the framework. This demonstration also helps refine the framework based on the few observed limitations such as the inputs collected and the features in the modelling technique. The demonstration of the framework along with ISCL highlights the important features of ISCL and the modules to be developed to enhance design support. In particular, support for design generation through standardized PSS ontology and the synthesis of different designs generated through Taguchi's factorial method would prove to be especially helpful for PSS designers. Additionally, the implementation of this framework using a step-by-step approach in software could greatly facilitate the design process by decreasing information load on PSS designers. An effective mechanism is required to support PSS designers to specify the conditions and consequences of each development design which are vital inputs for PSS simulation. Since the underpinning expectation that PSS will have a lower environmental impact is yet to be proven, the impact of PSS designs on environmental benefits should be studied separately. There is ongoing work in applying this framework to various case studies which involve other case companies who are in the process of refining the development of their PSS designs.

Competing interests
The authors declare that they have no competing interests.

Authors' contributions
GVAV carried out understanding industrial challenges, framework development, case study development, framework evaluation and manuscript preparation. HK participated in the framework development, Service CAD integration with the framework and manuscript preparation. RH participated in understanding industrial challenges and assisted in the framework development, case study development and framework evaluation. RR was the principal investigator of this project and assisted in all the phases of the work especially on the framework development and framework evaluation. TT supported in Service CAD integration with the framework. SE and AT participated in the framework development and framework evaluation. SW supported in the laser case study development and framework evaluation. All authors read and approved the final manuscript.

Acknowledgements
This research was funded by the Cranfield Innovative Manufacturing Research Centre (EPSRC Grant EP/E001874/1) as part of the research project 'Capability based conceptual PSS design'.

Author details
[1]Design, Manufacture and Engineering Management, University of Strathclyde, Glasgow G1 1XQ, UK. [2]Advanced Manufacturing Research Institute, National Institute of Advanced Industrial Science and Technology, Tsukuba, Ibaraki 305-8564, Japan. [3]School of Engineering, Cranfield University, Cranfield, Bedfordshire MK43 0AL, UK. [4]Manufacturing and Materials Department, Cranfield University, Cranfield, Bedfordshire MK43 0AL, UK. [5]Institute for Manufacturing, University of Cambridge, Cambridge CB3 0FS, UK. [6]Welding Engineering and Laser Processing Centre, Cranfield University, Cranfield, Bedfordshire MK43 0AL, UK.

References

1. Neely, A: Exploring the financial consequences of the servitization of manufacturing. Operation Management Research **1**, 103–118 (2008)
2. Annamalai Vasantha, GV, Roy, R, Lelah, A, Brissaud, D: A review of product–service systems design methodologies. J. Eng. Des. **23**(9), 635–659 (2012)
3. Mont, OK, Plepys, A: Customer satisfaction: review of literature and application to the product-service systems. Final report to the Society for Non-Traditional Technology, Japan. International Institute for Industrial Environmental Economics, Lund University, Lund, (2003)
4. Komoto, H, Tomiyama, T: Integration of a service CAD and a life cycle simulator. CIRP Ann. Manuf. Technol. **57**, 9–12 (2008)
5. Komoto, H, Tomiyama, T: Design of competitive maintenance service for durable and capital goods using life cycle simulation. Int. J. Automot. Techn. **3**(1), 63–70 (2009)
6. Umeda, Y, Nonomura, A, Tomiyama, T: Study on life-cycle design for the post mass production paradigm. Artificial Intelligence for Engineering Design, Analysis and Manufacturing **14**, 149–61 (2000)
7. Sakao, T, Shimomura, Y: Service engineering: a novel engineering discipline for producers to increase value combining service and product. J. Clean. Prod. **15**(6), 590–604 (2007)
8. Shimomura, Y, Hara, T, Arai, T: A service evaluation method using mathematical methodologies. CIRP Ann. Manuf. Technol. **57**(1), 437–440 (2008)
9. Shimomura, Y, Hara, T, Arai, T: A unified representation scheme for effective PSS development. CIRP Ann. Manuf. Technol. **58**(1), 379–382 (2009)
10. Maussang, N, Zwolinski, P, Brissaud, D: Product–service system design methodology: from the PSS architecture design to the products specifications. J. Eng. Des. **20**(4), 349–366 (2009)
11. Alonso-Rasgado, T, Thompson, G, Elfström, B: The design of functional (total care) products. J. Eng. Des. **15**(6), 515–540 (2004)
12. Alonso-Rasgado, T, Thompson, G: A rapid design process for total care product creation. J. Eng. Des. **17**(6), 509–531 (2006)
13. Zhao, Y, Harrison, A, Roy, R, Mehnen, J: Aircraft engine component deterioration and life cycle cost estimation. In: Hesselbach, J, Herrmann, C (eds.) Globalized Solutions for Sustainability in Manufacturing: Proceedings of the 18th CIRP International Conference on Life Cycle Engineering, pp. 657–662. Springer, Berlin (2011)
14. Neaga, EI, Henshaw, M, Yue, Y: The influence of the concept of capability-based management on the development of the systems engineering discipline. In: Proceedings of the 7th Annual Conference on Systems Engineering Research (CSER 2009), Loughborough University, Loughborough, 20–23. (2009)
15. Matzen, D: A systematic approach to service oriented product development. DTU Management Engineering, Technical University of Denmark, PhD dissertation (2009)
16. Tan, AR: Service-oriented product development strategies. PhD thesis, Department of Management Engineering, Technical University of Denmark. (2010)
17. Hussain, R, Lockett, H, Annamalai Vasantha, GV: A framework to inform PSS conceptual design by using system-in -use data. Computers in Industry, Special Issue: Product Service System Engineering: From Theory to Industrial Applications **63**, 319–327 (2012)
18. Tucker, A, Tischner, U: New Business for Old Europe - Product-Service Development. Competiveness and Sustainability, Greanleaf, Sheffield (2006)
19. NASA: NASA Systems Engineering Handbook: NASA/SP-2007-6105 Rev1. , NASA, Washington, D.C (2007)
20. Annamalai Vasantha, GV, Hussain, R, Cakkol, M, Roy, R, Evans, S, Tiwari, A: An ontology for product-service systems. In: Xu, Y (ed.) Decision Engineering Report Series. Cranfield University, Cranfield (2010)
21. Annamalai Vasantha, GV, Roy, R, Cakkol, M: Problem definition in designing product service systems. In: Proceedings of the 3rd CIRP Conference on IPS2, Braunschweig, 5–6 May 2011. ()
22. Roy, R: A Primer on the Taguchi Method. Society of Manufacturing Engineers, Dearborn. ()
23. Supply-Chain Council: Supply-Chain Operations Reference-Model: SCOR Version 7.0 Overview. Supply-Chain Council, Cypress. (2005)
24. Komoto, H, Tomiyama, T: Systematic generation of PSS-concepts using a service CAD tool. In: Sakao, T, Lindahl, M (eds.) Introduction of Product/Service-System Design, pp. 71–91. Springer, Berlin (2010)
25. ISCL software. (2005). http://staff.aist.go.jp/h.komoto/iscl_en.html (2011). Accessed 20 Aug 2013

A remanufacturing process library for environmental impact simulations

Nurul Hanna Ismail[1,2*], Guillaume Mandil[1,2] and Peggy Zwolinski[1,2]

* Correspondence: nurulhanna@hotmail.fr
[1]Université Grenoble Alpes, G-SCOP, F-38000 Grenoble, France
[2]CNRS, G-SCOP, F-38000 Grenoble, France

Abstract

'Closed loop' end-of-life strategies such as remanufacturing must be applied to create eco-efficient products. Remanufacturing may be a key element in reducing the environmental impact of products, but this remains to be proved. The aim of this study is to help designers evaluate the environmental impacts of their remanufacturing process during the design phase. The first task is to identify, list and classify the various remanufacturing processes (disassembly, cleaning, sorting and controlling, reconditioning, reassembly) by the type of process and then estimate the environmental impact for each process. These processes are then formalized by characterization and organized in a database. Using a simulator, the different processes can be aggregated to assess the environmental impacts of a remanufacturing line. An example is presented in the last part of this paper to illustrate the proposal.

Keywords: Remanufacturing process; Environmental impact; Simulator; Life cycle assessment

Background

Introduction

Technology currently has a considerable impact on production. Advances in tools such as machinery, computer technology and information technology make production easier, quicker and cheaper. However, this has indirectly created environmental problems such as more waste because large volumes of products reach the end of their lifetimes. To solve these problems, many programs have been introduced and governments are placing pressure on manufacturers to recover and exploit these products [1].

Several strategies exist for managing the end-of-life of a system. One of them is the closed loop end-of-life strategy or 'Cradle to Cradle' system. This strategy outlines a life cycle of a product or service, in which the 'good', once beyond effective use, is reintegrated in the manufacturing process to form an autonomous process (loop) that requires little or no new input to function. This strategy is believed to be a favourable option for the environment and will be reviewed in this study.

Figure 1 shows several options for the closed-loop end-of-life strategy. Reuse implies that items are used by a second customer without prior repair operations or as originally designed [2], whereas recycling is the process of taking a component material and processing it to make the same material or useful degraded material [2]. Current proposals for closed-loop end-of-life strategy focus on the material recycling process. While it is certainly successful in retrieving raw materials, it also results in the loss of

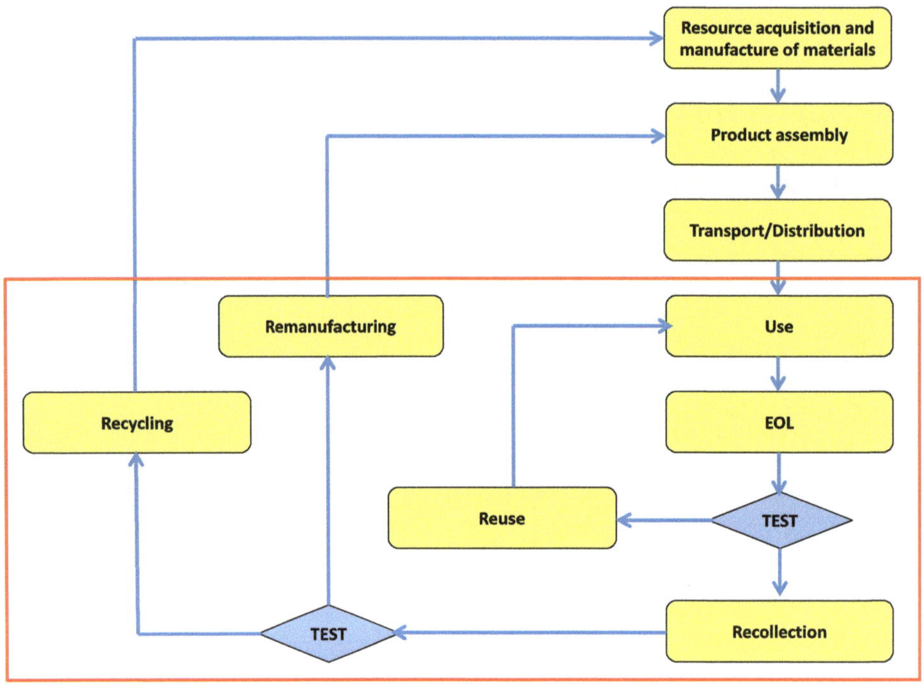

Figure 1 Closed-loop end-of-life product life cycle [23].

product shape [1]. This destroys most of the product's added value [3] and may represent a significant environmental impact [4].

Remanufacturing is defined as 'A process that brings a used product back to a new state through reuse, refurbishment and replacement of its components' [5]. It is a process of disassembling, cleaning, inspection and sorting, reconditioning or replacing parts (if applicable) and reassembling a used product in order to make it at least as good as, or better, than a brand new one. By applying the remanufacturing process, a product can be returned to service with a reasonably high degree of confidence that it will at least endure another full life cycle. The main purpose of remanufacturing is to prolong the product life cycle and at the same time preserve the added value of the material during the design and manufacturing phases. From the environmental point of view, remanufacturing ensures sustainable products that save energy and material resources. Remanufacturing also diverts material from landfills and creates jobs for skilled workers [6]. The decision to integrate the product remanufacturing process must be made at the earliest stages of the product life cycle [7,8] to ensure the outcome is successful.

To better understand this field, Lund [9] defined three different types of remanufacturing company. The first type is the Original Equipment Manufacturer (OEM) which remanufactures its own products. The second type is the Subcontractor Company which remanufactures products made by other companies (usually the OEMs). The last type is the Independent Remanufacturer (IR) which remanufactures products without much contact with the OEMs. They also have to buy or collect the core components for their process. Many studies have confirmed that remanufacturing is more profitable for OEMs [10,11].

Several studies have focused on proving the environmental benefits of remanufacturing. These studies include that of Kerr et al. [12] on the eco-efficiency gains derived from remanufacturing, based on a study of Xerox photocopiers in Australia. It attempted to quantify the overall life cycle environmental benefits of remanufacturing by comparing the remanufacturing of a traditionally designed photocopier and a photocopier designed to facilitate its own remanufacturing. In addition, Lindahl et al. [13] identified the general environmental advantages and disadvantages of remanufacturing by modelling a life cycle assessment of several households' appliances and products. An environmental comparison was performed of end-of-life scenarios (remanufacturing and recycling) for all of these appliances and products. Likewise with J. Amaya, who carried out a comparative assessment of remanufactured product life-cycles and classical life-cycle scenarios. This approach was illustrated by a case study of a truck injector [14].

Even though there have been many difficulties in defining and interpreting the system's boundaries, the measurement methods used and the result itself, this research has come to the conclusion that remanufacturing is a preferable option in comparison to new manufacturing and other end-of-life options [15].

Despite these advantages, remanufacturing lacks a tool for modelling the environmental impact of its process. Tools such as Remanufacturing Decision-Making Framework (RDMF) have been developed by researchers to promote remanufacturing in companies. It is a decision-making framework which helps management to make decisions on whether or not to remanufacture a product [7,8]. REPRO2 [16] is a tool that allows designers to monitor the evolution of design parameters for remanufacturing [1]. The CLOEE tool [17] has been developed at the G-SCOP laboratory. This tool is designed to assess the environmental impact of multiple end-of-life scenarios. It is particularly useful when many end-of-life strategies with different options have to be tested. However, CLOEE requires a certain quantity of environmental impact data as inputs in order to test each process or scenario. These data could be obtained using the remanufacturing process library proposed in this paper.

Nowadays, there are evermore methods and technologies that complete each remanufacturing phase. This phenomenon is positive for remanufacturers because they have a large variety of options available for setting up the remanufacturing system. Moreover, current technologies allow the remanufacturing of products that could not be remanufactured before. But having more choices leads to more confusion for designers when they are not equipped with the tools they need to highlight the environmental impacts of their process.

The absence of this kind of tool always leads designers to ask about the basic rules for choosing a method/technology capable of minimizing environmental impacts. The global impact of remanufacturing on the environment is also one of the aspects that still remain uncertain. All these issues show the importance of having tools to help designers answer these questions.

From the environmental standpoint, it is difficult to optimize the remanufacturing process without knowing its impacts on the environment for the entire product life cycle. Data from studies of existing remanufacturing systems have been limited and patchy [12]. Because of these problems, it is reasonable to propose a database and a tool intended to help remanufacturers and designers to evaluate and compare the environmental impacts of their remanufacturing process during the design phase. This study is aimed at helping remanufacturers and designers to compare environmental impacts and make decisions.

In this article, a global representation of the remanufacturing process is presented. This is followed by a detailed explanation of the construction of the remanufacturing process library, the calculation of the environmental parameters, and the creation of the environmental impact simulator. In the last part, the simulator is tested in a case study on an engine block and the results are discussed before the conclusion.

Methods

Global presentation of the remanufacturing process

It is essential to have a global picture of the different processes involved in order to create a remanufacturing library. Hence, this study started with research to obtain data on the variety of methods and technologies involved in performing each remanufacturing step. Normally, these data are available online and are often obtained from remanufacturers' and suppliers' websites. Once collected, they were sorted and arranged in a readable and structured format.

A mind mapping form was chosen to represent these data. It is a simple visual tool used to help structure information and better analyse, comprehend and synthesize ideas. The mind mapping diagrams always take the same basic format of a tree, with a single starting point in the middle that branches out and divides continuously. The tree is made up of words or short sentences connected by lines.

In the mind mapping task, the data were classified by types of processes. It started with a remanufacturing process which was divided into five essential steps: disassembly, cleaning, inspection and sorting, reconditioning and reassembly. Figure 2 shows these five steps arranged in a mind mapping form. For each step, the different methods and technologies currently used were listed. Since the cleaning and repair steps are the most critical steps in the remanufacturing process [18], their methods and technologies

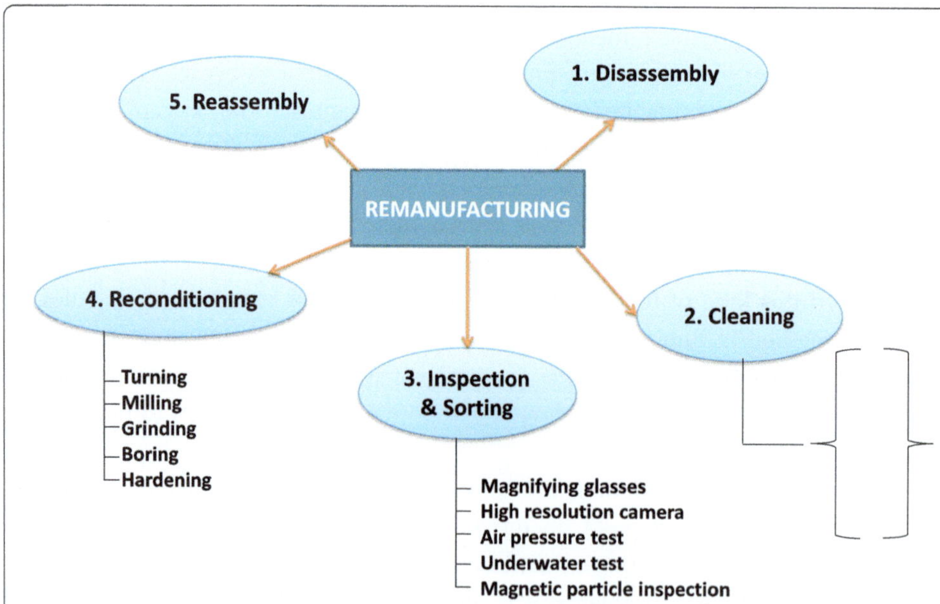

Figure 2 Remanufacturing mind map - first part.

were explored and listed until the deepest level (see Figure 3 [25-28]). It is essential to list them thoroughly, because every component is a variable that can change the environmental impact and must be taken into account during the design phase.

This global presentation is an attempt to identify as many existing technologies and methods as authors can. This list should be expanded as soon as new technologies or methods appear.

Standard characterization sheet (Remanufacturing process library)

After taking a global view of the remanufacturing process, designers require detailed information about the methods and technologies listed in the previous mind map. This detailed information is important to give designers a basic idea about such methods and technologies, and help them make their choice. This study offers essential and useful information about these methods and technologies, classified in a library of available processes for remanufacturing. In particular, it includes a sheet that contains the important parameters that differentiate one process from another from the environmental standpoint.

This characterization sheet was specially designed to be used with the remanufacturing simulator, in order to compute environmental impacts. Information and parameters were synchronized between this sheet and the simulator database. This sheet contains information about the process, i.e. its name, a useful description of the process, drawings, machine technology, durations of cycles, inputs and outputs of the process, parameters to be used by the simulator for modelling the process, parameters that affect the environment and consumable materials.

The characterization sheet works like a catalogue from which designers can choose methods and technologies suitable for their product and processes. The information

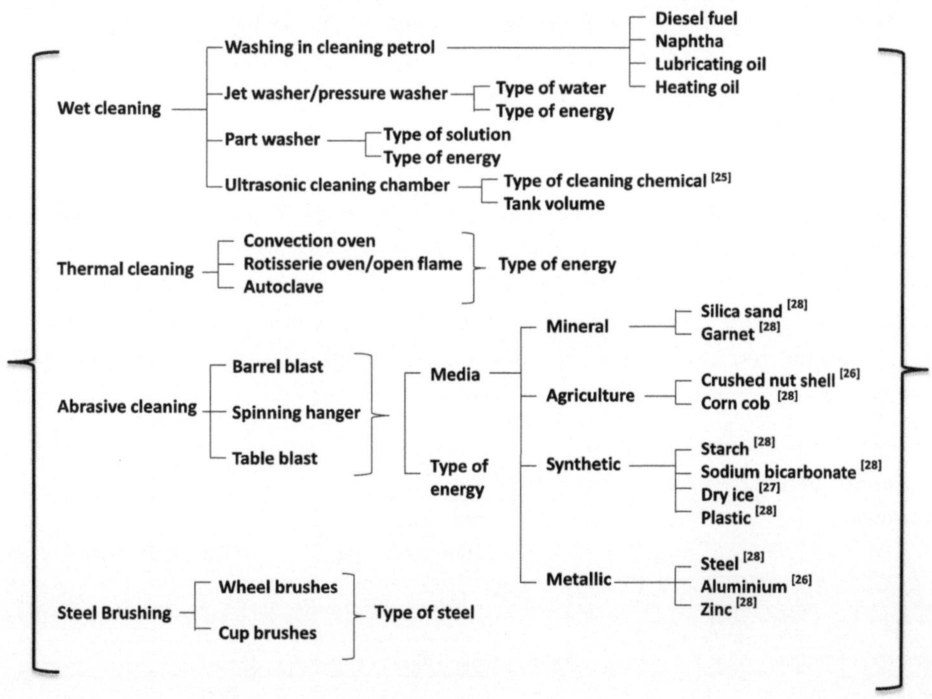

Figure 3 Remanufacturing mind map - second part.

gathered in this sheet can generally be obtained from remanufacturers' and suppliers' websites, and designers can also choose methods and technologies based on environmental objectives. The use of the data contained in this sheet helps designers to model their remanufacturing process in the environmental simulator.

Calculation of environmental parameters

In the remanufacturing process library, the most important data to store for the creation of the remanufacturing tool are the generic parameters. Firstly, all the physical parameters of the process are listed. Then, a single physical parameter that best describes the size of the machine or process is chosen. It will be called size parameter in the following section. Then, for all the input parameters of the process, the aim is to calculate the generic parameters to be listed in the process characterization sheet.

As for the other input parameters, they are listed in correlation with the size parameter. Supplier data are used to plot a graph of input parameters as a function of the size parameter. Then the generic parameters are calculated based on a linear slope that best fits all the points plotted. These data are important because they link the remanufacturing process library and the environmental impact simulator. The size parameter is the only data that users must know in order to assess the environmental impact of the corresponding process when using the simulator. Consequently, the table providing the correspondence between the size parameter and generic parameter must be stored in the characterization sheet of the process.

The importance of the generic parameter is that it gives a generic value for one input parameter of a process. It avoids the recalculation of an input parameter every time the user changes the characteristics of the machine, such as its size. Consequently, only one value has to be keyed in for one input parameter of a process.

In this section, an example of an ultrasonic cleaning process is used to explain how the generic parameter (see Table 1) is calculated. To understand more about the function of this parameter and on the relation between the remanufacturing process library and the environmental impact simulator, a case study was performed.

The main parameters used to design one remanufacturing process can be identified from the diagram of inputs and outputs (see Figure 4) found in the remanufacturing process library. The input parameters that generate a potential environmental impact were identified, and research into the machine was carried out using the constructor's data. The *environmental input parameter* section in the remanufacturing process library contained the research results. The information found was used to draw curves and linear slopes, and to calculate the generic parameters later on.

Table 1 Generic parameters for ultrasonic cleaning

Type of solution	Generic parameters	
	Solution weight (kg)	Electrical power (kW)
General purpose and medical instruments	0.05	19.23
Industrial metal brightening and rust remover	0.05	19.23
Industrial carbon remover	0.05	19.23
Industrial heavy oil and carbon remover	0.05	19.23
Aluminium and soft metal safe	0.05	19.23

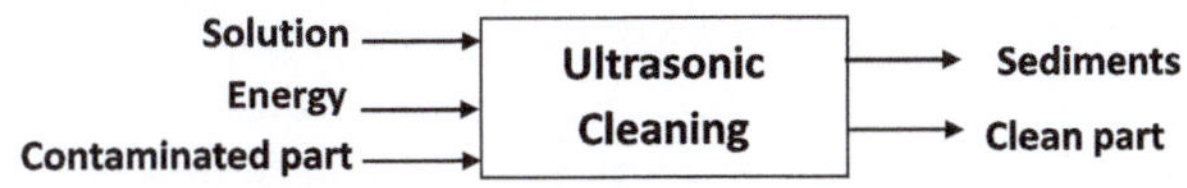

Figure 4 Inputs and outputs for ultrasonic cleaning.

Figures 5 and 6 show an example from an ultrasonic cleaning process. A table with the machine range (see Figure 5) is the constructor's data. It presents the volume as the size parameter, and the electrical power and solution weight as input parameters. In order to calculate the generic value for these input parameters, a power-vs.-volume curve was plotted using the information in the table (see Figure 6). From the curve, a power-vs.-volume slope that passes through zero and best fitted the previous points was drawn. In this case, a value of 19.23 (W/m^3) was found for this parameter. This value is a value of the generic parameter for the electrical power that was keyed in as information for the environmental impact simulator database. The same steps were performed to obtain the generic parameter for the solution weight and the other processes.

Environmental impact simulator

Once the remanufacturing process information was gathered and the essential generic parameters were calculated, it was then possible to create a remanufacturing simulator. In this study, a life cycle assessment (LCA) was used as it is an accepted environmental management tool that holistically and systematically quantifies the environmental burdens and their potential impacts over the whole life cycle of a product, process or activity from a multidisciplinary standpoint [19]. The LCA is a comparative tool and is used in the comparison of different ways of achieving a defined function [20] and as a decision-making tool in process selection, design, and optimization in order to identify clean technologies.

Environmental input parameters

Machine range

Enclosure size (cm)			Size parameter, Volume (l)	Power (kW)	Solution weight (kg)
Length	Width	Height			
15	13	10	1.75	55	0.0875
24	13	10	2.8	80	0.14
24	13	15	3.8	100	0.19
30	15	15	5.5	120	0.275
30	24	20	12	200	0.6
50	13	20	9	200	0.45
50	30	15	19	240	0.95
50	30	20	26	300	1.3
45	45	35	95	1000	4.75
61	45	35	125	2000	6.25
86	45	35	170	4000	8.5
84	61	30	208	4000	10.4
91	61	50	322	6000	16.1
122	66	50	416	8000	20.8

Cleaning durations :

- Transmission parts : 7-10 min
- Automobile parts : 10 min
- Electronic components : 3-10 min

Changeable parameters

Figure 5 Data collected from constructors for ultrasonic cleaning process.

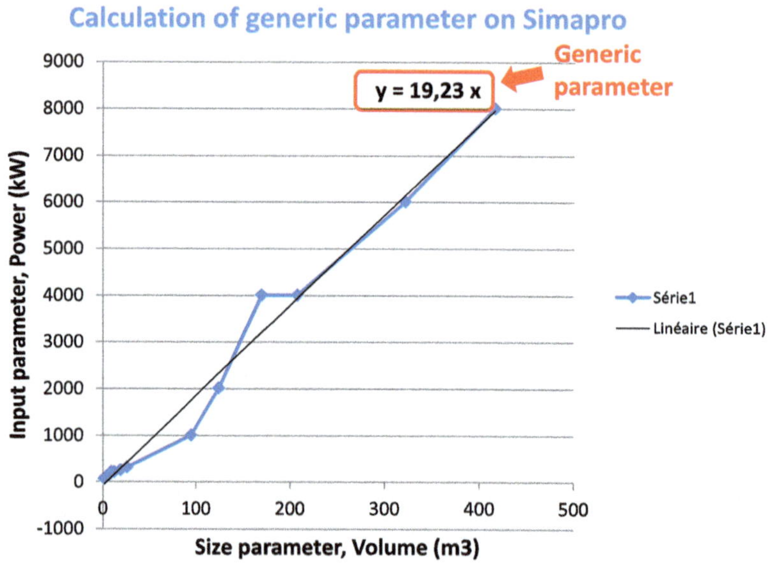

Figure 6 Graph showing calculation of electrical power *generic parameter* for ultrasonic cleaning process.

The Simapro software was chosen as it is probably the most used at present for LCA applications and has been described and used in a large number of published works [19]. Simapro includes several LCA methods such as Eco-indicator 99, EDIP 1997 and 2003 and EPS 2000. It allows viewing parts of the life cycle at different scales, and displaying their contributions to the total score [21].

A remanufacturing function was created in Simapro under the 'processing' category. In this function, all the remanufacturing process data found in the remanufacturing process library were listed and keyed in. The study focused on the 'cleaning' and 'reconditioning' steps, because the other steps consumed almost no energy or consumable material. It was created from scratch for the 'cleaning' process, while the 'reconditioning' process was created directly using data available in Simapro databases. A step-by-step user guide was also created to help designers use the simulator.

Below are some important pieces of information concerning the environmental impact simulator:

1. Number of processes performed: four (cleaning).
2. Number of process found in Simapro: three (reconditioning).
3. Some processes cannot be modelled (e.g. water jet cleaning).
4. The database used was 'Ecoinvent unit processes', located in the European Union (indicator U/RER). The unavailable data were noted in the characterization sheet.
5. The 'Impact 2002+' method was used to analyse the LCA results as it is universal and simple to use.

Basically, users simply have to choose the remanufacturing process, enter their product size parameters (weight, volume, etc.), choose the consumable material, enter the consumable material parameters, and change the duration of the cycle (depending on the type of product). The software performs the calculation and the results appear in a few minutes afterwards.

Case study on an engine block

A case study was carried out on a commercially remanufactured product, using the simulator to make ensure that it functioned as wished. This case study allowed detecting malfunctions of the system and the user's difficulties when they used it. It was also important in order to improve the simulator and its user guide. In this case study, a remanufacturing process was modelled for an engine block using the information found on the remanufacturer's website.

Engine block specification

Figure 7 shows the engine block used in this study. The specifications are given below:

Brand: Ford
Material: Aluminium
Weight: 58 kg
Volume: 191 l (dm^3)
Dimensions: 53 × 68 × 53 cm

Remanufacturing process

Once the engine block was defined, the remanufacturing process comprising more detailed information on the method and technology for each step was described using data obtained from promarengine.com.

The flowchart in Figure 8 shows a remanufacturing process for the engine chosen (see Figure 7). It can be seen that the engine block was subjected to two different inspection technologies, four cleaning methods and five reconditioning processes. These were the methods and technologies that will be entered in the environmental impact simulator using the information in the remanufacturing process library.

Field of study

The same methodology and phases used in the classical Life Cycle Assessment were applied to evaluate the previous remanufacturing process. According to the ISO 14040

Figure 7 Ford engine block [28].

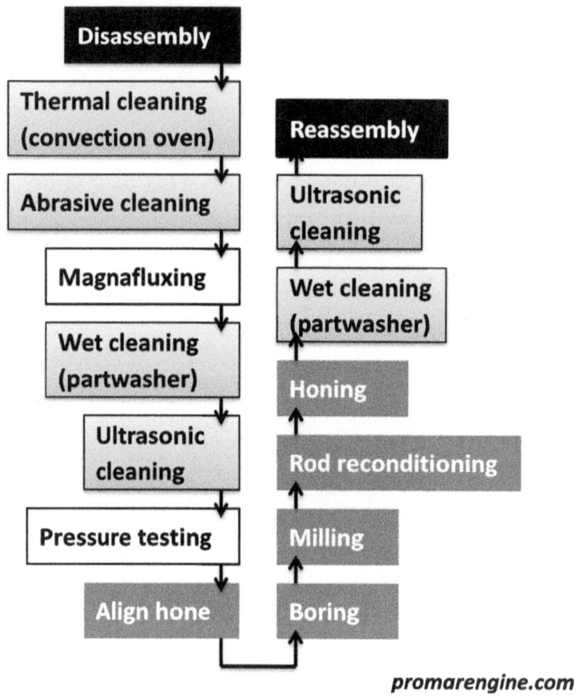

Figure 8 Remanufacturing process of an engine block [29].

and 14044 standards, a Life Cycle Assessment is carried out in four different phases, as listed below:

a) Definition of objective and scope: In this phase, a functional unit (which defines precisely what is being studied and quantifies the service delivered by the product system) of a product or service, the system boundaries, and any assumptions and limitations are defined. Normally, information research had been carried out beforehand to gather the information required.
b) Life cycle inventory: In this phase, all the inputs and outputs of a product or service are listed. The data must be related to the functional unit defined in the previous phase.
c) Impact assessment: The aim of this phase is to evaluate the significance of potential environmental impacts based on the Life cycle inventory (LCI) flow results. It consists of the selection of impact categories, category indicators, and characterization models.
d) Interpretation: The results from the inventory analysis and impact assessment are summarized during this phase. The outcome of the interpretation phase is a set of conclusions and recommendations for the study.

Definition of objective and scope

For this case study, the main objective was to test the environmental impact simulator in order to determine whether it performed the calculation or not. Once the results were obtained, the environmental impacts generated by several remanufacturing process scenarios were compared. Scenarios were created so that different input parameters could be filled in, making it possible to see the different impacts.

The functional unit used for this objective was the remanufacturing process of one engine block whose dimensions, weight and volume were those given in the previous section. This functional unit however cannot compare the scenario created in terms of functionality. For example, the same input weight has been used to perform the different types of material tests, even though their density and cleaning capacity were different. The geographical boundary of the system for this case study was that of the European Union. The study limitation was that only cleaning and reconditioning processes were included since it was assumed that the environmental impact of disassembly, inspection and sorting, and reassembly were negligible in comparison.

Life cycle inventory

Following the definition of the objective and scope, the life cycle inventory is provided in Table 2. The table shows the detailed process undergone by the engine block and the information keyed-in in the remanufacturing simulator. The 'step' column represents the process realized, while the 'simapro' column states the process name found in the remanufacturing simulator. The 'quantity' column indicates the estimated input quantity used in realizing the process. The 'remark' column is where the process is shortly described.

Once the result for the process in Table 2 is obtained, several tests using different scenarios were performed to determine whether the simulator was able to compare the environmental impacts. In every test that was performed, only one input was manipulated. Two remanufacturing scenarios are explained below and the results are shown in the later section.

Scenario 1: 'Abrasive cleaning' step: In this scenario, three different inputs, stainless steel, steel and zinc, were used as abrasive material.

Scenario 2: 'Ultrasonic cleaning' step: In this scenario, two different types of detergent are compared: 'aluminium-safe powder detergent, heavy contaminant' detergent and 'pH controlled and low foaming liquid detergent, heavy contaminant' detergent.

Table 2 Remanufacturing process inventory

Step	Simapro	Quantity	Remark
Thermal cleaning (convection oven)	Electric convection oven ED (compartment volume)	0.9 m^3	Oven with 1.8 m^3 chamber volume, divided into 2 as operation is in batches(two compartments per batch)
Abrasive cleaning	Stainless steel abrasive table blasting ED (part mass)	58 kg	Abrasive: Stainless steel
Wet cleaning (part washer)	Aluminium-safe powder detergent, heavy contaminant ED (part mass)	116 kg	Solution for aluminium, amount multiplied by 2 as two operations
Ultrasonic cleaning	Aluminium and soft metal safe ED (tank volume)	382 l	Solution for aluminium, amount multiplied by 2 as two operations
Milling	Milling, aluminium, dressing/ RER U	0.1 kg	Simapro data, assuming 100 g of removal material
Align honing + boring + rod conditioning + honing	Aluminium product manufacturing, average metal working/RER U	0.1 kg	Simapro data, assuming 100 g of removal material

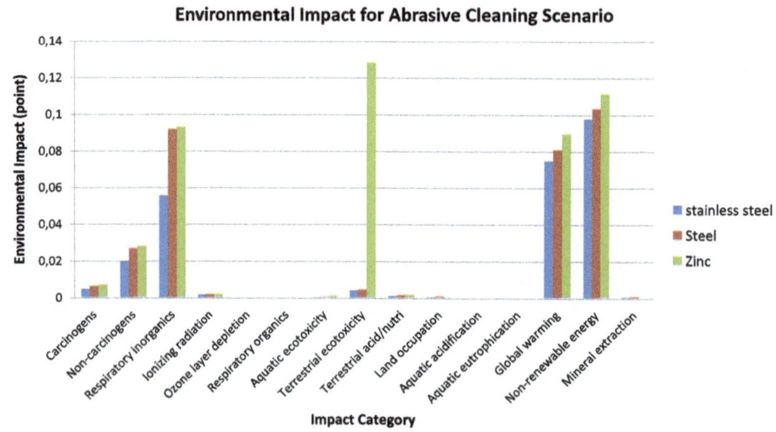

Figure 9 Result for abrasive cleaning scenario.

Results and discussion

Impact assessment

In this phase, the environmental impacts were assessed using the SimaPro software and the 'impact 2002+' [22] calculation method. The remanufacturing process library was introduced into the SimaPro process database. Figures 9 and 10 show some of the results for the different scenarios considered for the remanufacturing of the block engine.

The results showed the different environmental impacts for each scenario. The simulator performed the calculation and compared the environmental impact easily thanks to the creation of the remanufacturing process library.

Since the main objective of this case study was to test the simulator, it is not the authors' intention to interpret the results or give any recommendation concerning them. In order to do so, a further study must be carried out considering many aspects, such as the functionality, level of quality, and quantity used of each parameter chosen to carry out processes that can be compared in terms of functionality.

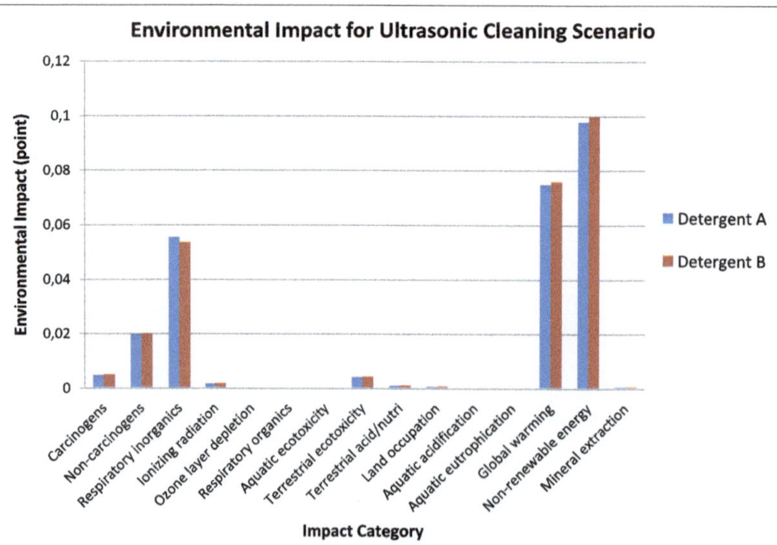

Figure 10 Result for ultrasonic cleaning scenario.

Conclusions and perspectives

The remanufacturing industry lacks tools to help designers choose their remanufacturing process. This study can help them to bridge this gap by providing an effective, comprehensive and simple tool that provides a global view of the remanufacturing process even in the design stage.

The generic data for a certain process was successfully synthesized in the remanufacturing process library and environmental impact simulator studies. Furthermore, the results from the engine block case study showed that the environmental impact simulator compares the remanufacturing scenarios well. With this result, this simulator could be used in the design phase as a design tool (in addition to classical design tool) in order to define the best product and remanufacturing scenario couple. It could help designers to design a remanufacturable product and its corresponding low-environmental impact remanufacturing processes.

The authors think that the outlook for improving this tool in the future is promising and that it could be used commercially. It will first be necessary to add data on different remanufacturing processes to the simulator database so that this simulator can serve the needs of designers more completely. Second, tests must be carried out to ensure that the results provided by the environmental impact simulator are useful for design activities. Such tests are important since the results of this study do not take this aspect into account.

Competing interests
The authors declare that they have no competing interests.

Authors' contributions
PZ defined the research aim and identified the problem. NHI collected data, carried out the literature review, conducted technical and environmental analyses and finally completed this manuscript. GM and PZ reviewed the technical and environmental analyses, participated in the discussion of the results and corrected the manuscript. All the authors read and approved the final manuscript.

Acknowledgements
Nurul Hanna Ismail sincerely appreciates the G-Scop Laboratory for the financial support during this Master thesis period.

References
1. Lopez Ontiveros, M: Intégration des contraintes de remanufacturabilité en conception de produits. *PhD Thesis*. Institut National Polytechnique de Grenoble (2004)
2. Ijomah, WL, Bennett, JP, Pearce, J: Remanufacturing: evidence of environmentally conscious business practice in the UK. In: Yoshikawa, H (eds.) Environmentally Conscious Design and Inverse Manufacturing, 1999. Proceedings. EcoDesign'99: First International Symposium On IEEE, Tokyo pp. 192–196. (1999)
3. Lund, R: Remanufacturing. In: Klein, J, Miller, J (eds.) The American Edge: Leveraging Manufacturing's Hidden Assets, pp. 228–229. McGraw-Hill, New York (1993)
4. Steinhilper, R: Remanufacturing. Fraunhofer IRB Verlag, Stuttgart (1998)
5. Amezquita, T, Hammond, R, Salazar, M, Bras, B: Characterizing the remanufacturability of engineering systems. In: Yoshimura, M, Eschenauer HA, Gilmore, B (eds.) ASME Advances in Design Automation Conference, 1995, American Society of Mechanical Engineersm, U.S. vol. 82, pp. 271–278, (1995)
6. Gray, C, Charter, M: Remanufacturing and product design. International Journal of Product Development **6**(3-4), 375–392 (2007)
7. Subramoniam, R, Huisingh, D, Chinnam, R: Remanufacturing for the automotive aftermarket strategic factors: literature review and future research needs. J. Clean. Prod. **17**, 1163–1174 (2009)
8. Subramoniama, R, Huisingh, D, Chinnam, R, Subramoniam, S: Remanufacturing decision-making framework (RDMF): research validation using the analytical hierarchical process. J. Clean. Prod. **40**, 212–220 (2013)
9. Lund, R: Remanufacturing: an American perspective. In: Responsive Manufacturing–Green Manufacturing (ICRM 2010), 5th International Conference, 11-13 Jan. 2010, pp. 1–6. (2010)
10. Hammond, R, Amezquita, T, Bras, B: Issues in the automotive parts remanufacturing industry: discussion of results from surveys performed among remanufacturers. Int. J. Eng. Des. Auto-Spec. Issue Environ. Conscious. Des. Manuf. **4**(1), 27–46 (1998)

11. Guide Jr, R, Daniel, V: Production planning and control for remanufacturing: industry practice and research needs. J. Oper. Manag. **18**, 467–483 (2000)
12. Kerr, W, Ryan, C: Eco-efficiency gains from remanufacturing a case study of photocopier remanufacturing at Fuji Xerox Australia. J. Clean. Prod. **9**, 75–81 (2001)
13. Lindahl, M, Sundin, E, Ostlin, J: Environmental issues within the remanufacturing industry. In: The 13th CIRP International Conference on Life Cycle Engineering. (2006)
14. Amaya, J, Zwolinski, P, Brissaud, D: Environmental benefits of parts remanufacturing: the truck injector case. In: Zhang, H-C (eds.) 17th CIRP International Conference on Life Cycle Engineering Proceedings, University of Technology Press, Hefei, China (2010)
15. Sundin, E, Lee, HM: In what way is remanufacturing good for the environment?. In: Matsumoto, M, Umeda, Y, Fukushige, S (eds.) Design for Innovative Value Towards a Sustainable Society, Springer, Netherlands pp. 552–557. (2012)
16. Online software REPRO2. http://repro2.inpg.fr/ang/indexa.php. Accessed 20 March 2013
17. Online software CLOEE. https://cloee.g-scop.grenoble-inp.fr/apps/WebObjects/CloeeApplication/. Accessed 20 March 2013
18. Sundin, E, Bras, B: Making functional sales environmentally and economically beneficial through product remanufacturing. J. Clean. Prod. **13**, 913–925 (2005)
19. Pieragostini, C, Mussati, M, Aguirre, P: On process optimization considering LCA methodology. J. Environ. Manag. **96**, 43–54 (2012)
20. Olsen, S, Christensen, F, Hauschild, M, Pedersen, F, Larsen, H, Torslov, J: Life cycle impact assessment and risk assessment of chemicals—a methodological comparison. Environ. Impact Assess. Rev. **21**, 385–404 (2001)
21. Simapro database manual. http://www.pre-sustainability.com/download/manuals/DatabaseManualMethods.pdf. Accessed 5 April 2013
22. Life cycle impact assessment methodology: Impact 2002+ user guide. http://www.quantis-intl.com/pdf/IMPACT2002_UserGuide_for_vQ2.21.pdf. Accessed 3 August 2013
23. Amaya Rivas, J: Evaluation des bénéfices environnementaux liés à des cycles de vie de produits en flux bouclés. *PhD Thesis*. Université de Grenoble (2012)
24. Supplier's website for ultrasonic cleaning chemical. http://www.ultrawave.co.uk. Accessed 3 January 2013
25. Supplier's website for abrasive cleaning media. www.kramerindustriesonline.com. Accessed 3 January 2013
26. Suppliers website for abrasive cleaning media. www.coldjet.com. Accessed 3 January 2013
27. Suppliers website for abrasive cleaning media. http://www.surfacepreparation.com. Accessed 3 January 2013
28. Photo of Ford engine block. http://www.autoblog.com/2009/06/13/ford-racing-introduces-new-boss-351-engine-block/. Accessed 3 October 2013
29. Remanufacturer's website for a remanufacturing process of an engine block. http://www.promarengine.com/engines.asp. Accessed 3 October 2013

3

Developing performance management systems for the green supply chain

Michael Mutingi[1,2*], Herbert Mapfaira[1] and Robert Monageng[1]

* Correspondence: mmutingi@gmail.com
[1]Mechanical Engineering Department, University of Botswana, P/Bag 0061 UB Post, Gaborone, Botswana
[2]Faculty of Engineering and the Built Environment, University of Johannesburg, Johannesburg, South Africa

Abstract

As 'green' issues continue to become a global concern in the manufacturing supply chain, developing appropriate performance measurement systems for specific supply chains is imperative. Various green supply chain management strategies have been proposed in different contexts. On the other hand, a number of performance management systems (PMS) have been proposed. However, given the variations in the contexts of the available green strategies and the performance measurement approaches, selecting or developing suitable performance measures and the ensuing PMS under a given supply chain context is not trivial. The purpose of this study is to develop a structured taxonomic approach to developing PMS under various green supply chain conditions, contexts, and business objectives. Therefore, we (i) explore extant empirical studies on green supply chain activities and environmental management, (ii) develop a taxonomy of green supply chain strategies, (iii) derive a structured approach to developing green performance management systems, and (iv) provide a taxonomic performance measurement framework consisting of environmental, economic, and social performance metrics. Unlike past studies, the taxonomic framework forms a practical platform to assist decision makers when developing a suitable set of performance measures and the ultimate PMS while considering the particular context of specific green strategies under which the PMS is supposed to operate.

Keywords: Green supply chain; Performance management systems; Performance measurement; Environmental management

Introduction

As environmental issues are rapidly becoming one of the most important topics in supply chain management, managers consider improvements in environmental or 'green' performance a basic competitive priority besides lower cost, short lead time, and high quality [1-3]. The ever-growing green concerns in the market place and the ensuing green movements have forced decision makers to manage their organizational performance from the ecological or environmental perspective. Business activities can have an impact on the natural environment through one or more of the following ways:

- Emissions to air, e.g., greenhouse gases, dust and particles, metal emissions to air, ozone depleting substances
- Emissions to water, e.g., waste water, metal emissions to water

- Emission to land, e.g., pesticides and fertilizers, metal emissions to land, radioactive waste
- Resource use, e.g., water use and abstraction, minerals, forestry

In various countries, regulatory bodies enforce recovery of product packaging [4]. A number of automotive industries have introduced product take back, recycling and re-use policies [5]. The introduction of green strategies for environmental management bears a significant impact on the choice and successful implementation of a performance measurement system. In this connection, it can be seen that linking specific green strategies to appropriate performance measurement systems is a serious challenge to most supply chain decision makers. Precisely, the implementation of green strategies requires changes in performance management in one or more of the following problem areas [1,6]:

- Procurement policies: New environmental performance requirements often demand significant modifications in materials and supply procedures;
- Product technology: The introduction of new environmental targets often demands new product structure in relation to design, components, and materials;
- Process technology: Environmental strategies often require new technologies, e.g., closed-loop and cleaner technologies with reduced environmental impact;
- Logistics and distribution: Environmental policies may require new ways of managing physical flows of goods and collaboration with other players in the supply chain.

Remarkable extant studies on the strategic importance of green issues and green performance management systems (PMS) are vast [1,2,7-10]. Most researchers and practitioners have focused their attention on two major study areas, namely (i) analysis of green strategies available to supply chain managers and ways of enhancing strategy formulation [1,2], and (ii) development of frameworks for assessing organizational environmental performance based on physical indicators [4], environmental costs [11], and life-cycle assessment methods [7]. Gunasekaran et al. [8] investigated issues concerned with current practices in supply chain performance measurement systems, providing insights into the future requirements in supply chain performance measurement. The main issues discussed include (i) the need for organization-wide and supply-chain-wide integration and coordination; (ii) the need for every metric to take a supply chain perspective in order to monitor performance effectively; (iii) the need for a common goal upon which each entity in the supply chain is measured and improved; and (iv) the need for designing additional new indicators, non-financial and financial, to cater for new developments in the supply chain context. As pointed out by Brewer and Speh [9], there are a number of concerns associated with the development and application of performance measurement systems across the entire supply chain. Some of these issues are outlined as follows [9]:

- Lack of understanding: Managers that are focused on internal systems find it difficult to understand the importance of multi-organizational performance measures or indicators;

- Lack of control: Most organizations and decision makers prefer to be evaluated on indicators they can control; without sufficient supply chain collaboration, the management and control of inter-organizational indicators is complex;
- Different goals and objectives: Organizations have different goals and objectives, therefore would argue on the selection of performance indicators;
- Lack of standardized performance indicators: In most supply chains, there are no agreed-upon performance measures in terms of units to use, structure, and format; and
- Difficulty in linking measures to stakeholder value: Linking measures to stakeholder or customer value and environmental issues is becoming more and more complex.

Besides the above-mentioned body of literature and the issues arising, there is continual research activity in the development and application of performance measurement and management systems, for both conventional and green supply chain systems. This demonstrates the ever-increasing importance of developing the right PMS system for the right supply chain. Surprisingly, most of these studies tend to propose these PMS from a general point of view [1,2,7]. On the contrary, the main benefit of this study is that it proposes green performance management frameworks that will be able to assist managers to know how to evaluate their organizational green performance taking into consideration both the constraints of their supply chain network, organizational pressures, and objectives of the business.

Also in spite of growing literature on these issues, little attention has been paid to the design and implementation of environmental performance measurement systems that take into account the specific context of green strategies, from a taxonomic view point. The concerns pointed out above stem from the fact that there are a number of green strategies and performance management tools and systems that can be applied in different contexts of the green supply chain strategies used. Moreover, different firms have different organizational objectives and capabilities. As such, the development of an appropriate performance system, given the right green strategy, is not a trivial task. In view of these issues, the design of a carefully defined framework for the development of effective performance measurement systems is imperative. Most of the frameworks in the green supply chain literature simply classify green strategies into reactive and proactive strategies, which is rather too broad [7,8,10].

Numerous extant studies dealt with traditional supply chain performance metrics without consideration of the environmental performance. The traditional supply chain only deals with indicators associated with cost, time, quality, and flexibility [8-10,12]. On the contrary, green supply chain performance management focuses on metrics concerned with green image and environmental issues [10]. With the advent of environmental concerns associated with manufacturing operations, the metrics and indicators for supply chain performance management have been expanded significantly. Given the wide scope of the possible metrics and the variety of supply chain strategies that can be implemented, a guiding framework is essential for effective understanding of performance measurement systems, specifically for the green supply chain.

In light of these issues, our approach seeks to provide a more detailed taxonomic analysis of performance measurement, deriving from our contributions on strategic green supply chain management (GSCM) and PMSs. As such, this study goes a step further to consider the context of specific green strategies and the appropriateness of specific performance measures. The study presents a taxonomic contingency framework which examines how different green strategies may be implemented and assessed in the context of specific PMSs. The specific objectives of this work are as follows:

(1) To perform a survey of case studies on green supply chain strategies and performance management systems;
(2) To develop a taxonomy of green strategies by analyzing the key drivers behind the adoption of specific green strategies;
(3) To develop an integrated framework for performance management for the green supply chain; and
(4) To suggest the most effective PMS metrics for the green supply chain, promoting understanding of green performance management systems.

The rest of the paper is organized as follows: The next section provides an overview of the research methodology and followed by the section which presents the results of literature search survey. An outline of green supply chain practices is presented in the fourth section. This is followed by a taxonomy of GSCM strategies in the fifth section. A framework for developing PMS is presented in the sixth section. Finally, concluding remarks and further research prospects are presented in the last section.

Research methodology

Green supply chain success indicators are quantifiers which are used in assessing the efficiency and effectiveness of green supply chain management practices. These indicators tend to vary between companies and industries, depending on their priorities and the focus of the performance criteria. The research methodology employed in this work considers existing empirical studies in the literature and seeks to coin a guiding platform for developing PMS in the context of specific green strategies. In this regard, the study consists of four phases, from literature survey through to framework development, as shown in Figure 1.

The first phase focuses on carrying out a literature search survey of real-world case studies on green practices and the implementation of relevant performance indicators. The aim is to figure out the drivers behind the implemented green practice and their influence on the performance indicators selected. The empirical studies were obtained through a wide search in academic studies, databases, and bibliographical lists so as to obtain the relevant information. Thus, the search was selected from published work in reputable journals, concerned with the implementation of green manufacturing strategies, such as Sustainability, Journal of Remanufacturing, International Journal of Production Economics, International Journal of Operations and Production Management, Supply Chain Management, Benchmarking: An International Journal, and Journal of Cleaner Production. The search criteria used included keywords such as 'green practices', 'green strategies', 'green supply chain', 'environmental issues', 'ecological', 'green supply chain performance measurement', 'performance indicators' and 'performance metrics'. The second phase highlights the critical areas of focus of each case study in order to determine the major driving elements behind the implementation of GSCM strategies. This would assist in answering questions as to why certain strategies are suitable for specific industrial contexts, and what was the performance measurement focus for each

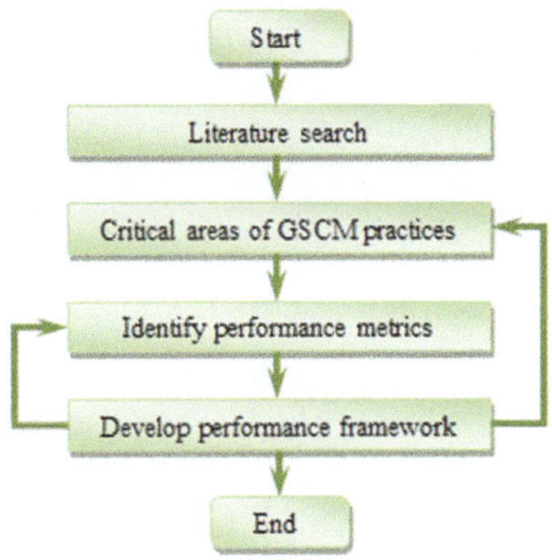

Figure 1 Research methodology.

industrial context. The third phase seeks to highlight green performance metrics chosen and the drivers behind the metrics chosen under different contexts of the case studies. The fourth and final phase aims to develop a generic framework for selecting the most appropriate PMS metrics for specific GSCM strategy.

In the next section, a summary of the outcome of the literature search survey is outlined.

Results of the search survey

The literature search survey indicated that some of the major contributors in the literature relevant to GSCM and PMS were Hart [13,14] and Porter and Van der Linde [15]. Other researchers in the area include Azzone and Noci [1], Kumar and Yamaoka [5], and Azzone et al. [6]. Following the literature search process, further perusal of selected publications also indicated that a number of organizations have embarked on introducing green practices. Some of these practices are green procurement, green product design, green production or manufacturing processes, green distribution, recycling and remanufacturing. For instance, Wal-Mart adopted green procurement of biodegradable and/or recyclable packaging. Automotive companies such as Toyota and Ford require ISO 14000 certification for their suppliers. Furthermore, it was realized that a number of firms have invested in recycling and reuse practices, for instance, Dell, Hewlett Packard, Toshiba and other electronics industries [16,17]. A number of legislations and regulatory systems are being put in place by governments and concerned stakeholders to account for environmental performance. In Western Europe, there is an obligation for 100% collection on 'white goods' [18]. The general acceptance of green activities has led to increasing empirical studies on the external and internal factors leading to the uptake of green practices and their impact on organizational performance [19]. Interesting results were also obtained with regard to the common trends in the development of performance measurement and management systems.

Table 1 provides a summary of the main areas of focus identified from the existing empirical studies investigated in this study. A closer look at the results of the literature search survey shows that there are three basic categories of areas of focus in green supply chains, namely (1) improving environmental performance, (2) improving economic performance, and (3) improving the firm's green image by promoting green initiatives that contribute towards the firm's social responsibility over the well-being of the society and its environment. The focus areas explain the need for developing performance indicators associated with each area. To measure, manage and control these areas of focus, it is crucial to establish appropriate performance indicators, and to put in place the necessary green strategies to enhance achievement of the desired goals within the respective areas. In retrospect, we can somewhat conveniently classify the performance indicators into the following:

(1) Environmental performance
(2) Economic or financial performance
(3) Social performance

Further explanations to the performance indicators in each of the above-mentioned category will be discussed in the 'Developing green PMS metrics' section. However, it is important to take a closer look at the green practices behind the focus areas.

Green supply chain management practices

In as much as several researchers have investigated the concepts of greening the environment at strategic and operational levels, many related researchers have studied greening practices, such as green product design [39,40], green process design [15,41], green purchasing [42] and green manufacturing practices at large [16,43,44]. From these greening concepts, the environmentally conscious practices, and the motive for organizational competitiveness, GSCM can be defined as follows:

$$\text{GSCM} = \text{Green Purchasing} + \text{Green Manufacturing} + \text{Green Distribution} + \text{Green Reverse Logistics}$$

Figure 2 shows the green practices in a closed loop green supply chain. Supply chain players such as used parts warehouses/storages, waste collectors, recycle plants, final

Table 1 Identified empirical studies on the GSCM performance

Area of focus	Performance metrics	Reference source
1. Environmental performance	Air emissions	[1,20-24]
	Waste water or water pollution	[7,8,11,21,22,24-26]
	Solid waste or hazardous materials	[21,22,27-32]
	Energy usage or consumption	[21,22,29,33,34]
2. Economic performance	Reverse logistics costs	[24,35-38]
	Sustainability costs	[8,11,23,26,37]
	Energy consumption costs	
3. Social responsibility performance	Green image	[8,23,26,27,29,30]
	Customer perspective	
	Percent recycling	[8,11,17,23,26,27,30,32,36]
	Scrap rate	

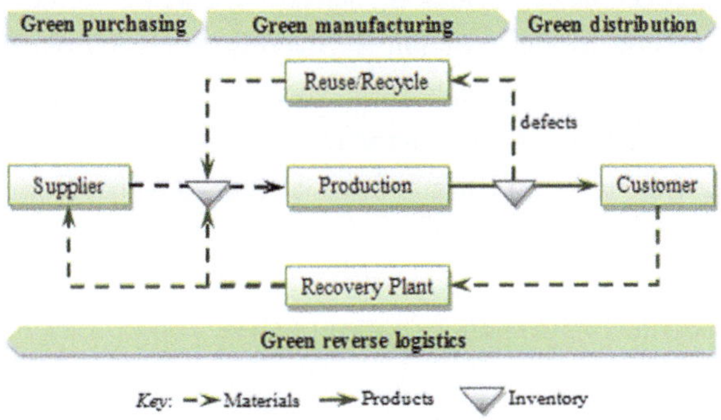

Figure 2 Activities in green closed-loop green supply chain.

treatment (landfill) and other stakeholders are involved in carrying out these green activities.

Green purchasing practices

These are concerned with purchasing environmentally friendly materials and products, certification of potential suppliers, total quality environmental management (pollution prevention, internal performance measurement and auditing), environmentally friendly packaging and distribution [42]. Material recovery, reuse, and recycling are put in place so that all purchasing procedures satisfy environmental quality standards [45-47].

Green manufacturing practices

Green manufacturing defines efficient product design and production processes that use inputs with low environmental impacts, capable of producing minimal or no waste nor pollution [39,40]. Green product design ensures that the number of components in a product is minimized, assembly and disassembly processes are simplified, and material recovery, reuse and recycling are enhanced. In addition, green manufacturing consists of environmentally friendly processes that minimize raw material usage, energy usage, wastes, and emissions. The end result of green (re)manufacturing is reduced material costs, energy costs, occupational safety expenses, as well as improved production efficiency and corporate image.

Green distribution practices

Green distribution involves the application of green packaging practices that encourage the use of standard downsized packaging, recyclable materials, returnable packaging methods, and promotion programs on recycling and reuse [46,47]. In addition, green distribution encourages the use of energy-efficient distribution methods that minimize energy usage, carbon emissions, and transportation costs.

Green reverse logistics

Green reverse logistics is the process of retrieving the product materials or parts from the end user (consumer) in order to recapture value or dispose the materials in an environmentally friendly manner [46-48]. Green reverse logistics include activities such as waste collection, parts collection, inspection, selection, sorting, direct recovery, reprocessing, redistribution, and disposal [49,50]. Recovering products, refurbishing goods, and salvaging parts such as precious metals that can be recycled or reused can bring a huge benefit to the environment and to the manufacturer [50,51].

In light of the above, the choice and the success of these green practices depend on the strategies and capabilities of an organization. We present a taxonomic analysis of green strategies in the next section.

GSCM strategies

The adoption of green strategies has a far-reaching and multi-dimensional impact on performance [1]. In particular, the introduction of GSCM strategies may result in (i) management complexities due to changes in logistics for product recovery and recycling, leading to (ii) multiple changes in performance measurement, and in the management of operations at large. It is therefore imperative to identify the conditions under which these strategies are best implemented, and to analyze the performance measurement systems in light of the specific green strategies [45]. A question then arises as to what influences the selection and implementation of green strategies.

Drivers behind green supply chain strategies

The environmental awareness of an organization will influence the perspective from which green strategies are developed [45]. In the absence of environmental awareness, an organization adopts a naive perspective where the enterprise takes no initiatives towards improving environmental performance. In this work, two categories of GSCM drivers are identified: internal and external drivers. Internal drivers are as follows:

- Eco-efficiency perspective: Green strategies are developed with the objective of reducing wastes and resource usage, while maximizing environmental performance [52];
- Competitive advantage perspective: Green strategies are used as means of attaining a competitive advantage through product and process technology innovation [53];
- Ecological and ethical perspective: Green strategies are driven by the desire to reduce the industrial impact on the natural environment, rather than by the desire to improve on economic performance [51,54].

Various researchers have identified total quality control, cost control, communities and environmental regulations as the main drivers behind an organization's environmental perspective [45-48,55,56]. However, from a broader view, the influence of such factors depends on a combination of factors at industry, country and even global levels. In addition to the above-mentioned internal drivers, we further identify the following external drivers:

- Compliance perspective: The need to comply with current environmental regulations forces adoption of green strategies;

- Supply chain relationship: Supply chain collaboration and relationship will always influence the company's environmental behavior. For example, automotive companies require ISO 14000 certification for their suppliers.
- Green pressures: 'Green' movements arising from customers and other stakeholders lead to a widespread adoption of green strategies across the supply chain.

The drivers (internal and external) highlighted above are vital for the analysis of green supply chain strategies. In our investigation, it was observed that the selection of green strategies depends largely on a combination of internal and external drivers. For instance, a firm which is highly concerned with ecological and ethical responsibilities, located in a collaborative supply chain environment tends to choose closed-loop supply chain strategies (see Figure 3).

The green supply chain strategies

In view of the above internal and external drivers, we distinguish four major categories of GSCM strategies:

(1) Compliance-cantered,
(2) Lean-based,
(3) Innovation-cantered, and
(4) Closed-loop strategies.

Of the four green strategies, compliance-centered is the most basic with the closed-loop strategy being the most complex. The adoption of the compliance-centered strategy is mostly due to external pressures while the implementation of the remaining three is mostly due to internal drivers. Organizations can thus start adoption of GSCM practices starting from the compliance-centered strategy and moving on to adopt the remaining strategies in turn until finally adopting the closed-loop strategy. Should any organization choose to fully implement a higher level green strategy, some of the green practices pertaining to lower strategies have to be implemented, as the starting point. In terms of performance measurement of each strategy, the higher level strategies contain all the performance indicators from the lower level strategies. In other words, adoption of the lean-based strategy allows the firm to also meet the requirements of the compliance-centered strategy.

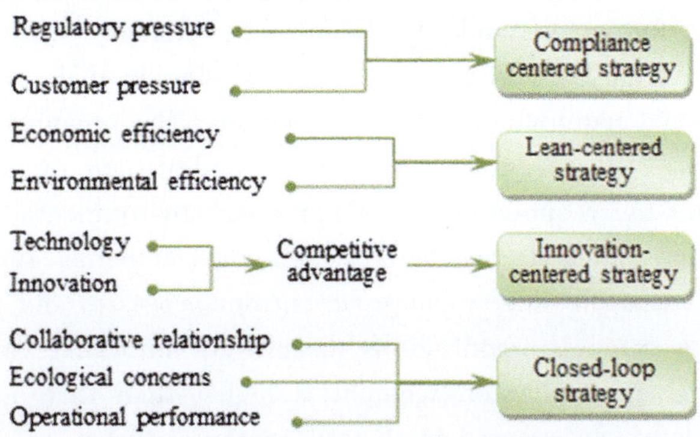

Figure 3 **The main drivers behind closed loop supply chain strategies.**

Compliance-centered strategy

This strategy is adopted in response to environmental regulations, pressures from customers and other stakeholders such as government and green pressure groups [57]. Companies may be obliged to adopt international standard systems such as ISO 14001 [50], and purchasing contracts with suppliers aimed at meeting certain regulatory requirements. The strategy is mainly adopted by companies whose main objective is to comply with the requirements of their stakeholders, as a way of getting permission to do business (regulatory compliance) and qualifying to do business with relevant organizations (inter-organizational compliance). Companies adopting compliance-centered strategy are reactive to events and tend to view changes needed for compliance as a burden. Typically, these companies' investment to the greening of the supply chain is minimal [48]. The compliance-centered strategy is the most basic of the GSCM strategies and it is therefore the natural starting point for GSCM strategy implementation, and it requires minimal customer-supplier engagement.

Lean-based strategy

This is a more complex GSCM strategy normally adopted by organizations that focus on eco-efficiency or lean and green approach to GSCM. With this strategy, additional to meeting the requirements of the compliance-centered strategy, suppliers are required to satisfy certain operation-based efficiency targets. The strategy thus links environmental performance with operational efficiency through waste minimization and optimal resource usage [19,44]. The environmental performance benefits from this strategy arise from the adoption of specific manufacturing practices which are focused on resource use reduction and waste minimization, and in so doing also provide secondary environmental performance benefits. The lean-based strategy requires higher levels of customer-supplier engagement and it ties environmental performance to operational processes in the supply chain. To ensure economic benefits, all firms participating in the supply chain have to adopt lean-based operational process. This strategy is widely implemented within the automotive industry where automakers expect their suppliers to adopt sustainable lean-based operational processes, which are focused on waste minimization and resource use reduction.

Innovation-centered strategy

This is adopted by companies whose objective is to keep up-to-date with changes in environmental regulations through specialized technologies, product designs, processes and strict green performance standards [16,50]. The strategy requires organizations to adopt a product life-cycle perspective when developing their products, and give stricter environmental requirements to their suppliers. The organizations adopting this strategy routinely support their suppliers to modify their operational processes in order to meet the requirements of the newest environmental regulations. Participating organizations integrate specific relevant green activities, such as green design and green purchasing in order to improve current processes and product development. The strategy compels organizations to cultivate innovative capabilities to GSCM practices. Customer-supplier engagement is higher than that in lean-based strategy. The strategy is widely adopted in the automotive industry, an example being Chrysler Corporation [58].

Closed-loop strategy

This strategy is adopted by firms in highly collaborative supply chains where companies can keep abreast with complex requirements of closed-loop supply chain and to integrate with suppliers, from design to product take-back [59]. Material recovery, recycling and remanufacturing improve environmental performance of the entire supply chain [24,36]. This strategy is the most complex and collaborative type of GSCM strategy. It targets eco-efficiency achieved by the management of all processes in the supply chain that focuses on customer service level, which covers from raw materials to product delivery to customers, through design, procurement, production, distribution, use, reuse and disposal, product take-back, reverse logistics, including aspects such as reduction of time and resource use, costs and environmental impacts.

The strategy thus links the environmental performance to the entire supply chain activities and allows the integration of business goals pertaining to economic and operational performance, social performance and environmental performance. Organizations adopting this strategy require high levels of control over the capture and return of used materials. Of the four GSCM strategies, the closed-loop strategy requires the highest level of customer-supplier engagement. An example of companies adopting this strategy is Toyota Motor Corporation and its suppliers [60]. As an example, to increase its vehicle recycling performance, Toyota formulated the Toyota Recycling Vision in 2003, whose target is to achieve a vehicle recovery rate of 95% by 2015. The company achieved the target ahead of time in 2007, and went on to achieve a vehicle recycling rate of 99% in 2011 [61].

Developing PMS for the green supply chain

To develop an effective green PMS for the green supply chain, the decision maker needs a guide or framework for developing the PMS and for identifying appropriate green performance indicators. There are four basic requirements that should be considered when developing an effective performance measurement system that can support green supply chain performance [1]:

- Strategic focus: the ability of the PMS system to account for the long-term impacts;
- Measurability: the ease of measurement, to enable assessment of green efficiency;
- Completeness: ability of the PMS to account for all relevant performance indicators;
- Timeliness: the amount of time taken by the PMS to collect and analyze the performance related data.

To effectively meet these requirements, there are critical areas of concern which supply chain decision makers should focus on. As identified from various empirical studies in the literature, critical success factors for every PMS in a green supply chain include (i) top management commitment, (ii) the availability of diagnosis tools, (iii) the ability to formulate an appropriate action plan deriving from the findings of the diagnosis, (iv) the availability of suitable performance measurement indicators, and (v) the availability of a performance management system for performance monitoring and improvement. The commitment of top management is imperative; this essentially calls for an establishment of a performance management function consisting of members in the top management, not forgetting the concerned operational managers. In addition, the availability of appropriate diagnosis tools is essential for quantitative and qualitative measurement of the supply chain performance. Deriving from the findings from diagnosis and analysis activities, appropriate action plans can then be formulated. However, to enable the supply chain decision maker to evaluate and monitor performance and effectiveness of the action plans taken, the availability of a set of efficient and effective performance measures/indicators is crucial. Above all, the availability of a performance management system is critical in order to monitor and improve the performance of the green supply chain. Based on these observations, a green performance management framework can be formulated.

PMS development framework

Figure 4 provides a framework for the development of green performance management systems, consisting of five phases: (1) developing a performance management function, (2) diagnosis and analysis, (3) developing an action plan, (4) developing a performance measurement system, and (5) developing a performance management system. These are explained in the following sections.

Phase 1: GSCM performance function

Proper management of the performance of the green supply chain requires commitment and focus from top management. A dedicated organizational structure must be set up to facilitate the performance improvement effort. Depending on the supply chain's needs, size, and characteristics, this structure can be in two basic forms: (a) a performance management unit, headed by an environmental performance manager who reports directly to senior management or (b) a cross-functional team comprising performance coordinators appointed from the concerned operational units.

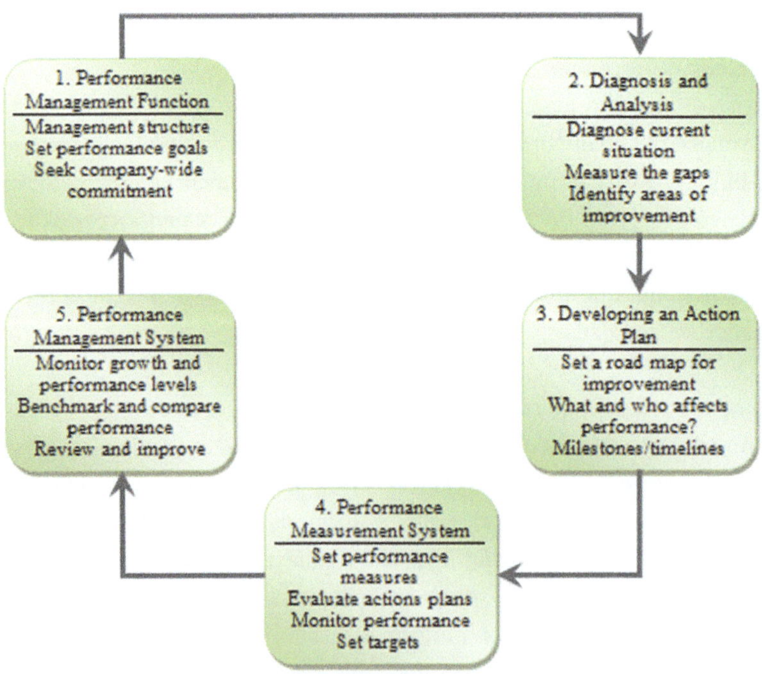

Figure 4 Green performance management framework.

Phase 2: diagnosis and analysis

For any green performance-related intervention to be effective, one should have a thorough understanding of the current situation of the supply chain. This is done through a performance diagnosis to assess the supply chain's position, pinpoint its strengths and weaknesses, and recommend areas for improvement. A green performance diagnosis covers (a) a qualitative assessment of the supply chain performance in relation to the green levers and (b) a quantitative assessment of the supply chain performance based on certain key indicators or measures that are linked to the various selected green levers.

Phase 3: developing an action plan

After completion of the diagnosis step, the next step is to develop an action plan or a road map based on the findings obtained. The road map helps to direct specific activities towards the desired green performance goals in a systematic manner. The performance action plan is expected to address the following questions: What affects green performance? Who affects green performance? When should the activities be undertaken?

Phase 4: performance measurement system

Performance improvement initiatives must be complemented by a sound measurement system. The measurement system forms an integral part of the supply chain's performance management system. A sound measurement system consists of performance indicators that are useful for (i) monitoring environmental, economic, and social performance, (ii) setting performance targets, and (iii) developing appropriate supply chain strategies, or improving existing strategies.

Phase 5: performance management system

A performance measure informs the supply chain decision maker on how and why the supply chain is performing at a specific rate, and what it should do in view of its current performance. The next step is to use these indicators to manage green performance. Overall, performance management covers two main areas: (a) activities to monitor performance and (b) activities to reinforce performance across the supply chain. However, it is important to emphasize that the effectiveness of these activities hinge on the underlying system of performance metrics developed in the previous phase. As such, the next section deliberates on developing a system of green performance metrics.

Developing green PMS metrics

It has been noted that, when developing a system of green performance indicators, one should take into account the critical requirements, including [1] (i) strategic focus, the ability of the system of indicators to account for long-term investment decisions and their impacts; (ii) measurability, the ease with which measurement and assessment of green efficiency can be done; (iii) completeness, the ability of the performance system to account for all the relevant performance indicators for effective assessment of the firm's PMS; (iv) timeliness, so as to enable evaluation of the time taken by the PMS to analyze the collected data; and (v) cost of analysis, concerned with the cost of the resources needed to implement the PMS chosen.

Based on the characteristics of specific green strategies outlined in the taxonomy, PMS models can be developed to enhance the strategic focus and its measurability, while considering a judicious trade-off between completeness, timeliness, and cost of the performance measurement activities.

Environmental performance indicators (measures) are the core requirements of a PMS for effective evaluation of the environmental performance of processes, activities, and services of the supply chain. The taxonomy of green strategies assists the supply chain managers in identifying the most effective performance measures and in designing the most suitable PMS structure. Performance indicators range from air emissions to energy recovery and recycling. Possible PMS designs include (i) models which use physical indicators to describe environmental impact resulting from the strategy adopted, (ii) models based on economic indices linked with economic efficiency, (iii) models based on life-cycle analysis methods indicating the economic and physical impact of a given strategy throughout the product life cycle, and (iv) models based on an integrated perspective that indicates the effect of the adopted green strategy on shareholders' value, physical indices, and economic indices. The following section analyses performance metrics for the green strategies identified in this study [4,5,11]. To assess the impact of industrial processes, three categories of indicators are utilized:

(1) Physical indicators - to measure, in physical terms, how the supply chain activities affect the natural environment, or
(2) Economic or financial indicators - to measure the variation of the life-cycle costs associated with the product in question, or
(3) Social indicators - to measure the green image of a firm based on customer or stakeholder's evaluation.

In the most ideal case, effective PMSs should basically include both physical and economic indicators in an integrated framework. While physical indicators provide a timely assessment of the activities of the supply chain, economic or financial indicators indicate how the adopted supply chain activities affect the individual organization's profitability [4]. Therefore, an efficient PMS supporting the implementation of a green strategy should consider the physical environmental indicators, which express the organization's environmental performance. These indicators can be aggregated into the following categories [5]:

- Volume indices: These assess the amount of solid wastes resulting from a firm's production operations, such as percentage of scrap, and percentage of recyclable products;
- Process efficiency indices: These assess the environmental efficiency of a company's operations regarding waste water, air emissions and energy consumption.
- Design efficiency indices: These measure the 'green' design features defining the 'green' products, such as assembly time, number of materials or parts in a product, and number of levels in the bill of materials.

In order to monitor the firm's contribution to economic or financial value creation in the supply chain, a set of carefully selected economic features need to be considered. These features can be classified into the following categories:

- Operational efficiency: costs of green manufacturing operations, materials, (non) manufacturing overheads, direct labor, and environmental compliance audits;
- Distribution costs: costs due to transportation, waste disposal and product take-back;
- Price of the product as well as demand for the product.

Suggested indicators and their descriptions are provided in the next sections.

Indicators for compliance-based strategies

The compliance-based strategy normally arises from the introduction of environmental regulations, governmental regulations, and extreme customer pressure [57]. The strategy seeks to identify the key areas of concern in line with the organization's capabilities, leading to the adoption of operations that ensure that the organization meets the required environmental performance. In this connection, the adopted PMS should offer the following:

- Timely measurement of physical indices to highlight the firm's performance;
- Accurate monitoring of physical indices in order to monitor the trend of the performance

Air emissions, waste water, solid wastes and energy consumption indicators should be included in the PMS in accordance with regulations. Table 2 shows the suggested indicators and their descriptions. For environmental performance assessment, the most effective way is to benchmark the current performance with the regulation standards. When selecting among different potential solutions or programs, an aggregate indicator, derived from the above indicators, can be used to evaluate the solutions. Thus, four indicators derived from these indicators can be used to describe the trends in environmental performance. Specialized indices can be aggregated into these four basic categories.

The identified indicators can be used to evaluate a decrease of various costs: (a) costs associated with material usage, energy usage, raw material usage and (b) costs associated with fees paid for waste discharge, waste treatment, including transportation and distribution. These are concerned with process efficiency indices.

The compliance-based PMS is most suitable to organizations that choose to follow a passive approach to environmental performance with the objective of introducing green concepts in order to meet customer pressure, or to comply with some limiting

Table 2 Compliance-based performance indicators

No.	Indicator	Description
1	Air emissions	Percentage of gas emissions into the air, e.g., chlorinated pollutants
2	Waste water	Percentage of pollutants in water, e.g., total nitrogen
3	Solid waste	Percentage of materials sent for disposal
4	Energy	Amount of energy consumed, e.g., electricity, oil

environmental regulations. One notable negative impact of this perspective is that it offers limited competitive advantage in the presence of aggressive innovations in environmental performance. Due to its reactive nature, the passive approach often lags behind the target performance required, resulting in loss of market share. In this regard, timeliness of the PMS is crucial, especially in the presence of fast-changing regulations and customer pressures.

Indicators for lean-based strategies

Lean-based strategies go beyond regulatory compliance through the requirement for the suppliers to meet operation-based targets. This strategy maximizes on economic performance while simultaneously providing secondary environmental performance benefits through waste and resource use reductions. The main objective of this strategy is to gain dual environmental and economic performance benefits. Therefore, when developing performance indicators for lean-based strategies, the following categories of measures should be included, if the strategies are to succeed [53,54]:

- Physical indicators and
- Economic indicators

The physical indicators correspond to the compliance-based indicators listed in Table 2 in the previous section. In addition to this, lean-based economic performance indicators suggested in this study are summarized in Table 3.

Indicators for innovation-based strategies

The innovation-based strategies often lead to the introduction of new and complex green product and process technologies. These complex initiatives need complete and timely assessment techniques, usually with less attention on cost disadvantages, if any. Innovation-centered indicators seek to measure such factors as life cycle of product components, and number of assemblies or number of parts in a product. In this view, the set of performance indicators suitable for this strategy should include the following categories:

- Physical indicators,
- Economic indicators, and
- Innovation-specific indices

Further to the indicators defined in Tables 2 and 3, the PMS should also include indicators concerned with product life, product green efficiency, and green image. Indicators

Table 3 Lean-based economic indicators

No.	Indicator	Description
1	Material costs	Decrease of materials purchasing costs
2	Energy	Decrease of energy consumption costs
3	Inventory costs	Decrease of inventory (storage) costs
4	Waste treatment	Decrease of fees paid for waste treatment
5	Waste discharge	Decrease of fees paid for waste discharge
6	Transportation	Decrease of transportation related costs

defining product life show how long a product, or its components, can be used. This evaluation is crucial for long-term forecasting of energy and material usage. Table 4 presents a summary of innovation-based indicators.

Indicators for closed-loop strategies

Performance indicators for the closed-loop supply chain range from air emissions to energy recovery and recycling indices. Possible designs include (i) physical indicators to describe environmental impact resulting from the green manufacturing activities, (ii) economic indices linked with economic efficiency, (iii) life cycle analysis methods indicating the economic and physical impact of the supply chain throughout the product life cycle.

In a collaborative supply chain environment characterized by integrated relationships, innovative initiatives such as product take-back and take-back of recyclable end-of-life materials can easily be supported through involvement of suppliers [62]. In this vein, life cycle assessment is an appropriate model that should be implemented. This enables the organization to measure the environmental impacts of the product over its entire lifecycle, right from collection of raw materials and design to recovery and/or disposal. The life cycle analysis model can be used to assess the impact of the industrial processes using: physical, economic, and social indicators. Hence, in addition to the compliance-based indicators (Table 2), the lean-based indicators (Table 3), and the innovation-based indicators (Table 4), the closed-loop strategy should include the social or ecological performance indicators as well.

Table 5 provides a list of suggested indicators for measuring social performance of a supply chain. These indices include green image, which is a measure of the customers' perspective due to the firm's green practices such as product take-back, recycling, and product recovery. Such activities have a positive impact on the firm's green image from the perspective of the society.

Conclusions and further research

In every supply chain context, with specific green strategies, it is essential to select or develop appropriate PMS, if the system is to be effective. This research proposed a structured approach to guide supply chain managers who intend to develop effective PMS for specific GSCM strategies.

Managerial implications

The motivation for the adoption of GSCM practices varies across different organizations. Most organizations adopt GSCM practices due to external pressures, and in most cases, the level of adoption is at the baseline level, that is, they only meet the minimum requirements [63]. This problem is mostly brought about by the difficulty in justifying

Table 4 Innovation-based performance indicators

No.	Indicator	Description
1	Product Life	Life cycle of different product components or sub-assemblies
2	Product green efficiency	Increase of green design features in a product; number of parts, number of materials, etc
3	Green image	Increase in customer goodwill due to addition of innovative greening activities

Table 5 Indicators for social performance

No.	Indicator	Description
1	Percent recycling	Increase in recycled material compared to material disposal
2	Product green efficiency	Increase of green design features in a product; number of parts, etc
3	Green image	Increase in customer goodwill due to greening activities

the economic benefits for the adoption of GSCM practices [29]. Due to a lack of robust performance measurement systems, it is hard to quantify the tangible benefits derived from the adoption of different GSCM practices. This study provides a systematic way of developing performance indicators to allow the consistent quantification of the benefits of adoption of different GSCM practices, across the four green strategies. The taxonomic approach and the measurement techniques presented in this study provide operational guidelines for devising a set of performance indicators that are suitable for the selected green strategy. In this regard, the approach is applicable across various supply chain domains. In addition, the framework can be used to evaluate the possible available green options when changing from one green strategy to the other. For instance, when changing from compliance-centered to innovation-centered strategy, new indicators that pertain to the contribution of the enterprise's innovative product and process technologies should be designed.

Since the adoption of higher level green strategies requires that an organization satisfactorily meets the requirements of lower level strategies, the GSCM framework can be used as a basis of a maturity model for GSCM practices. Organizations can thus use the model to assess their level of adoption of GSCM practices. For example, are they at the stage of compliance-centered or lean-based?

The GSCM framework, through the four strategies, also provides a basis for benchmarking performance against other firms, and enhances inter-organizational relationship [62,63]. Additionally, the framework can be used as a basis for supplier selection - a firm can decide to work with organizations at a specific level of GSCM strategy adoption, e.g. lean-based strategy.

Further research

Further research directions include the validation of the application of green performance indicators for each PMS identified in this research. This will help to establish the validity of the indicators or the metrics for green supply chain performance measurement across different categories of green strategies. Furthermore, new comprehensive or hybrid indicators may be designed for integrated assessment of both environmental and economic performance.

Competing interests

The authors declare that they have no competing interests.

Authors' contributions

MM contributed in developing the concept of context-specific performance measurement systems under various supply conditions and objectives, derived the structured approach to developing green performance management systems, and made a draft of the paper. HM contributed in further analysis of case studies and the performance metrics used in specific contexts. RM contributed in further literature search survey, critical revision of the paper for practical relevance. All authors read and approved the final manuscript.

Acknowledgements

The authors would like to acknowledge the anonymous reviewers for their comments on the earlier version of the paper.

References

1. Azzone, G, Noci, G: Identifying effective PMSs for the deployment of "green" manufacturing strategies. Int. J. Operations Produc. Manag. **18**(4), 308–335 (1998)
2. Azzone, G, Bertele, U: Exploiting green strategies for competitive advantage. Long Range Plan. **27**, 69–81 (1994)
3. Bloom, GS, Morton, MS: Hazardous waste is every managers' problem. Sloan Manage. Rev., 75–84 (1991).
4. Noci, G: Supporting decision making for recycling based investments. Bus. Strateg. Environ. **4**(2), 62–71 (1995)
5. Kumar, S, Yamaoka, T: System dynamics study of the Japanese automotive industry closed loop supply chain. J. Manuf. Technol. Manag. **1**(2), 115–138 (2007)
6. Azzone, G, Bianchi, R, Mauri, R, Noci, G: Defining operating environmental strategies: programmes and plans within Italian industries. Environ. Manag. Health **8**(1), 1–15 (1997)
7. Hervani, AA, Helms, MM, Sarkis, J: Performance measurement for green supply chain management. Benchmarking Int. J. **12**(4), 330–353 (2005)
8. Gunasekaran, A, Patel, C, Tirtiroglu, E: Performance measures and metrics in a supply chain. Int. J. Operations Product. Manag. **21**, 71–87 (2001)
9. Brewer, PC, Speh, TW: Adapting the balanced scorecard to supply chain management. Supply Chain Manage Rev. **5**(2), 48 (2001)
10. Beamon, BM: Measuring supply chain performance. Int. J. Operations Product Manag. **19**, 275–292 (1999)
11. Gray, R: Accounting and environmentalism: an exploration of the challenge of gently accounting for accountability, transparency and sustainability. Acc. Organ. Soc. **17**(5), 399–425 (1992)
12. Chan, FTS, Qi, HJ: An innovative performance measurement method for supply chain management. Supply Chain Manag. **8**, 209–223 (2003)
13. Hart, SL: A natural-resource-based view of the firm. Acad. Manag. Rev. **20**(4), 986–1014 (1995)
14. Hart, SL: Beyond greening: strategies for a sustainable world. Harv. Bus. Rev. **75**(1), 66–76 (1997)
15. Porter, M, Van Der Linde, C: Green and competitive: ending the stalemate. Harv. Bus. Rev. **28**(6), 120–134 (1995)
16. Hu, AH, Hsu, CW: Empirical study in the critical factors of green supply chain management (GSCM) practice in the Taiwanese electrical and electronics industries. IEEE Int. Conf. Manage. Innov. Tech., 853–857 (2006)
17. Hu, AH, Hsu, CW: Critical factors for implementing green supply chain management practice-an empirical study of electrical and electronics industries in Taiwan. Manage. Res. Rev. **33**(6), 586–608 (2010)
18. Vlachos, D, Gaorrgiadis, P, Iakovou, E: A system dynamics model for dynamic capacity planning of remanufacturing in closed-loop supply chains. Comput. Oper. Res. **34**, 367–394 (2007)
19. Simpson, D, Samson, D: Developing strategies for green supply chain management. Decision line. **39**(4), 12–15 (2008)
20. Agarwal, G, Vijayvargy, L: Green supplier assessment in environmentally responsive supply chains through analytical network process. Proceed. Int. Multi. Conference Eng. Computer Sci. Hong Kong. **2**, 14–16 (2012). 1-6
21. Kainuma, Y, Tawara, N: A multiple attribute utility theory approach to lean and green supply chain management. Int. J. Prod. Econ. **101**(1), 99–108 (2006)
22. Zhu, Q, Sarkis, J, Lai, K: Confirmation of a measurement model for green supply chain management practices implementation. Int J Product Econ **111**, 261–273 (2008)
23. Zhu, Q, Sarkis, J, Lai, K: Initiatives and outcomes of green supply chain management implementation by Chinese manufacturers. J. Environ. Manag. **85**, 179–189 (2007)
24. Olugu, EU, Wong, KY, Shaharoun, AM: A comprehensive approach in assessing the performance of an automobile closed-loop supply chain. Sustainability **2**, 871–889 (2010)
25. Chia, A, Goh, M, Hum, S: Performance measurement in supply chain entities: balanced scorecard perspective. Benchmarking Int. J. **16**(5), 605–620 (2009)
26. Tsoulfas, GT, Pappis, CP: A model for supply chains environmental performance analysis and decision making. J. Clean. Prod. **16**, 1647–1657 (2008)
27. Olugu, EU, Wong, KY, Shaharoun, AM: Development of key performance measures for the automobile green supply chain. Resour. Conserv. Recycl. **55**(6), 567–579 (2011)
28. Eltayeb, TK, Zailani, S, Ramayah, T: Green supply chain initiatives among certified companies in Malaysia and environmental sustainability: Investigating the outcomes. Resour. Conserv. Recycl. **55**, 495–506 (2011)
29. Li, Y: Research on the performance measurement of green supply chain management in China. Int. J. Sustain. Dev. **4**(3), 101–107 (2011)
30. Rao, P, Holt, D: Do green supply chains lead to competitiveness and economic performance? Int. J. Operations Product Manage. **25**, 898–916 (2005)
31. Rao, P: Greening of the supply chain: a new initiative in South East Asia. Int. J. Operations Product Manage. **22**, 632–655 (2002)
32. McIntyre, K, Smith, H, Henham, A, Pretlove, J: Environmental performance indicators for integrated supply chains: the case of Xerox Ltd. Supply Chain Manag. **3**, 149–156 (1998)
33. Wu, J, Dunn, S, Forman, H: A study on green supply chain management practices among large global corporations. J. Supply. Chain Operations Manag. **10**(1), 182–194 (2012)
34. Wang, HF, Gupta, SM: Green supply chain management-A product life cycle approach. McGraw-Hill Prof. Med/Tech., (2011)
35. Choudhary, M, Seth, N: Integration of green practices in supply chain environment - the practices of inbound, operational, outbound and reverse logistics. Int. J. Eng. Sci. Technol. **3**(6), 4995–4993 (2011)
36. Mondragon, AEC, Lalwani, C: Measures for auditing performance and integration in closed-loop supply chain. Supply Chain Manage. Int. J. **16**(1), 43–56 (2011)
37. Diabat, A, Govindan, K: An analysis of the drivers affecting the implementation of green supply chain management. Resour. Conserv. Recycl. **55**(6), 659–667 (2011)

38. Olugu, EU, Wong, K: A study on the validation of green supply chain performance measures in the automotive industry. Commun. IBIMA, **2011**, 1–14 (2011)
39. Gupta, M: Environmental management and its impact on the operations function. Int. J. Operations Product. Manag. **15**(8), 34–51 (1995)
40. Allenby, B: Supporting environmental quality: developing an infrastructure for design. Total Qual. Environ. Manag. **2**(3), 303–308 (1993)
41. Klassen, RD, McLaughlin, CP: The impact of environmental management of firm performance. Manag. Sci. **42**(8), 1199–1215 (1996)
42. Handfield, R, Walton, SV, Sroufe, R, Melnyk, SA: Applying environmental criteria to supplier assessment: a study in the application of the analytical hierarchy process. Eur. J. Oper. Res. **141**, 70–87 (2002)
43. Winsemius, P, Guntram, U: Responding to the environmental challenge. Bus. Horiz. **35**(2), 12–20 (1992)
44. Mutingi, M, Mbohwa, C: A taxonomic framework for developing green supply chain strategies. Proceed. 2012 Int. Conference Ind. Eng. Operations Manag. Istanbul Turkey **6**, 326–335 (2012)
45. Korchi, AE: Conditions of emergence of OEM's reverse supply chains. J. Remanufactur. **4**(3), (2014). doi:10.1186/2210-4690-4-3
46. Srivastava, S: Green supply-chain management: a state-of-the-art literature review. Int. J. Manag. Rev. **9**(1), 53–80 (2007)
47. Tibben-Lembke, RS: Life after death: reverse logistics and the product life cycle. Int. J. Phys. Distr. Log. Manag. **32**, 223–244 (2002)
48. Van Hoek, RI: From reversed logistics to green supply chains. Supply Chain Manag. **4**, 129–134 (1999)
49. Schot, J, Fischer, K: Environmental Strategies for Industry: International Perspectives on Research Needs and Policy Implications. Island Press, Washington D.C (1993)
50. Kollman, K, Prakash, A: EMS-based environmental regimes as club goods: examining variations in firm-level adoption of ISO 14001 and EMAS in U.K., U.S., and Germany. Policy. Sci. **35**(1), 43–67 (2002)
51. CRR: The Centre for Remanufacturing and Reuse. (2014). http://www.remanufacturing.org.uk/. Accessed 15 July 2014
52. Ninlawan, C, Seksan, P, Tossapol, K, Pilada, W: The implementation of green supply chain management practices in electronics industry. Proceed Int. Multi-Conference Eng. Computer Sci. **3**, (2010)
53. Azzone, G, Noci, G: Measuring the environmental performance of new products: an integrated approach. Int. J. Prod. Res. **34**(11), 3055–3078 (1996)
54. Sarkis, J, Rasheed, A: Greening the manufacturing function. Bus. Horiz. **38**(5), 17–27 (1995)
55. Diabat, A, Kannan, D, Mathiyazhagan, K: Analysis of enablers for implementation of sustainable supply chain management: a textile case study. J. Clean. Prod. **14**(5), 1–13 (2014)
56. Zhu, Q, Sarkis, J: An inter-sectoral comparison of green supply chain management in China: drivers and practices. J. Clean. Prod. **14**, 472–486 (2006)
57. Zhu, Q, Sarkis, J, Geng, Y: Green supply chain management in China: pressures, practices and performance. Int. J. Operations Product Manag. **25**(5), 449–468 (2005)
58. Braun, J, Guthrie, M, McCampbell, E, Sit, V: Chrysler Corporation: Innovations In Supply Chain Management, Case Study. University of Michigan. (1997). Available from: http://www-personal.umich.edu/~afuah/cases/case3.html. Accessed 03 August 2014
59. Beamon, BM: Designing the green supply chain. Logist. Inf. Manag. **12**(4), 332–342 (1999)
60. Lopes, CP, Lobato, AM, Akabane, GK, Mendonca, C, Soares, WLP, Barreto, RM: Toyota motor European (TME) sustainable logistics: An example to Brazil. POMS 21st Ann. Conf. Vancouver Canada (2010)
61. Toyota Motor Corporation: Vehicle Recycling, Sustainability Report. (2014). Available from: http://www.toyota-global.com/sustainability/report/vehicle_recycling/pdf/vr_all.pdf. Accessed 07 August 2014
62. Lind, S, Olsson, D, Sundin, E: Exploring inter-organizational relationships in automotive component remanufacturing. J. Remanufactur. **4**, 5 (2014)
63. Zhu, Q, Sarkis, J, Kai, K: Green supply chain management: practices and performance within the Chinese automobile industry. J. Clean. Prod. **15**, 1041–1052 (2007)

4

Conditions of emergence of OEM's reverse supply chains

Akram El Korchi[1*] and Dominique Millet[2]

* Correspondence:
a.elkorchi@uiz.ac.ma
[1]ENSA AGADIR, Université Ibn ZOHR, Agadir 80000, Morocco
Full list of author information is available at the end of the article

Abstract

Although academic research and some industrial experiences show that a reverse supply chain (RSC) based on remanufacturing offers the possibility to transform the constraint of environmental regulations on product recovery into an opportunity for value creation, few companies have managed to set up their own RSCs. Five research propositions on the condition of emergence of RSCs follow from a literature review and are applied to a case study from a global player in the electrical and electronic equipment (EEE) industry. The findings support the importance of an integrated approach for the design and implementation of a RSC. This integrated approach responds to the traditional questions of OEMs, 'Why' set up a RSC?; 'What' main conditions must be taken into account in the three segments of a RSC (procurement, production, and distribution)?; and 'How' can these conditions be satisfied?

Keywords: Remanufacturing; Reverse supply chain; Conditions of emergence

Introduction

In some industries, the value created by the forward supply chain (FSC) can no longer ensure the firm's viability. Recently, some industrial sectors (e.g., the industry of photocopiers) have seen the emergence of a new form of value creation called the reverse supply chain (RSC). This new chain is intended to complement, not to replace, the chain of value creation initiated by the FSC. A RSC can be considered as an organizational innovation that enables the firm to have a double chain of value creation. The RSC can be defined as 'the series of activities required to retrieve a used product from a customer and either dispose of it or reuse it. Indeed, for a growing number of manufacturers in industries ranging from carpets to computers, reverse supply chains are becoming an essential part of business' [1].

With the enforcement of waste electrical and electronic equipment (WEEE) regulations in Europe, manufacturers of electronic and electrical products have become responsible for the recovery and recycling of their end-of-life products. To meet this statutory requirement, most manufacturers use the services of approved reverse logistics providers. Given the complexity and heterogeneity of the recovered flows, the recovery channels implemented by reverse logistics service providers are primarily recycling-oriented. Generally, these channels do not create economic value for original equipment manufacturers (OEMs) and represent an additional burden on industry. This is why OEMs prefer to outsource most of these channels. Unlike a recovery channel based on recycling, a RSC creates tangible value for the company. Although academic research [2,3] and some industrial experiences (Xerox, IBM, etc.) show that RSC-oriented remanufacturing offers the possibility of transforming this extra burden into an opportunity for value creation, few OEMs have established their own RSCs. According to Hauser and Lund [4], among the 2,000 companies in the USA which claim to be 'remanufacturers', only 6% are OEMs. The remanufacturing industry is represented mainly by SMEs creating value by remanufacturing OEM products.

Some proactive companies have already managed to implement RSCs in an environment characterized by high uncertainty. Simply imitating these companies' models does not guarantee success for other companies [5]. However, most companies interested in creating RSCs tend to copy existing models in their industry. DiMaggio and Powell [6] assume that the ubiquity of certain structures is rather due to the universality of mimetic processes than to any concrete evidence showing the effectiveness of the adopted models.

This tendency to copy existing RSCs is due to the lack of models adopting a holistic vision (systemic) that take into account all aspects of the RSC [7-9]. The remanufacturing industry lacks a comprehensive framework that includes all the major strategic factors necessary for making effective remanufacturing decisions earlier in the conceptual stage of product development [10,11]. Most research focuses on treating some particular problem of the RSC and trying to make local optimizations (e.g., inventory management, production planning) [7]. Given their lack of knowledge about RSC design, managers first need general models that demonstrate and explain the conditions to be met at the first stage of RSC design. These conditions represent the common basis that underlies all RSCs; elements that differentiate RSCs, such as the model of inventory management, are involved in the second stage of the design process (detailed design).

Thus, the conditions of RSC emergence must be considered right from the first phase of design (foundations of the RSC). Satisfying these conditions can reduce the uncertainty and risk perceived by management [12], thus strengthening management decisions to invest in establishing a RSC. To date, few contributions have investigated the strategic conditions of development and implementation of RSCs [13,14]. The aim of this paper is to explore the strategic conditions of emergence of OEM's RSCs through a literature review and to assess these conditions by using a case study in the context of an EEE manufacture.

The paper is structured into three parts. In the first part, the emergence of a RSC is defined and five research propositions related to the proposed conditions of emergence of a RSC are derived from a literature review. Next, a case study from a global EEE company is analyzed to explore the research propositions in practice. Finally, in the third part, the revised theoretical propositions are incorporated in a proposed model.

Literature

Literature is relating to our research focuses on identifying the main conditions of RSC emergence. These conditions are classified into three categories: the 'Whys', the 'Whats', and the 'Hows', representing respectively the traditional questions asked by OEMs interested in a RSC project: Why set up a RSC?; What main conditions must be taken into account in the three segments of a RSC (procurement, production, and distribution)?; and How can these conditions be satisfied?

The 'Whys'

With the emergence of sustainable development and life cycle management, the notion of sustainability is attracting more and more companies. Every viable system should be designed bearing in mind its economic, environmental, and social dimensions. Several studies have explored these three dimensions in the context of remanufacturing [15,16]. In the following paragraphs, we will examine the main economic, environmental, and social reasons that drive companies to set up RSCs.

Economic drivers

Originally, companies' interest in remanufacturing stemmed from economic motivations. Guide and Van Wassenhove [7] state that the remanufacturing industry in the USA is well organized and economically profitable; its development is based on profit rather than environmental concerns. Our literature analysis revealed the economic incentives shown in Table 1.

Environmental drivers

Several industrial experiences (Xerox, Caterpillar, etc.) showed that firms' environmental responsibility is not inconsistent with the creation of economic value. Besides bringing economic benefits, remanufacturing generates an environmental gain greater than that generated by recycling [15,24]. Remanufacturing appears to be one of the best solutions for combining corporate environmental responsibility with profitability [3]. The main environmental motivations appearing in the literature are synthesized in Table 2.

Social drivers

In the academic field, few studies have examined the social drivers regarding remanufacturing [15]. In the practical field, several mentions of the social advantages of remanufacturing could be found. The example of the remanufacturing industry in the USA shows that products are recovered, remanufactured, and reconsumed within the USA. Simply entering the words 'remanufacturing locally' in an Internet search engine reveals how widespread this social dimension is in the remanufacturing industry. In the USA, this industry supports the idea that local remanufacturing is cheaper than exporting remanufacturing to a low-cost country in view of the logistics costs involved. Remanufacturing may also prompt some companies to relocate their production activities to the country of origin (i.e., the USA) if they can capitalize on the synergy between the production of new products and remanufacturing. Our analysis of the literature and some remanufacturers' websites shows that the main social motivations are as follows:

Table 1 The main economic incentives

Common focus	Main economic incentives	Sources
Economic incentives for remanufacturing and RSC	Cost reduction	[7,13,15,17-23]
	Increased profit margin	
	Acquisition of a green image	
	Imitation of green competitors	
	Improved market value	
	Control of the secondary market	

Table 2 The main environmental incentives

Common focus	Main environmental incentives	Sources
Environmental incentives to remanufacturing and RSC	Compliance with environmental regulations	[3,7,13-15,17-24]
	Anticipation of changes in regulation	
	Reducing the environmental impact of product throughout its life cycle	

- Employment creation
- Saving local jobs
- Relocation of production to countries of origin
- Access of low-income households to low-cost products

To summarize, the existing bodies of knowledge identified in the literature review suggest that the emergence of RSCs has various economic, environmental, and social drivers. Thus, we can formally state the following proposition:

Proposition 1: The emergence of a RSC is conditioned by the fulfillment of sufficient economic, environmental, and social gains for the OEM.

This proposition is a combination of the three following sub-propositions:

Proposition 1-a: A RSC emerges if remanufacturing promises to generate satisfactory economic gain.
Proposition 1-b: A RSC emerges if remanufacturing promises to generate satisfactory environmental gain.
Proposition 1-c: A RSC emerges if the remanufacturing promises to generate satisfactory social gain.

The 'Whats'

Having addressed the first question: 'why' designing a RSC? we now turn to a second level in which the firm raises the question of what to consider in designing a RSC. By analogy with the FSC, a RSC comprises three main processes: procurement, production, and distribution. Considering that these three components give interested managers a holistic view of a RSC project, the following sub-sections discuss the main conditions to be satisfied in each RSC component.

Distribution chain: market orientation of remanufacturing

For Singh [25], market orientation represents the set of organizational activities coordinated to generate customer satisfaction through high-performance products, while remaining competitive in the market. If market reaction cannot be ignored when introducing a new product, it is essential to take this dimension into account when introducing a remanufactured product [26]. Indeed, the response to the customer expectations regarding remanufactured products is a strategic decision that requires managers to understand the market, the customers, and the potential influence of remanufacturing on strategy and the firm's distinctive competencies [13]. Among the influences on business strategy, introducing remanufactured products could involve the risk of market cannibalization [8]. This is why in addressing the area of market orientation in the context of remanufacturing, we focus primarily on customer needs and the risk of cannibalization.

Customer needs Customers are the source of the product's returns, and at the same time, they are the potential buyers of remanufactured products. Guide and Van Wassenhove [1] state that the expectations of customers who buy remanufactured products are different from those of customers who buy new products. Thus, a company interested in remanufacturing must first identify its customers' needs. Our literature review identifies four factors influencing the customer decision when buying a remanufactured product:

a. Price of remanufactured products
   Consumers typically value remanufactured products less than the corresponding new products [27]. Debo et al. [17] have explored the impact of customers' profiles on the potential of remanufacturing. Based on empirical findings, they showed that customers value remanufactured goods less than new products and they are therefore not willing to pay the same price for remanufactured products. This difference between the price of new and remanufactured products is due to the uncertainty associated with the quality of remanufactured products; it is probably also due to customers' lack of knowledge about remanufacturing [26].
b. Quality of remanufactured products
   Remanufacturing is often confused with recycling [21,28]. Given customers' lack of knowledge concerning the concept of remanufacturing, the image of its quality is relatively impacted by the image of recycling. Bronstad and Evans-Correia [29] mention that a major barrier in the business of recycled materials is the poor perception of quality. So, to be considered as remanufactured, products must meet the same standards of quality as new products [13]. To exploit this quality image, OEMs must invest in marketing to convince customers that the quality of remanufactured products and new products is similar [30]. Customers who purchase remanufactured products because of this argument will be particularly demanding of product quality.
c. Environmental impact of remanufactured products
   Environmental aspects are an added aspect of customer needs. However, this need is not directly expressed by customers [31], may not be 'what they really want.' When the environmental aspects cause a price increase for the customer, they often constitute an obstacle to the act of purchase [31]. However, this problem does not arise in the context of remanufacturing because remanufactured products are generally cheaper than new ones. Furthermore, the 'green' customer segment clearly expresses environmental needs and favors the purchase of remanufactured products over new ones [32].
d. Upgradability of remanufactured products
   Upgradability is the updating of remanufactured product functions to the current level of technology. Umeda et al. [33] classify the reasons why consumers reject products into two categories. The first is 'physical obsolescence' of the product (performance, scratches, etc.), and the second is 'value obsolescence', meaning the obsolescence of product functionality. For example, many computers and cell phones are discarded due to obsolescence of their functional value. The upgradability of product features in order to meet new functional expectations may extend product lifetime [28].

Market cannibalization by remanufacturing The interrelationship between the demands on two products can be described as independent, complementary, or substitutable. In the case of substitution or cannibalization, reducing the price of product 'A' results in a reduction in quantity demanded for product 'B' [34].

Despite some company managers' interest in remanufacturing, they often find their efforts hampered internally because of fears of cannibalization of the market for new products [26]. Many managers expressed that market cannibalization is a real obstacle for remanufacturing projects [8].

Based on the literature review, the following proposition regarding the market orientation of remanufacturing is suggested.

Proposition 2: The emergence of a viable RSC starts with a market orientation of remanufacturing.

This proposition is a combination of the two following sub-propositions:

Proposition 2-a: To ensure customer purchase of remanufactured products, the following needs should be satisfied:

- *A lower price than the new product*
- *A lower environmental impact than the new product*
- *A quality equal to that of the new product*
- *Upgradability to new product technology level*

Proposition 2-b: Remanufacturing should avoid the market cannibalization of new products.

Production chain: design for remanufacturing

The purpose of remanufacturing is to bring used products to quality standards equivalent to those of new products. Used products are disassembled and inspected thoroughly to replace obsolete or faulty modules [30]. The remanufacturing is organized as an industrial process in order to achieve the benefits of mass production [35]. To facilitate the production process and to ensure high quality, products must be designed following the standards of design for remanufacturing. This design facilitates the remanufacturing process (inspection, cleaning, disassembly, etc.) [21,28]. A better design for remanufacturing also increases the economic and environmental benefits that can be obtained from the remanufacturing industry [28]. Thus, the following proposition can be stated:

Proposition 3: The emergence of a viable RSC is dependent on a product design facilitating remanufacturing.

Procurement chain: the uncertainties

The particular feature of the RSC is that it has two sources of supply. The first source is the supply of new non-reusable modules. This is handled in much the same way as the supply of modules for new products. The second source is the supply of used modules recovered from end-of-life products. This second source is referred to in the literature as 'reverse logistics network'. A reverse logistics network is how products are collected from the end user and returned to a facility for repair, remanufacturing, or recycling [36]. A large portion of reverse logistics is concerned with core acquisition management. This area is responsible for core acquisition and ensuring an adequate supply of cores for remanufacturing [36]. This paper is particularly interested in core acquisition because it is the weakest link in the procurement chain.

The procurement chain of used modules is marked by high uncertainty about time, quality, and quantity of product returns [21,37]. In traditional production-distribution systems, supply is an endogenous variable in the sense that timing, quality, and quantity of inputs are controlled according to the system requirements. In contrast, supply is an exogenous variable in product recovery systems and it is difficult to control [38]. Therefore, the following proposition can be suggested:

> *Proposition 4: The emergence of a viable RSC is conditioned by the reduction of uncertainties concerning time, quality, and quantity in the procurement chain.*

The 'Hows'

Pfeffer and Salancik [39] argue that with increased resource dependence, firms tend to increase vertical coordination. In the case of higher uncertainty, firms try to increase the level of vertical coordination between the focal firm and its partners [40]. In the context of RSCs, Carter and Ellram [9] emphasize that a high level of vertical coordination with customers and suppliers can lead to heightened adoption of new technologies, which also includes the adoption of reverse logistics.

Integrating internal and external supply chain actors into RSC design is a key factor if a RSC is to emerge. Genchev's study [41] about a large computer wholesale company shows a general change in perspective regarding the development of reverse logistics programs and emphasizes the importance of integrating the different stakeholders into the RSC design process. This section discusses how it is essential to coordinate decisions among actors of both internal and external supply chains to satisfy the 'whats' in the various RSC components.

Distribution chain: coordinating for better market orientation

As it was seen previously, the market orientation of remanufacturing consists of meeting customer needs and reducing the risk of cannibalization.

Coordination to satisfy customer needs In order to avoid product market failure, any design process should begin by defining customer expectations. The customer's voice defines the whole of the subsequent process [42]. RSC design must also comply with this logic starting by defining customer needs; indeed, all stakeholder decisions must focus on meeting those needs.

In order to favor purchasing of remanufactured products over new products, marketing teams must assess the conditions imposed by customers, namely price, quality, environmental impact, and upgradability. These conditions should be integrated into both product and RSC designs. For example, with regard to quality (it must be equivalent to a new product), designers should focus on product quality as well as the whole process of reverse logistics. In addition, the design team must identify the interactions among all the design parameters. Identifying these interactions shows how product design affects RSC design and vice versa [43-45].

Coordination to avoid market cannibalization At the launch of the platform of used goods in 2001, Amazon CEO Jeff Bezos said that he was 'pleased with the start' of the used product initiative and added that 'sellers love the platform'. Bezos said that he was not worried that Amazon's used business would hurt sales of new products (although the company said the slowing sales growth was due in part to sales of used products that are not recorded as sales; Amazon only takes a commission) [46].

To reduce the effect of market cannibalization, it is essential to coordinate the decisions of internal and external stakeholders. Concerning external actors of the supply chain, marketing must coordinate with distributors to develop a common strategy to promote remanufactured products. In some cases, if the profit margin from the sale of remanufactured products is high enough, retailers can promote the sale of remanufactured products over the sale of new ones [47]. Concerning internal actors, in companies where the production operations of new and remanufactured products are managed by two different divisions, the performance of each division is generally evaluated independently [27]. In such cases, internal coordination is essential in order to establish a strategy for allocating costs and benefits between the two divisions [27].

Production chain: coordination for better design for remanufacturing

To facilitate the process of remanufacturing (inspection, sorting, disassembly, etc.), products must be designed according to the standards of design for remanufacturing. Better design for remanufacturing increases the economic and environmental benefits that can be obtained from remanufacturing [28].

Redesigning products to comply with the standards of design for remanufacturing can influence other design parameters. Indeed, changing product design can affect production costs, environmental impact, quality, and modularity. According to Toktay and Wei [27], design for remanufacturing can result in higher production costs for new products; therefore, the additional production costs should be allocated to the cost of remanufacturing. Moreover, modifying the modularity of the product may change the modularity of the supply chain [48]. Thus, better design for remanufacturing requires the coordination of actors' decisions to take into account various issues related to any changes in product design

Procurement chain: coordination to reduce uncertainty

Because of uncertainties in the procurement chain, it is difficult to plan production and predict the amount of products that can be remanufactured. Uncertainty about the quantities to be 'remanufactured' is mainly due to two factors: firstly, uncertainty as to the time of returns and secondly, uncertainty about the quality of recovered products [49]. These two variables specific to the remanufacturing context can be added to other variables defining the quantity to be produced in a conventional production system, namely the quantity of supply (quantity of returns) and the quantity of demand. Figure 1 shows the interrelationships among different variables that define the quantity to be remanufactured.

Controlling uncertainty about the quantity to be remanufactured depends on controlling uncertainty about related variables upstream. It is possible to control these variables if decisions made by supply chain actors are better coordinated. For example, the development of options for acquiring products from customers (leasing, buyback, acquisition premiums, etc.) can give more predictable information about the time and quantity of returns; similarly, communicating market forecasts to the production

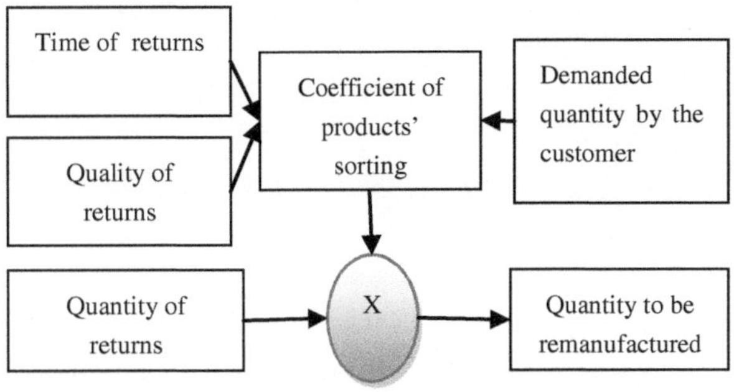

Figure 1 Variables determining the quantity of remanufactured products.

department improves forecasting of quantities demanded. Thus, we can formally state the following proposition:

> *Proposition 5: The emergence of a viable RSC involves coordinating the decisions of internal and external supply chain actors around market orientation of remanufacturing, design for remanufacturing, and reducing uncertainties in the procurement chain.*

The case study

Case methodology

The five propositions were explored in a single case study of a global EEE company. The choice of this case is valid given that the company met the selection criteria derived from the research propositions [50]. Firstly, it was predisposed to set up a RSC for remanufacturing its end-of-life products. Secondly, access to the company enabled us to explore the RSC from a cross-functional perspective using key informants; this factor also supports construct validity [51]. Thirdly, the focus of the case study was restricted to the French market which is clearly driven by a leasing business model favorable for remanufacturing [11].

Given that the main purpose of our own study is to explain a hitherto unstudied phenomenon and possibly to develop theoretical propositions [52], this study can be considered as inductive research. A strong advantage of the case study method is that we can investigate the conditions of emergence of a RSC without removing it from its social 'real life' context [53]. Furthermore, this methodological choice allowed us to access critical business data such as bills of materials and production costs.

Data collection

The case study presented in this paper is part of a wider research project which aimed at developing a methodology for designing remanufacturing systems by combining eco-design and RSC. This case study is about a French producer of EEE machines interested in remanufacturing the company's end-of-life products. In view of data confidentiality, we will call this company 'Alpha'. The company produces electrical and electronic equipment for professional use and is ranked second internationally in its field. Its products are subject to WEEE regulation on end-of-life product recovery. The object of our intervention in this company was to transform an existing recovery channel of products at end of lease (Figure 2), qualified as partially industrial, to a RSC based on an optimized industrial process. This research comprised two main steps: the first was to collect data to analyze and understand the existing situation and the second was to co-design the new remanufacturing system in collaboration with managers.

Given the exploratory nature of our research, the focus was on the collection of qualitative data that was highly relevant to our research subject. Regarding the study population, the sample was limited to key stakeholders involved in the design of the RSC. These belong to both internal and external supply chains of the company.

Six semi-structured interviews were conducted with heads of various departments inside the company involved in the remanufacturing project, two non-structured interviews with the providers of collection and recycling services, and several non-structured interviews with the quality, health, safety, and environment (QHSE) and eco-design managers who were our two main contacts within the company (Table 3). The questions in the interview guide were adapted to the function of each interviewee. No phone interviews were held. We also held six multidisciplinary meetings with the heads of the department involved in the remanufacturing project (Table 3). This specific study focuses on a single instrumental case [54]. These six meetings were organized to facilitate the investigation of the parameters to be taken into account in the RSC design and to present the progress of the project. Data collection of multidisciplinary meetings was carried out over 1 year.

Other secondary sources of data such as field observations and internal company documents were also used. The company agreed to make available many sources of data on the condition of strict confidentiality. All interviews and group meetings were recorded and transcribed. Coding and content analysis were conducted using NVivo 8 software. The plan for case study follows the suggested guidelines developed by Ellram

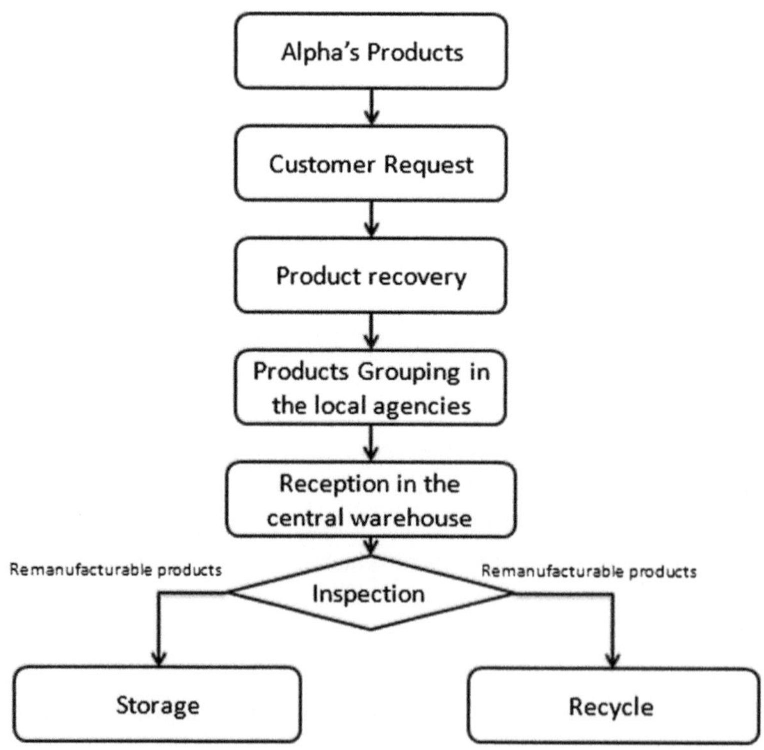

Figure 2 Alpha existing recovery channel.

Table 3 Detail of interviews and multidisciplinary meetings

	Function	Mission	Interview type	Interview duration	Participated in the 6 meetings
Internal supply chain	Marketing	Director	Semi-structured	1 h	Yes
	Production	Director	Semi-structured	1 h	Yes
	Supply chain	Director	Semi-structured	1 h	Yes
	R&D	Director	Semi-structured	1 h	Yes
	Procurement	Top manager	Semi-structured	1 h	
	Methods	Top manager	Semi-structured	1 h	Yes
	QHSE	Top manager	Non-structured	Several	Yes
	Eco-design	Top manager	Non-structured	Several	Yes
External supply chain	Collection provider	QHSE Director	Non-structured	2 h	
	Recycling provider	Top manager	Non-structured	2 h	

[51], and a case study research protocol is listed in the Appendix. The details of the research protocol are available from the authors upon request.

Findings

The results are now discussed in two sections: firstly, the role of RSC in the business of this company and secondly, a summary analyzing the case and comparing the findings with the research propositions.

The role of the RSC in the company business

Alpha's willingness to address the RSC competence stemmed from an acute awareness of current problems with managing product end-of-life. Even as a global player present all over the world, the company saw the French market unequivocally as highly attractive and critical for successful further expansion into European markets. The company had experienced strong growth in Europe seen in the high performance of European subsidiaries, the rise of leasing services, and the acquisition of distributors in Switzerland, Sweden, Norway, and Germany. Nevertheless, despite the strategic importance of the European market, customer requirements were beginning to change more rapidly, competition was fluid, and environmental regulations were becoming more stringent. The main strategic objective was thus to transform these constraints into a competitive advantage through remanufacturing. This involves managing two distinct but related processes: 'manufacturing' new products and 'remanufacturing' end-of-life products. Multidisciplinary meetings enabled us to identify the salient issues concerning remanufacturing. Interestingly, the problems were located throughout the whole of the RSC, from product recovery to redistribution. This is a good illustration of the notion of systemic approach to RSC design.

A common problem across the RSC was the need for a disciplined process. An in-depth analysis revealed bottlenecks within the current cross-functional processes. There is a time-consuming and cumbersome company-wide process for the recovery of end-of-lease products. This process took more than 12 weeks overall. The two most time-consuming activities were company agents' recovery of products (4 weeks) and storage in local agencies before shipping to the central warehouse (8 weeks). On the marketing side, knowledge about customer needs and buying behaviors was still patchy. Marketers were overwhelmed by the specifics of the remanufacturing market, starting from a highly active secondary market to tightly controlled sales channels. Furthermore, Alpha managers stressed that remanufacturing parameters should be integrated earlier in the product design process. Everyone agreed that in order to streamline the remanufacturing process, the commitment of all functions to reduce the time-sensitive steps was essential.

Case analysis

The purpose of the case study was to explore in practice the five propositions derived from the literature review.

Regarding the first proposition, the findings from Alpha show that motivations for developing a RSC are mainly economic and environmental. At the time of the study, Alpha's economic targets were, respectively, to reduce manufacturing costs, to increase profit margin, to access new market segments, and to imitate competitors engaged in remanufacturing. Alpha's managers were confident that they felt comfortable that their RSC approach would support the achievement of these objectives. Furthermore, our findings reveal that acquiring a green image and controlling the secondary market appear not to be priorities for the company. Regarding the environmental drivers, complying with regulatory requirements and reducing products' environmental impact seem to be very important because they contribute to improve perceived customer value. The results also highlight the strong correlation between environmental and economic drivers.

Proposition 2 states that the emergence of a viable reverse supply chain starts with a market orientation of remanufacturing. The case study shows that market orientation is not only supported by the RSC but can also trigger the development of RSC competence. As with many companies today, satisfying customer needs is a strategic priority for Alpha. The analysis focused on which needs had to be satisfied: Alpha should remanufacture products priced lower than new products, with a quality equivalent to new products and upgradability to new product levels of technology; this is particularly true for software. The variable concerning environmental impact was not been verified. Alpha managers think that a big decrease in price could have a negative effect on customers' perceived quality of the remanufactured product. With regard to market cannibalization, data analyses show this risk as a major factor blocking the development of remanufacturing. Discourse analysis allowed us to classify market cannibalization into three categories, namely market cannibalization of new products by remanufactured ones of the same generation, market cannibalization of new products by the remanufactured ones of an older generation, and market cannibalization by competitors. The first and second categories are cannibalizations that are 'negative' for the company, while the third is 'positive'. Managers' fear is mainly related to negative cannibalization. Alpha managers think that this negative cannibalization could be avoided by selling remanufactured products at the same price as new ones.

Regarding proposition 3, results show that one of the key points of the RSC project is to integrate the principles of 'design for remanufacturing' into product design in order to obtain a 'remanufacturable' product. Indeed, design for remanufacturing should facilitate the remanufacturing process (disassembly, cleaning, etc.) in order to reduce costs and increase the quality of the treatment process. A modular design appears to be an encouraging route to achieving this end.

The results related to Proposition 4 show that managers perceive the uncertainties about time, quality, and quantity of return flows as a real constraint. However, they do not consider this constraint as a factor that might block the development of Alpha's remanufacturing program. There are two main reasons for this: first, there is a considerable stock of end-of-lease machines available in the company's central warehouse. Second, the company's business model is based on a leasing system allowing some control over the time, quality, and quantity of returns. In contrast, findings highlight a new uncertainty: that of the cost of remanufacturing as an obstacle to implementing the RSC. In fact, managers attach great importance to assessments of economic gain of remanufacturing compared to the production of new products. They argue that traditional accounting methods are unsuited to carrying out such evaluations. This means that new accounting methods need to be developed if possible gains generated by remanufacturing are to be demonstrated.

Regarding Proposition 5 which covers the coordination of decisions, our interpretation of managers' perceptions confirms the importance of coordinated decision making to give remanufacturing a market orientation while integrating the principles of design for remanufacturing into product design and reducing uncertainties in the procurement chain. Furthermore, our results highlight the emergence of two moderating variables for the emergence of the RSC. The first is a business model that offers a global concept, taking into account the main aspects of the remanufacturing system and demonstrating its viability. The second is overcoming the constraints imposed by the decentralized company structure which is not favorable to coordination. This can be done by appointing a project manager with sufficient authority to deploy the strategy of remanufacturing throughout the company.

Summary - towards a model of RSC design

On the basis of the propositions derived from the literature review and the case study findings, the theoretical propositions are revised for further research and combined into a model that integrates the conditions for RSC emergence (see Figure 3).

P 1': The emergence of a reverse supply chain is contingent on a satisfactory combination of economic and environmental benefits derived from remanufacturing.

P 1-a': The reverse supply chain emerges if the remanufacturing promises to generate a satisfactory economic gain.

P 1-b': The reverse supply chain emerges if the remanufacturing promises to generate a satisfactory environmental gain.

P 2': The emergence of a viable reverse supply chain starts with a market orientation and the avoidance of negative market cannibalization.

2-a': To ensure that customers purchase remanufactured products, the latter must meet customer needs according the following criteria: priced lower than the new product, quality equivalent to the new product, and upgradability to the level of technology of the new product.

P 2-b': The remanufactured products should avoid negative market cannibalization.

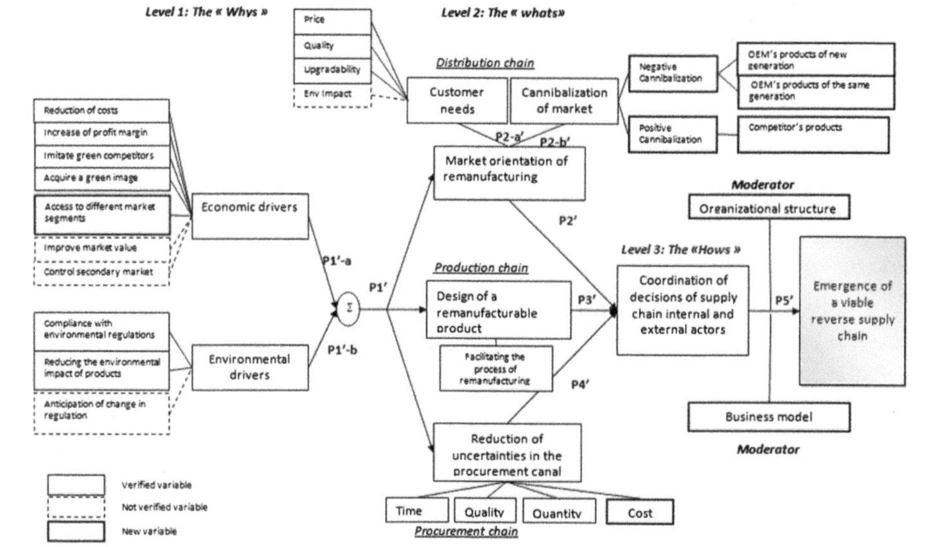

Figure 3 General model of emergence of a RSC.

P 3': The emergence of a viable reverse supply chain is dependent on a product design facilitating remanufacturing.

P 4': The emergence of a viable reverse supply chain is conditioned by the reduction of uncertainties about time, quality, quantity, and cost in the procurement chain.

P 5': The emergence of a viable RSC involves coordinating decisions of internal and external supply chain actors around a market orientation of remanufacturing, a design for remanufacturing, and the reduction of uncertainties in the procurement chain.

Conclusion

The constraints imposed by the business environment are increasingly pushing the emergence of RSCs. An integrated approach (as opposed to a fragmented approach) to the design and implementation of RSC is essential [37,45]. According to Stock [55], this integrated approach should follow flow mapping in the RSC. This is why we followed the RSC from product recovery to product reintroduction on the market in order to identify the key strategic conditions for RSC emergence.

Our case study followed the process of a company attempting to implement a RSC project. This case study allowed us to experience the design of the RSC as part of the project team. It demonstrates that setting up a RSC requires a proactive approach that necessitates organizational and cultural change within the company. It shows that the success of a RSC project depends on the involvement of internal and external actors of the supply chain.

The model and theoretical propositions raise a range of questions regarding managerial and organizational implications. Nevertheless, our findings contribute to our understanding of the dynamics involved in the RSC design. While our framework and theoretical propositions are a starting point providing directions on the relationships between the main concepts, more research is needed to develop hypotheses which further specify the causal links.

Appendix

Case study research protocol

I. Research questions/issues covered in the interviews and meetings.
- Why set up a RSC?
  - What are the economic drivers for setting up a RSC?
  - What are the environmental drivers for setting a RSC?
  - What are the social drivers for setting up a RSC?
- What are the main conditions to be taken into account in the procurement chain, the production chain, and the distribution chain?
- How can these conditions be satisfied?
  - What is the impact of coordination on adopting a market orientation?
  - What is the impact of coordination on adopting the design for remanufacturing?
  - What is the impact of coordination on reducing uncertainties in the procurement chain?

II. Methodology/case study design

A. Single case study design
- Case study multidisciplinary meetings organized as co-development discussions
- Propositions will be explored from a practitioner perspective
- Different mindsets of functional representatives will be captured

B. Case study selection

Company meets the relevant validity criteria of the research:
- Predisposition to the concept of RSC and remanufacturing
- Access ensures that RSC can be explored from a cross-functional perspective using key informants
- The focus on the French market which is driven by a leasing business model favorable for remanufacturing

C. Interview protocol

The final interview protocol was developed with the assistance of two practitioners who were in charge of the RSC project within the company targeted for this study. These individuals provided valuable corrections to the original interview protocol. The interview protocol was designed to be easy to understand and easy to answer. Also, anonymity was offered to respondents. Respondents were selected based on their direct knowledge of the subject of remanufacturing and, therefore, their ability to answer the questions properly.

III. Data collection - sources of evidence

1. Interviews with head of departments - key source of information
2. Multidisciplinary meetings
3. Direct observations
4. Documentation
   - Internal reports and documents
   - Bill of materials and product design structure
   - Market reports and sale volumes by range of products

IV. Data analysis

1. Pattern matching
   - Look for patterns based on drivers for RSC implementation
   - Patterns based on uncertainties in the procurement chain

- Patterns based on design for remanufacturing
- Patterns based on market orientation of remanufacturing
- Associations based on how coordination is important for reducing uncertainties in the procurement chain
- Associations based on how coordination is important for facilitating design for remanufacturing
- Associations based on how coordination is important for market orientation of remanufacturing

2. Explanation building
   - Single case study: risk of misrepresenting the associated findings
   - No intention to generalize the findings, but to describe the causal powers between the concepts investigated

Competing interests

The authors declare that they have no competing interests.

Acknowledgements

The authors wish to thank the CETIM Foundation and ADEME for sponsoring this research. Also, the authors would like to express sincere gratitude to the managers of Alpha company for their collaboration.

Author details

[1]ENSA AGADIR, Université Ibn ZOHR, Agadir 80000, Morocco. [2]SUPMECA, Toulon 83957, France.

References

1. Guide Jr, VDR, Van Wassenhove, LN: The reverse supply chain. Harv. Bus. Rev. **80**, 25–26 (2002)
2. Fleischmann, M, Van Nunen, J, Grave, B, Gapp, R: Reverse logistics - Capturing Value in the Extended Supply Chain In: ERIM Report Series Reference No. ERS-2004-091-LIS (2004)
3. Geyer, R, Jackson, T: Supply loops and their constraints: the industrial ecology of recycling and reuse. Calif. Manag. Rev. **46**, 55–73 (2004)
4. Hauser, W, Lund, R: Remanufacturing: Operating Practices and Strategies. Boston University, Boston (2008)
5. Fernanndez, I, Kekale, T: The influence of modularity and industry clockspeed on reverse logistics strategy: implications for the purchasing function. J. Purch. Supply Manage. **11**, 193–205 (2005)
6. DiMaggio, P, Powell, WW: The New Institutionalism in Organizational Analysis. University of Chicago Press, Chicago (1991)
7. Guide Jr, VDR, Van Wassenhove, LN: The evolution of closed-loop supply chain research. Oper. Res. **57**, 10–18 (2009)
8. Prahinski, C, Kocabasoglu, C: Empirical research opportunities in reverse supply chains. Omega **34**, 519–532 (2006)
9. Carter, CR, Ellram, LM: Reverse logistics: a review of the literature and framework for future investigation. J. Bus. Logist. **19**, 85–102 (1998)
10. Subramoniam, R, Huisingh, D, Chinnam, RB, Subramoniam, S: Remanufacturing decision-making framework (RDMF): research validation using the analytical hierarchical process. J. Clean. Prod. **40**, 212–220 (2013)
11. Sundin, E, Bras, B: Making functional sales environmentally and economically beneficial through product remanufacturing. J. Clean. Prod. **13**, 913–925 (2005)
12. March, JG, Shapira, Z: Managerial perspectives on risk and risk taking. Manag. Sci. **33**, 1404–1418 (1987)
13. Dowlatshahi, S: A strategic framework for the design and implementation of remanufacturing operations in reverse logistics. Int. J. Prod. Res. **43**, 3455–3480 (2005)
14. Subramoniam, R, Huisingh, D, Chinnam, RB: Remanufacturing for the automotive aftermarket-strategic factors: literature review and future research needs. J. Clean. Prod. **17**, 163–1174 (2009)
15. Lund, RT: The Remanufacturing Industry: Hidden Giant. Boston University, Boston (1996)
16. Seitz, M: A critical assessment of motives for product recovery: the case of engine remanufacturing. J. Clean. Prod. **15**, 1147–1157 (2007)
17. Debo, LG, Toktay, LB, Van Wassenhove, LN: Market segmentation and product technology selection for remanufacturable products. Manag. Sci. **51**, 1193–1205 (2005)
18. Carter, CR, Rogers, D: A framework of sustainable supply chain management: moving toward new theory. Int. J. Phys. Distrib. Logistics Manage. **38**, 360–387 (2008)
19. Lebreton, B: Strategic Closed-Loop Supply Chain Management. Springer, Berlin (2007)
20. Hart, SL: Beyond greening: strategies for a sustainable world. Harv. Bus. Rev. **75**, 66–76 (1997)
21. De Brito, MP: Managing reverse logistics or reversing logistics management? PhD thesis. Erasmus University Rotterdam (2004)
22. Tibben-Lembke, R, Rogers, DS: Differences between forward and reverse logistics in a retail environment. Supply Chain Manage. **7**, 271–282 (2002)
23. Ummenhofer, M: La logistique dans une perspective d'écologisation : vers l'éco-logistique intégrée. Thèse de doctorat en sciences de gestion, université de la Méditerranée (Aix-Marseille II) (1998)

24. Quariguasi Frota Neto, J, Bloemhof-Ruwaard, JM: The environmental gains of remanufacturing: evidence from the computer and mobile industry. (2009). Research Paper ERS-2009-024-LIS Revision ERIoME ed
25. Singh, S: Market Orientation, Corporate Culture and Business Performance. Ashgate, Burlington (2004)
26. Ferrer, G, Whybark, DC: From garbage to goods: successfully manufacturing systems and skills. Bus. Horiz. **43**, 55 (2000)
27. Toktay, LB, Wei, D: Cost allocation in manufacturing-remanufacturing operations. Working paper I ed, Fontainebleau, France (2005)
28. Sundin, E: Product and process design for successful remanufacturing. PhD dissertation. Linköpings Universitet, Linköping, Sweden (2004)
29. Bronstad, G, Evans-Correia, K: Green purchasing: the purchasing agent's role in corporate recycling. In: Conference Proceedings of the National Association of Purchasing Management, pp. 117–121. (1992)
30. Thierry, M, Salomon, M, Van Nunen, J, Van Wassenhove, L: Strategic issues in product recovery management. Calif. Manag. Rev. **37**, 114–135 (1995)
31. Yim, H, Herrmann, C: Eco-voice of consumer (VOC) on QFD. In: Proceedings of EcoDesign2003: Third International Symposium on Environmentally Conscious Design and Inverse Manufacturing. Tokyo, Japan (2003). 8–11 Dec
32. Atasu, A, Sarvary, M, Van Wassenhove, LN: Remanufacturing as a marketing strategy. Manag. Sci. **54**, 1731–1746 (2008)
33. Umeda, Y, Daimon, K, Kondoh, S: Life cycle option selection based on the difference of value and physical lifetimes for life cycle design. In: International Conference on Engineering Design. ICED'07, Paris, France (2007). 28–31 Aug
34. Kerin, RA, Harvey, MG, Rothe, JT: Cannibalism and new product development. Business Horizons **21**, 25 (1978)
35. Steinhilper, R, Rosemann, B, Freiberger, S: Product and process assessment for remanufacturing computer controlled automotive components. In: 13th CIRP International Conference on Life Cycle Engineering. LCE2006, Leuven, Belgium (2006)
36. Guide Jr, VDR: Production planning and control for remanufacturing: industry practice and research needs. J. Oper. Manag. **18**, 467–483 (2000)
37. Fleischmann, M, Beullens, P, Bloemhof-Ruwaard, JM, van Wassenhove, LN: The impact of product recovery on logistics network design. Prod. Op. Manage. **10**, 156–173 (2001)
38. Fleischmann, M, Krikke, H, Dekker, R, Flapper, SDP: A characterisation of logistics networks for product recovery. Omega **28**, 653–666 (2000)
39. Pfeffer, J, Salancik, GR: The External Control of Organizations: A Resource Dependence Perspective. Harper & Row, New York (1978)
40. Achrol, R, Reve, T, Stern, L: The environment of marketing channel dyads: a framework for comparative analysis. J. Mark. **47**, 55–67 (1983)
41. Genchev, SE: Reverse logistics program design: a company study. Stud. Bus. Horiz. **52**, 139–148 (2009)
42. Akao, Y: New product development and quality assurance – quality deployment system. Standardization and Quality Control **25**, 7–14 (1972)
43. Umeda, Y, Nonomura, A, Tomiyama, T: Study on life-cycle design for the post mass production paradigm. In: AIEDAM (Artificial Intelligence for Engineering Design, Analysis and Manufacturing), pp. 149–161. (2000)
44. Krikke, H, Bloemhof-Ruwaard, J, Van Wassenhove, LN: Concurrent product and closed-loop supply chain design with an application to refrigerators. Int. J. Prod. Res. **41**, 3689–3719 (2003)
45. El Korchi, A, Millet, D: Designing a sustainable reverse logistics channel: the 18 generic structures framework. J. Clean. Prod. **19**, 588–597 (2011)
46. Milliot, J: Low prices, more used product for Amazon.com's book group. Publishers Weekly **248**(31), 9 (2001)
47. Giuntini, R, Gaudette, K: Remanufacturing: the next great opportunity for boosting US productivity. Bus. Horiz. **46**, 41–48 (2003)
48. Fine, CH: Clockspeed: Winning Industry Control in the Age of Temporary Advantage. Perseus, Cambridge (1998)
49. Van der Laan, E, Salomon, M, Dekker, R, Van Wassenhove, LN: Inventory control in hybrid systems with remanufacturing. Management Science. **45**, 733–747 (1999)
50. Mentzer, J, Flint, D: Validity in logistics research. J. Bus. Logist. **18**, 199–216 (1997)
51. Ellram, L: The use of case study method in logistics research. J. Bus. Logist. **17**, 93–138 (1996)
52. McCutcheon, D, Meredith, J: Conducting case research in operations management. J. Oper. Manag. **11**, 239–256 (1993)
53. Bonoma, T: Case research in marketing: opportunities, problems, and a process. J. Mark. Res. **22**, 199–208 (1985)
54. Yin, RK: Case Study Research: Design and Methods, 3rd edn. Sage, London (2003)
55. Stock, JR: Development and Implementation of Reverse Logistics Programs. Council of Logistics Management, Oak Brook, Illinois (1998)

5

An integrated fuzzy approach for evaluating remanufacturing alternatives of a product design

Xiaojun Wang[1*] and Hing Kai Chan[2]

* Correspondence: xiaojun.wang@bristol.ac.uk
[1]Department of Management, University of Bristol, Bristol BS8 1TN, UK
Full list of author information is available at the end of the article

Abstract

Remanufacturing has emerged as a competitive strategy for manufacturers to tackle environmental and economic challenges. In this paper, an integrated fuzzy approach is developed for the evaluation of remanufacturing alternatives. Then, importance weights of main remanufacturing processes and evaluation criteria are obtained through fuzzy extent analysis. Fuzzy hierarchical TOPSIS is then applied to evaluate the alternatives. A case study is presented to demonstrate the applicability of the proposed approach. The analysis results show that it is a viable approach and can be used as an effective tool for design evaluation from the remanufacturing point of view. Finally, conclusions are discussed and future research directions are suggested.

Keywords: Remanufacturing; Hierarchical model; Life cycle assessment; Fuzzy extent analysis; TOPSIS

Background

In the last two decades, environmental concerns diffuse into almost all aspects of the manufacturing industry and all phases of products' life cycles. This is simply because resources consumed during the course of manufacturing and production are enormously high, and hence, the amount of waste generated from those processes is also notorious [1]. One of such key areas is the end-of-life treatment [2]. Remanufacturing is one of many end-of-life strategies.

Remanufacturing is not a new topic but had not been considered as an important strategic area until the recent decade. In the past, remanufacturing activities focus mainly on recapturing economical values from used products or have been driven by regulatory pressure [3]. Typical activities include recycling of materials and reuse of parts or components, among others, to produce close-to-new refurbished products. Figure 1 shows a flowchart of a typical remanufacturing process. Nevertheless, the processing procedures may vary depending on the nature of the product being remanufactured [4]. Obviously, there are lots of uncertainties in remanufacturing [5]. With the backdrop of increasing environmental awareness, remanufacturing is one of many ways to mitigate environmental impacts by reducing the consumptions of virgin materials, resources in primary production and etc. This has been becoming popular in the last decade [6]. The contemporary school of thought considers that remanufacturing can not only (re-)gain financial benefits, but also reduce the environmental burdens [5]. This is a typical multi-objective problem. Remanufacturing is now referred to as a value-adding process and has emerged

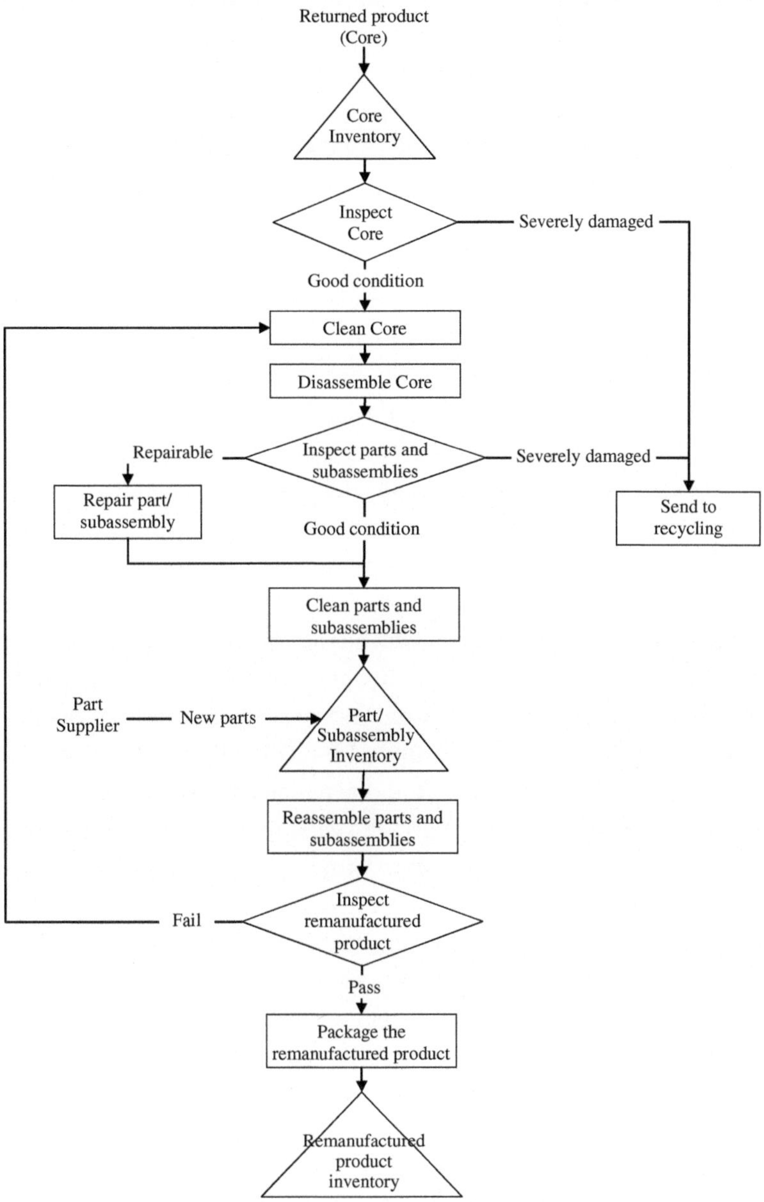

Figure 1 Remanufacturing process [4].

as part of closed-loop supply chains [7]. This trend implies the importance of developing decision-making models when remanufacturing activities are involved.

Life cycle assessment (LCA) provides the basic modelling framework for evaluating the environmental load and impact throughout the entire product life cycle [8]. It is an effective, comprehensive and practical tool in assessing environmental impact of products [9]. For example, Chan et al. [10] adopted the concept of LCA and proposed a comprehensive framework for the selection of green product designs. The life cycle concept is also applicable to remanufacturing process. For instance, Schau et al. [11] conducted an LCA study of remanufactured alternators. Three designs were considered and the associated environmental impacts were evaluated. However, the major obstacle is that remanufacturing activities are not well structured, so applying LCA to evaluate all design options would be time-consuming, if not impractical. Therefore, it is important to provide designers/engineers a more efficient 'screening' approach to assess the environmental and economic performance of alternative designs.

Evaluating the environmental and economic impact of a product or process is essentially a multi-criteria decision-making (MCDM) problem. LCA, for example, considers multiple inputs and multiple outputs, and they are not homogenous in most cases. Saaty [12] developed a groundbreaking tool, called analytic hierarchy process (AHP), to deal with MCDM problems. The merit of AHP is that both qualitative and quantitative factors can be considered in a hierarchical model. Since then, applications of AHP are numerous, with a trend to integrate with other methods [13]. One strand of such integrated approaches is to combine the method with fuzzy theory, which was developed by Zadeh [14] and can handle imprecise information. This characteristic supplements the pairwise comparisons in standard AHP so that a higher degree of uncertainty can be included in the decision-making process. The fuzzy AHP approach provides such practical solution, which is simple and less demanding upon the resources needed to make a decision by converting uncertain variables into linguistic variables. In other words, the process can be simplified in that sense. Nevertheless, it is still very easy to have over a hundred pairwise comparisons in order to make a design selection decision, which relies heavily on subjective decisions and is therefore not effective in terms of computational complexity. This research confronts this challenge through integration of fuzzy extent analysis and fuzzy hierarchical Technique for Order of Preference by Similarity to Ideal Solution (TOPSIS) for conducting effective evaluation of design alternatives from the remanufacturing perspective.

Fuzzy extent analysis, developed by Chang [15], stems from the AHP method that is used routinely to estimate comparative weights with a view in solving MCDM problems. Studies that apply fuzzy extent analysis leverage the benefits of fuzzy set theory and make use of linguistic terms (e.g. high, very high) or a fuzzy number in lieu of a precise numerical value when conducting pairwise comparison e.g. [16]. It has been widely applied in different problem environments in the literature: Kahraman et al. [17] developed an analytical selection tool to measure the customer satisfaction in catering firms in Turkey, Celik et al. [18] developed fuzzy AHP methodology based on Chang's extent analysis to model shipping registry selection, and Wang et al. [19] applied fuzzy extent analysis to develop a risk assessment model that enabled a structured analysis of aggregative risk in the food supply chain. The trends in utilizing fuzzy extent analysis in fuzzy AHP evident in the literature have been continued in many of the operational disciplines due to its ease of use and computational simplicity.

Fuzzy TOPSIS [20,21] is derived from the TOPSIS technique proposed by Hwang and Yoon [22] to evaluate the performance of alternatives. TOPSIS ranks the alternatives according to their distances from the ideal and the negative ideal solution. The positive ideal solution maximizes the benefit criteria and minimizes the cost criteria, while the negative ideal solution maximizes the cost criteria and minimizes the benefit criteria. The most preferred alternative is then derived as the shortest distance from the positive ideal solution and the longest distance from the negative ideal solution. Despite its popularity and simplicity in concept, TOPSIS is often criticized for its inability to deal with uncertainty and imprecision inherent in the process of mapping the perceptions of experts [23]. To address the limitation, scholars have made use of fuzzy TOPSIS (combination of fuzzy logic with TOPSIS) for expert systems in areas such as plant location selection [21], supplier selection and evaluation [20], analysis of network uncertainty [24] and assessment of green supply chain initiatives [25]. Fuzzy hierarchical TOPSIS will benefit from both the superiority of the hierarchical structure and easiness of implementation of TOPSIS in a fuzzy environment.

In this article, a decision support model is proposed based on fuzzy synthetic extent analysis method and hierarchical fuzzy TOPSIS method to make quick selection decisions regarding remanufacturing alternatives. This is an effective modelling approach for such evaluation, which is the major contribution of this paper. In addition, this paper also makes practical contributions as shown in the case study which demonstrates the operations of the proposed model. The rest of the paper is organized as follows: The 'Methods' section presents the details of the model. It is then followed with a case study of a real-life example, which is obtained from a published study, in order to demonstrate the applicability of the proposed methodology. Finally, some concluding remarks and directions for future researches are presented in the 'Conclusions' section.

Methods

The proposed methodology consists of a hierarchical evaluation model, fuzzy extent analysis and fuzzy hierarchical TOPSIS techniques. In the hierarchical model, the critical aspects for sustainable remanufacturing are first defined and the criteria under each aspect are identified. Fuzzy extent analysis is then used to determine the relative importance weights of evaluation criteria. Finally, the fuzzy hierarchical TOPSIS is applied to assess alternative design options. Details of the proposed method are elaborated in each of the following sections.

The hierarchical model

The proposed model can be broadly divided into four parts as illustrated in Figure 2. The first level is a collection of main remanufacturing processes. Then, the corresponding criteria within each remanufacturing process are identified and presented in the second level. The third level includes the performance measures employed to evaluate the remanufacturing alternatives provided at the final level. In other words, the decision is made based on the relative importance of each process against each performance measure, and then an aggregated score can be computed in order to help make a decision.

The aim of identifying the first level process is to break down the whole remanufacturing operation into a number of processes so that the importance of each process with respect to the remanufacturing operation of different alternatives can be evaluated. This is analogous to the life cycle phases mentioned in the 'Background' section. However, the objective is not the same as life cycle assessment as only remanufacturing is considered in this paper. Due to its unstructured nature, there is no generic process for handling remanufacturing processes either. As a consequence, a number of common processes are observed from the existing literature. With reference to a number of studies [26-29], the following remanufacturing processes are identified:

- C_1. Disassembly
- C_2. Detoxification
- C_3. Machine operations

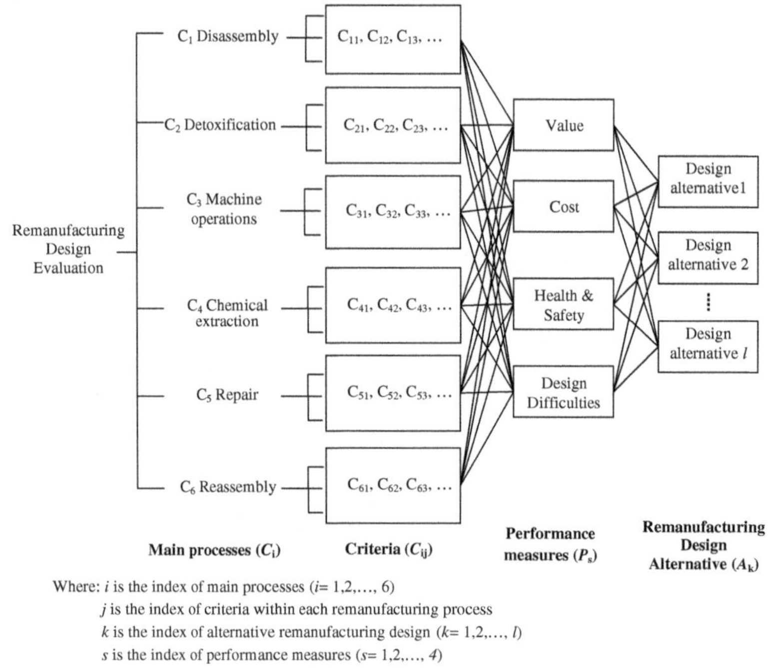

Figure 2 A hierarchical model for the evaluation of product designs from the remanufacturing perspective.

- C_4. Chemical extraction
- C_5. Repair
- C_6. Reassembly

Then, associated criteria within each remanufacturing process should be identified and listed at the second level for further evaluation. These criteria could be generic criteria associated with individual remanufacturing processes or be more specific with respect to particular products. This will not undermine the usefulness of the model because this is not a restriction to use pre-defined processes and their associated criteria in the model. Construction of the hierarchical model will be varied dependent on the actual cases, and any new processes can be added accordingly.

At the third level, the performance measures are provided to evaluate the remanufacturing alternatives. Hatcher et al. [30] recently conducted a case study regarding the barriers and challenges for remanufacturing, which can be adopted in this model. They are value (e.g. rare metal content, competition between imitated products, environmental impacts), cost involved, employee health and safety, and design difficulties (e.g. supplier relationship, technological advancement which discourage the use of old components). Finally, all the remanufacturing alternatives are presented at the fourth level for the evaluation.

Fuzzy extent analysis

Here, the fuzzy synthetic extent analysis method is introduced to calculate the synthetic extent value of the pairwise comparison. An extent analysis adaptation to fuzzy AHP was proposed in order to obtain a crisp priority vector from a triangular fuzzy comparison matrix [15]. The triangular fuzzy scale of preferences is given in Table 1, $M_z = (m_{z1}, m_{z2}, m_{z3})$, where $z = 1, 2,..., 9$. Triangular fuzzy numbers (TFNs) M_1, M_3, M_5, M_7 and M_9 are

Table 1 Linguistic classification of triangular fuzzy numbers

Rating level	Linguistic values	TFNs
1	Equal	(1, 1, 1)
3	Moderately more important	(2, 3, 4)
5	Fairly more important	(4, 5, 6)
7	Much more important	(6, 7, 8)
9	Absolute more important	(9, 9, 9)
2, 4, 6, 8	Midpoint preference values lying between above values	(1, 2, 3), (3, 4, 5), (5, 6, 7), (7, 8, 9)

used to represent the pairwise comparison of decision variables in the range from 'Equal' to 'Absolute more important', when these are employed as descriptive terms attached to the level of importance of paired variables. M_2, M_4, M_6 and M_8 represent the midpoint preference values lying between them.

Next, let $P = \{p_1, p_2,..., p_n\}$ be an object set and $Q = \{q_1, q_2,..., q_m\}$ be a goal set. Here, m equals the number of criteria identified in the whole remanufacturing process multiplied by the number of performance measures. According to the method of extent analysis [15], each object is taken and extent analysis is performed for each goal respectively. Therefore, the m extent analysis values for each object are obtained as follows: $M^1_{g_i}, M^2_{g_i},..., M^m_{g_i}$, $i = 1, 2,..., n$, where all the $M^j_{g_i}$ $(j = 1, 2,..., m)$ are TFNs. The following is a summary of the procedures with reference to the study conducted by Chan and Wang [16].The value of fuzzy synthetic extent with respect to the ith object is defined as

$$S_i = \sum\nolimits_{j=1}^{m} M^j_{g_i} \otimes \left[\sum\nolimits_{i=1}^{n}\sum\nolimits_{j=1}^{m} M^j_{g_i}\right]^{-1}, \tag{1}$$

and $\left[\sum\nolimits_{i=1}^{n}\sum\nolimits_{j=1}^{m} M^j_{g_i}\right]^{-1}$ can be calculated as

$$\left[\sum\nolimits_{i=1}^{n}\sum\nolimits_{j=1}^{m} M^j_{g_i}\right]^{-1} = \left(\frac{1}{\sum\nolimits_{i=1}^{n} m_{3i}}, \frac{1}{\sum\nolimits_{i=1}^{n} m_{2i}}, \frac{1}{\sum\nolimits_{i=1}^{n} m_{1i}}\right). \tag{2}$$

The degree of possibility of $M_1 \geq M_2$ is defined as

$$V(M_1 \geq M_2) = \sup_{x \geq y}[\min(u_{M_1}(x), u_{M2}(y))]. \tag{3}$$

When a pair (x, y) exists, such that $x \geq y$ and $u_{M_1}(x) = u_{M2}(y) = 1$, then we have $V(M_1 \geq M_2) = 1$. Since M_1 and M_2 are convex fuzzy numbers, we have that

$V(M_1 \geq M_2) = 1$ if $m_{12} \geq m_{22}$,

$$V(M_1 \geq M_2) = hgt(M_1 \cap M_2) = u_{M_1}(d) \tag{4}$$

where d is the ordinate of the highest intersection point D between u_{M_1} and u_{M_2} (see Figure 3). When $M_1 = (m_{11}, m_{12}, m_{13})$ and $M_2 = (m_{21}, m_{22}, m_{23})$, then the ordinate of D is computed by

$$V(M_2 \geq M_1) = hgt(M_1 \cap M_2) = \frac{m_{11} - m_{23}}{(m_{22} - m_{23}) - (m_{12} - m_{11})}. \tag{5}$$

To compare M_1 and M_2, both the values of $V(M_1 \geq M_2)$ and $V(M_2 \geq M_1)$ are required. The degree of possibility for a convex fuzzy number to be greater than k convex fuzzy

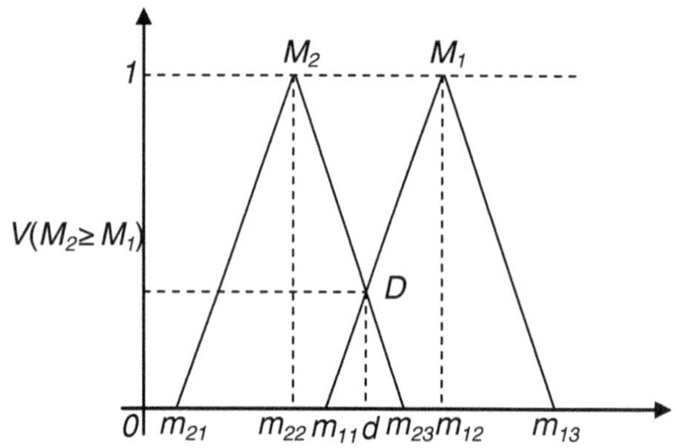

Figure 3 Membership functions of the set of importance ratings.

numbers M_i ($i = 1, 2,..., k$) can be defined by

$$\begin{aligned} &V(M \geq M_1, M_2, ..., M_k) \\ &= V[(M \geq M_1) \text{ and } (M \geq M_2) \text{ and}, ..., \text{and } (M \geq M_2)] \\ &= \min \ V(M \geq M_i), i = 1, 2, ..., k \end{aligned} \tag{6}$$

if

$$d(X_i) = \min \ V(S_i \geq S_k). \tag{7}$$

For $k = 1, 2,..., n$; $k \neq i$, then the rating vector is given by

$$W' = (d(X_1), d(X_2), ..., d(X_n))^T \tag{8}$$

where X_i ($i = 1, 2,..., n$) are n different criteria. Via normalization, the normalized rating vectors are

$$W = (R(X_1), R(X_2), ..., R(X_n))^T \tag{9}$$

where W is a non-fuzzy number that provides priority weights of an evaluation criterion over others.

For the accuracy of the method, the consistency measure is performed to screen out inconsistency between responses. Since M_i is a triangular number, it has to be defuzzified into a crisp number to compute the consistency ratio (CR). The centre of area (COA) approach is used here for defuzzifying M_i. TFN $M_i(m_{i1}, m_{i2}, m_{i3},)$ can be defuzzified into a crisp value by

$$P(Mi) = [(m_{i3} - m_{i1}) + (m_{i2} - m_{i1})]/3 + m_{i1}. \tag{10}$$

Therefore, the CR of each judgement can be calculated and checked to ensure that it is lower than or equal to 0.1.

Fuzzy hierarchical TOPSIS

To evaluate remanufacturing alternatives, four fuzzy decision matrixes, $\tilde{D}_s$, are constructed with respect to four performance measures. Assume there are l alternative designs A_k ($k = 1, 2,..., l$) and n main remanufacturing processes. Each remanufacturing process has N_i criteria where the total number of criteria is equal to $\sum_{i=1}^{n} N_i$. $\tilde{x}_{kij}$ represents the value of the jth criterion within ith remanufacturing process of the kth alternative, which can be crisp data or appropriate linguistic variables which can be further represented by fuzzy numbers (e.g. $\tilde{x}_{kij} = (a_{kij}, m_{kij}, b_{kij})$). A hierarchical MCDM problem can be concisely expressed in a fuzzy decision matrix as

$$\tilde{D}_s = \begin{array}{c} \\ \\ A_1 \\ A_2 \\ \vdots \\ A_K \end{array} \begin{array}{c} \begin{array}{ccccccccc} & C_1 & & & \cdots & & C_n & & \\ C_{11} & C_{12} & \cdots & C_{1N_1} & \cdots & C_{n1} & C_{n2} & \cdots & C_{nN_n} \end{array} \\ \begin{bmatrix} \tilde{x}_{111} & \tilde{x}_{112} & \cdots & \tilde{x}_{11N_1} & \cdots & \tilde{x}_{1n1} & \tilde{x}_{1n2} & \cdots & \tilde{x}_{1nNn} \\ \tilde{x}_{211} & \tilde{x}_{212} & \ddots & \tilde{x}_{21N_1} & \cdots & \tilde{x}_{2n1} & \tilde{x}_{2n2} & \cdots & \tilde{x}_{2nNn} \\ \vdots & \vdots & & \vdots & \cdots & \vdots & \vdots & \ddots & \vdots \\ \tilde{x}_{k11} & \tilde{x}_{K12} & \cdots & \tilde{x}_{K1N_1} & \cdots & \tilde{x}_{kn1} & \tilde{x}_{kn2} & \cdots & \tilde{x}_{knNn} \end{bmatrix} \end{array} \tag{11}$$

$k = 1, 2, \cdots, l; i = 1, 2, \cdots, n; S = 1, 2, 3, 4.$

where $\tilde{x}_{kij}$ is the fuzzy evaluation score of alternative A_k with respect to criterion C_{ij}. N_i is the number of criteria within the remanufacturing process C_i. s is the number of performance measures.

In general, the evaluation criteria can be classified into two categories: benefit and cost. The benefit criterion means that a higher value is better, while for the cost criterion, the opposite is valid. The data of the decision matrix $\tilde{D}_s$ comes from different sources. Therefore, it is essential to normalize it in order to transform it into a dimensionless matrix, which allows the comparison of the various criteria. Here, the normalized fuzzy decision matrix is denoted by $\tilde{R}$ which is shown as

$$\begin{aligned} &\tilde{R} = [\tilde{r}_{kij}]_{l\times n}, \\ &k = 1, 2, \cdots, l; i = 1, 2, \cdots, n; j = 1, 2, \cdots, N_i. \end{aligned} \tag{12}$$

The normalization process can then be performed by the following operations:

$$\tilde{r}_{kij} = \begin{cases} \left(\dfrac{a_{kij}}{u_{ij}^{+}}, \dfrac{m_{kij}}{u_{ij}^{+}}, \dfrac{b_{kij}}{u_{ij}^{+}}\right), \forall_{ij}, \tilde{x}_{ij} \text{ is a benefit criterion} \\ \left(\dfrac{u_{ij}^{-}}{a_{kij}}, \dfrac{u_{ij}^{-}}{m_{kij}}, \dfrac{u_{ij}^{-}}{b_{kij}}\right), \forall_{ij}, \tilde{x}_{ij} \text{ is a cost criterion} \end{cases} \tag{13}$$

where u_{ij}^{+} and u_{ij}^{-} present the largest and the lowest value of each criterion, respectively. The weighted fuzzy normalized decision matrix is shown as

$$\tilde{V} = [\tilde{v}_{kij}]_{k\times n}, k = 1, 2, \cdots, l;\ i = 1, 2, \cdots, n;\ j = 1, 2, \cdots, N_i \tag{14}$$

where $\tilde{v}_{kij} = \tilde{r}_{kij} \otimes W_{ij}$.

Here, W_{ij} is the final weight score for each criterion which is the product of the main remanufacturing process weight score and the criterion weight score with respect to the corresponding process as follows:

$$W_{ij} = w_{C_i} \otimes w_{C_{ij}} = w_i \otimes \begin{bmatrix} w_{i1} \\ w_{i2} \\ \vdots \\ w_{iN_i} \end{bmatrix}, \quad (i = 1, 2, \cdots, n) \tag{15}$$

where w_{C_i} and $w_{C_{ij}}$ denote the weight score of the ith main remanufacturing process and the criterion C_{ij}, respectively. Both w_{C_i} and $w_{C_{ij}}$ are obtained through the fuzzy extent analysis method discussed in the 'Fuzzy extent analysis' section. The calculation results of Equation 14 can be summarized as

$$\tilde{V}_s = \begin{array}{c} \\ \\ A_1 \\ A_2 \\ \vdots \\ A_K \end{array} \begin{array}{c} \begin{array}{ccccccccc} & C_1 & & & \cdots & & C_n & & \\ C_{11} & C_{12} & \cdots & C_{1N_1} & \cdots & C_{n1} & C_{n2} & \cdots & C_{nN_n} \end{array} \\ \begin{bmatrix} \tilde{v}_{111} & \tilde{v}_{112} & & \tilde{v}_{11N_1} & \cdots & \tilde{v}_{1n1} & \tilde{v}_{1n2} & \cdots & \tilde{v}_{1nN_n} \\ \tilde{v}_{211} & \tilde{v}_{212} & \cdots & \tilde{v}_{21N_1} & \cdots & \tilde{v}_{2n1} & \tilde{v}_{2n2} & \cdots & \tilde{v}_{2nN_n} \\ \vdots & \vdots & \ddots & \vdots & \cdots & \vdots & \vdots & \ddots & \vdots \\ \tilde{v}_{k11} & \tilde{v}_{k12} & \cdots & \tilde{v}_{K1N_1} & \cdots & \tilde{v}_{kn1} & \tilde{v}_{kn2} & \cdots & \tilde{v}_{knN_n} \end{bmatrix} \end{array} \tag{16}$$

Subsequently, the fuzzy addition principle is used to aggregate the values within each remanufacturing process as follows:

$$\tilde{v}'_{ki} = \sum\nolimits_{j=1}^{C_i} \tilde{v}_{kij}, k = 1, 2, \cdots, l; i = 1, 2, \cdots, n. \tag{17}$$

The matrix $\tilde{V}$ is thus converted into the final weighted normalized fuzzy decision matrix $\tilde{V}'$,

$$\tilde{V}'_s = \begin{array}{c} \\ A_1 \\ A_2 \\ \vdots \\ A_l \end{array} \begin{array}{c} \begin{array}{cccc} C_1 & C_2 & \cdots & C_n \end{array} \\ \begin{bmatrix} \tilde{v}'_{11} & \tilde{v}'_{12} & \cdots & \tilde{v}'_{1n} \\ \tilde{v}'_{21} & \tilde{v}'_{22} & \cdots & \tilde{v}'_{2n} \\ \vdots & \vdots & \ddots & \vdots \\ \tilde{v}'_{l1} & \tilde{v}'_{l2} & \cdots & \tilde{v}'_{ln} \end{bmatrix} \end{array} . \tag{18}$$

Again, the fuzzy addition principle is used to aggregate the values of performance measures as follows:

$$\tilde{Y} = \Sigma_{s=1}^{4} \ \tilde{V}'_s, s = 1, 2, 3, 4 \tag{19}$$

and

$$\tilde{Y} = \begin{array}{c} \\ A_1 \\ A_2 \\ \vdots \\ A_l \end{array} \begin{array}{c} \begin{array}{cccc} C_1 & C_2 & \cdots & C_n \end{array} \\ \begin{bmatrix} \tilde{y}_{11} & \tilde{y}_{12} & \cdots & \tilde{y}_{1n} \\ \tilde{y}_{21} & \tilde{y}_{22} & \cdots & \tilde{y}_{2n} \\ \vdots & \vdots & \ddots & \vdots \\ \tilde{y}_{l1} & \tilde{y}_{l2} & \cdots & \tilde{y}_{ln} \end{bmatrix} \end{array} . \tag{20}$$

The addition operation is important as the hierarchical structure can be reflected only when aggregation of the weighted values within each main remanufacturing process and four performance measures is conducted.

Now, let A^+ and A^- denote the fuzzy positive idea solution (FPIS) and fuzzy negative ideal solution (FNIS), respectively. According to the aggregated fuzzy decision matrix, we have

$$\begin{aligned} A^+ &= \left(\tilde{y}_1^+, \cdots, \tilde{y}_i^+, \cdots \tilde{y}_n^+\right) \\ A^- &= \left(\tilde{y}_1^-, \cdots, \tilde{y}_i^-, \cdots \tilde{y}_n^-\right) \end{aligned} \tag{21}$$

where $\tilde{y}_i^+$ and $\tilde{y}_i^-$ are the fuzzy numbers with the largest and the smallest generalized mean, respectively. For each column i, the greatest generalized mean of $\tilde{y}_i^+$ and the lowest generalized mean of $\tilde{y}_i^-$ can be obtained, respectively. Consequently, the FPIS (A^+) and the FNIS (A^-) are derived. Then, the distances (d^+ and d^-) of each alternative from A^+ and A^- can be calculated by the area compensation method as

$$\tilde{d}_k^+ = \sum\nolimits_{i=1}^{n} d\left(\tilde{y}_{ki}, \tilde{y}_i^+\right), \quad k = 1, 2, \cdots, l; i = 1, 2, \cdots, n \tag{22}$$

$$\tilde{d}_k^- = \sum\nolimits_{i=1}^{n} d\left(\tilde{y}_{ki}, \tilde{y}_i^-\right), \quad k = 1, 2, \cdots, l; i = 1, 2, \cdots, n \tag{23}$$

$$d(\tilde{A},\tilde{B}) = \sqrt{\frac{1}{3}\left[(a_1-b_1)^2 + (a_2-b_2)^2 + (a_3-b_3)^2\right]}. \quad (24)$$

By combining the difference distances d^+ and d^-, the relative closeness index is calculated as follows:

$$\check{\varphi}_k = \frac{\tilde{d}_k^-}{\tilde{d}_k^+ + \tilde{d}_k^-}. \quad (25)$$

According the index value, the remanufacturing alternatives can be ranked from the most preferred to the least preferred feasible options.

Case study

In this section, a case study is presented to illustrate how the proposed approach can be applied to support decision-making for remanufacturing alternative evaluation. The product used in the case study is an automotive alternator. The remanufactured alternators can be used again in the vehicle. According to Kim et al. [31], the alternator has the highest remanufacturing rate. Schau et al. [11] presented a case study of remanufactured alternators, in which they applied life cycle sustainability assessment (LCSA) to three different alternator designs. However, their LCSA approach and derived results are also dependent on the underlying assumptions and data availability. In this study, the authors make reference to the case to demonstrate how the proposed model can facilitate and simplify the evaluation process from a sustainable remanufacturing design perspective. With reference to the case [11], the main remanufacturing processes are defined and the associated evaluation criteria within each process are identified as illustrated in Table 2. The total number of remanufacturing processes for evaluation is not necessary to be restricted to six as shown in Table 2. Since there is no generic guideline for handling remanufacturing processes, it varies between individual products. To be clear, relevant data, e.g. remanufacturing processes and bill of materials, has to be collected to construct a similar hierarchical structure. Here, three different alternative designs are examined through the proposed method. Design 1 is a conventional alternator with belt fitting, fan and steel bearings and cast iron housing. Design 2 is a lightweight alternator with a plastic fan and aluminium housing. Design 3 is an ultra-lightweight alternator where also the belt fitting and bearings are replaced by lightweight parts (aluminium and plastic, respectively).

After constructing the hierarchical model, it is essential to know how important one process (or its associated criterion) is over another for remanufacturing purpose. In other words, decision-makers have to determine the weights between the remanufacturing processes and the associated criteria. The different weights were calculated using the fuzzy extent analysis discussed in the 'Fuzzy extent analysis' section. Using the main remanufacturing processes as an example, the fuzzy comparison matrix of five phases is constructed as described in Table 3.

The importance weights through the pairwise comparison of the five processes with respect to the remanufacturing operation are expressed by TFNs. The different values of fuzzy synthetic extent with respect to the five main processes are denoted by S_1, S_2, S_3, S_4, and S_5, respectively. By applying Equation 2, we have

Table 2 A hierarchical structure for the evaluation of three alternative remanufacturing designs

Remanufacturing processes	Criteria	Performance measures	Alternative designs
C_1. Disassembly	C_{11}. Number of components		
	C_{12}. Joint type of components		
	C_{13}. Disassembly directions		
C_2. Detoxification	C_{21}. Brushing		
	C_{22}. Washing with chemicals		
	C_{23}. Cleaning	P_1. Value	
C_3. Machine operations	C_{31}. Crushing	P_2. Cost	Design 1 (A_1)
	C_{32}. Separation		Design 2 (A_2)
	C_{33}. Polishing	P_3. Health and Safety	Design 3 (A_3)
	C_{34}. Surface grinding	P_4. Design difficulties	
C_4. Repair	C_{41}. Parts repair		
	C_{42}. Parts replacement		
	C_{43}. Testing of parts		
C_5. Reassembly	C_{51}. Hand tools		
	C_{52}. Manual labour		
	C_{53}. Testing of finished products		

$$S_1 = (4.3, 5.5, 7.0) \otimes (1/39.5, 1/29.3, 1/21.3) \\ = (0.11, 0.19, 0.33)$$

$$S_2 = (6.0, 10.0, 14.0) \otimes (1/39.5, 1/29.3, 1/21.3) \\ = (0.15, 0.34, 0.66)$$

$$S_3 = (4.3, 6.5, 9.0) \otimes (1/39.5, 1/29.3, 1/21.3) \\ = (0.11, 0.22, 0.42)$$

$$S_4 = (3.7, 4.0, 5.0) \otimes (1/39.5, 1/29.3, 1/21.3) \\ = (0.09, 0.14, 0.24)$$

$$S_5 = (2.9, 3.3, 4.5) \otimes (1/39.5, 1/29.3, 1/21.3) \\ = (0.07, 0.11, 0.21)$$

The degree of possibility of S_i over S_j $(i \neq j)$ can be determined by Equations 3, 4, 5.

$$V(S_1 \geq S_2) = 0.54,$$

$$V(S_1 \geq S_3) = 0.87,$$

Table 3 Synthetic pairwise comparison matrix for remanufacturing processes

	C1	C2	C_3	C_4	C_5
C_1	(1, 1, 1)	(1/3, 1/2, 1)	(1, 1, 1)	(1, 1, 1)	(1, 2, 3)
C_2	(1, 2, 3)	(1, 1, 1)	(1, 2, 3)	(1, 2, 3)	(2, 3, 4)
C_3	(1, 1, 1)	(1/3, 1/2, 1)	(1, 1, 1)	(1, 2, 3)	(1, 2, 3)
C_4	(1, 1, 1)	(1/3, 1/2, 1)	(1/3, 1/2, 1)	(1, 1, 1)	(1, 1, 1)
C_5	(1/3, 1/2, 1)	(1/4, 1/3, 1/2)	(1/3, 1/2, 1)	(1, 1, 1)	(1, 1, 1)

Note: CI/RI = 0.073.

Table 4 Summary of comparative weightings of remanufacturing processes and their associated criteria

Remanufacturing processes	W_l	Criteria	$W_{l\,c}$	Final weights
C_1	0.197	C_{11}. Number of components	0.371	0.073
		C_{12}. Joint type of components	0.415	0.082
		C_{13}. Disassembly directions	0.214	0.042
C_2	0.366	C_{21}. Brushing	0.052	0.019
		C_{22}. Washing with chemicals	0.567	0.208
		C_{23}. Cleaning	0.381	0.140
C_3	0.255	C_{31}. Crushing	0.033	0.008
		C_{32}. Separation	0.231	0.059
		C_{33}. Polishing	0.416	0.106
		C_{34}. Surface grinding	0.320	0.081
C_4	0.106	C_{41}. Parts repair	0.409	0.043
		C_{42}. Parts replacement	0.409	0.043
		C_{43}. Testing of parts	0.182	0.019
C_5	0.076	C_{51}. Hand tools	0.219	0.017
		C_{52}. Manual labour	0.219	0.017
		C_{53}. Testing of finished products	0.561	0.043

$$V(S_1 \geq S_4) = 1,$$

$$V(S_1 \geq S_5) = 1.$$

Similarly,

$$V(S_2 \geq S_1) = 1, \cdot V(S_2 \geq S_3) = 1, \cdot V(S_2 \geq S_4) = 1, \cdot V(S_2 \geq S_5) = 1;$$

$$V(S_3 \geq S_1) = 1, V(S_3 \geq S_2) = 0.69, V(S_3 \geq S_4) = 1, V(S_3 \geq S_5) = 1;$$

$$V(S_4 \geq S_1) = 0.71, V(S_4 \geq S_2) = 0.29, V(S_4 \geq S_3) = 0.60, V(S_4 \geq S_5) = 1;$$

$$V(S_5 \geq S_1) = 0.58, V(S_5 \geq S_2) = 0.21, V(S_5 \geq S_3) = 0.49, V(S_5 \geq S_4) = 84;$$

Based on Equation 7, we obtain

$$\begin{aligned} d(L_1) &= \min \ V(S_1 \geq S_2, S_3, S_4, S_5) \\ &= \min(0.54, 0.87, 1, 1) \\ &= 0.54 \end{aligned} .$$

Table 5 Linguistic classification of performance measures and the corresponding TFNs

Rating level	Linguistic values	TFNs
1	Extremely high	(0, 0, 1/6)
2	Very high	(0, 1/6, 2/6)
3	High	(1/6, 2/6, 3/6)
4	Medium	(2/6, 3/6, 4/6)
5	Low	(3/6, 4/6, 5/6)
6	Very low	(4/6, 5/6, 1)
7	Extremely low	(5/6, 1, 1)

Table 6 The aggregated fuzzy decision matrix with respect to four performance measures

		A_1	A2	A_3
C_1	P_1	(0.07, 0.10, 0.13)	(0.07, 0.10, 0.13)	(0.07, 0.10, 0.13)
	P_2	(0.10, 0.14, 0.17)	(0.07, 0.10, 0.13)	(0.04, 0.07, 0.11)
	P_3	(0.08, 0.11, 0.14)	(0.09, 0.12, 0.15)	(0.09, 0.12, 0.15)
	P_4	(0.09, 0.12, 0.16)	(0.07, 0.11, 0.14)	(0.05, 0.09, 0.12)
C_2	P_1	(0.12, 0.18, 0.24)	(0.12, 0.18, 0.24)	(0.12, 0.18, 0.24)
	P_2	(0.11, 0.17, 0.23)	(0.15, 0.21, 0.27)	(0.18, 0.24, 0.30)
	P_3	(0.11, 0.17, 0.23)	(0.13, 0.19, 0.25)	(0.19, 0.25, 0.31)
	P_4	(0.13, 0.19, 0.25)	(0.12, 0.18, 0.24)	(0.12, 0.18, 0.24)
C_3	P_1	(0.11, 0.15, 0.20)	(0.08, 0.13, 0.17)	(0.06, 0.10, 0.14)
	P_2	(0.06, 0.10, 0.15)	(0.08, 0.13, 0.17)	(0.11, 0.15, 0.19)
	P_3	(0.09, 0.14, 0.18)	(0.08, 0.13, 0.17)	(0.08, 0.12, 0.16)
	P_4	(0.14, 0.18, 0.22)	(0.08, 0.13, 0.17)	(0.03, 0.07, 0.12)
C_4	P_1	(0.04, 0.05, 0.07)	(0.04, 0.06, 0.08)	(0.03, 0.05, 0.06)
	P_2	(0.06, 0.08, 0.10)	(0.04, 0.05, 0.07)	(0.01, 0.02, 0.04)
	P_3	(0.04, 0.05, 0.07)	(0.04, 0.05, 0.07)	(0.04, 0.05, 0.07)
	P_4	(0.06, 0.08, 0.10)	(0.03, 0.05, 0.06)	(0.02, 0.04, 0.05)
C_5	P_1	(0.03, 0.04, 0.05)	(0.03, 0.04, 0.05)	(0.03, 0.04, 0.06)
	P_2	(0.03, 0.05, 0.06)	(0.03, 0.04, 0.05)	(0.02, 0.04, 0.05)
	P_3	(0.03, 0.04, 0.05)	(0.03, 0.04, 0.05)	(0.03, 0.04, 0.05)
	P_4	(0.05, 0.06, 0.07)	(0.03, 0.04, 0.05)	(0.01, 0.02, 0.04)

Similarly,

$$d(L_2) = 1, d(L_3) = 0.69, d(L_4) = 0.29, d(L_5){-}0.21\ .$$

Therefore, $W' = (0.54,\ 1{,}0.69,\ 0.29,\ 0.21)$ after the normalization process, so the weight vector with respect to the five main remanufacturing processes - C_1, C_2, C_3, C_4 and C_5 - can be expressed as

$$W = (0.197, 0.366, 0.255, 0.106, 0.076).$$

Using the same approach, the weights of identified evaluation criteria with respect to their associated remanufacturing processes can be derived, and the results are summarized in Table 4. The final weight scores for evaluation criteria were obtained by calculating the product of criteria weight scores and the weight scores of its associated

Table 7 The final aggregated fuzzy decision matrix

	A_1	A_2	A_3
C_1	(0.34, 0.47, 0.60)	(0.29, 0.42, 0.55)	(0.25, 0.38, 0.51)
C_2	(0.47, 0.71, 0.96)	(0.51, 0.76, 1.00)	(0.61, 0.85, 1.10)
C_3	(0.40, 0.57, 0.74)	(0.34, 0.51, 0.68)	(0.28, 0.44, 0.61)
C_4	(0.20, 0.27, 0.34)	(0.14, 0.21, 0.28)	(0.09, 0.16, 0.23)
C_5	(0.13, 0.18, 0.23)	(0.11, 0.16, 0.21)	(0.09, 0.14, 0.19)

Table 8 The relative closeness index of alternative remanufacturing designs along with the final ranking

	d^+	d^-	$\check{\varphi}_k$	Rank
A_1	0.138	0.369	0.727	1
A_2	0.285	0.223	0.439	2
A_3	0.369	0.138	0.273	3

remanufacturing process. At the same time, the consistency ratio of each judgement was calculated and checked to ensure that it is lower than or equal to 0.1.

After that, fuzzy hierarchical TOPSIS is employed for the evaluation of three remanufacturing alternatives. Performance was rated to the three alternative designs with respect to the four proposed remanufacturing performance measures against all the evaluation criteria. The qualitative explanation of rating levels and its corresponding TFNs are described in Table 5. The results were then used to constitute a hierarchical decision-making matrix $\tilde{D}$ as shown in the Appendix. The hierarchical decision-making matrix was then normalized using Equation 13. By computing the product of the normalized hierarchical decision matrix $\tilde{D}$ and the final weight scores for each evaluation criterion, the weighted normalized fuzzy decision matrix $\tilde{V}$ is obtained. By aggregating the values belonging to each remanufacturing process by fuzzy addition principle, the weighted normalized fuzzy decision matrix $\tilde{V}'$ is acquired as illustrated in Table 6. By grouping the four performance measure outputs, the final fuzzy decision matrix $\tilde{Y}$ is obtained as shown in Table 7.

The largest generalized mean and the smallest generalized mean of each main criterion could then be selected constituting the FPIS (A^+) and the FNIS (A^-). Now, the difference distances of each of the alternatives (d_k^+ and d_k^-) can be calculated as in Equations 22, 23, 24. Finally, combining the difference distances, the relative closeness index for each alternative can be obtained. The results are presented in Table 8, together with the corresponding rankings based on the index values. Among the three alternative designs, the conventional alternator design (A_1) has the highest relative closeness index and therefore should be recommended.

Using the relative closeness index, design 1 (A_1) tops the ranking list among the three remanufacturing alternatives. It is followed by design 2 (A_2) and design 3 (A_3). In order to

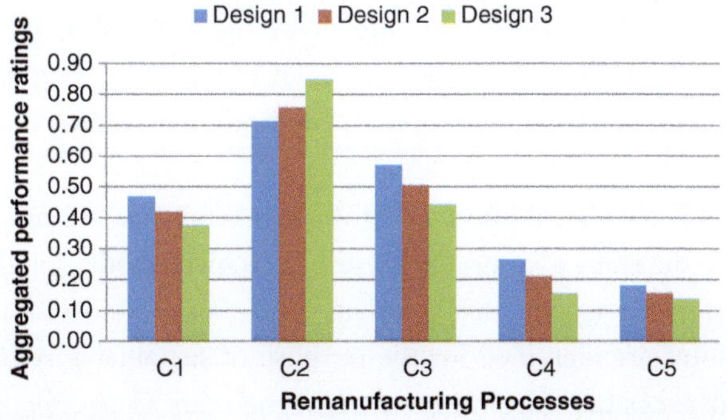

Figure 4 Weighted performance measures of three alternative designs with respect to main remanufacturing processes.

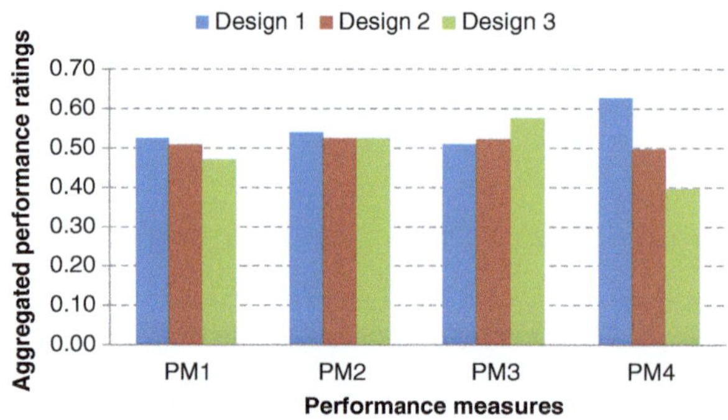

Figure 5 Performance ratings of three alternative designs with respect to different measures.

provide insights of this decision, further analysis was conducted. The analysis result displayed in Figure 4, shows the performance ratings of the three alternative designs with respect to the main remanufacturing processes. Overall the detoxification process (C_2) contributes most to the whole remanufacturing operation followed by the machine operations process (C_3). Although design 1 (A_1) has a slightly lower performance than the other designs in the detoxification process, crucially, it performs better in the other remanufacturing processes compared to the other two designs. This is one of the key reasons that design 1 stands out among the alternative remanufacturing designs. This is further proven in the ratings of three alternative designs with respect to different performance measures as illustrated in Figure 5. The conventional alternator design (A_1) tops the list in three out four performance measures including P_1 (value), P_2 (cost) and P_3 (design difficulties). Nevertheless, this does not underline the significance of other remanufacturing processes or performance measures, among which, design 1 (A_1) was scored lower than the other two designs (A_2 and A_3). In fact, it is important for designers to take a balanced approach when evaluating design options for sustainable remanufacturing purpose.

Conclusions

Remanufacturing is increasingly playing an important role in moving towards a more sustainable economy. The concept of remanufacturing can be deemed as a competitive strategy for manufacturers to satisfy diverse requirements from customers as well as policy makers. This paper proposed a hierarchical framework for evaluating alternative designs from the remanufacturing perspective. In addition to the evaluation framework, fuzzy extent analysis is used to calculate the importance weights of remanufacturing processes and associated evaluation criteria, and fuzzy hierarchical TOPSIS is applied to evaluate alternative product designs. A case study of remanufacturing an automotive alternator was presented to validate the proposed methodology and to demonstrate its effectiveness for remanufacturing design evaluation.

This article makes three key contributions. First, from a remanufacturing point of view, this research specifically develops a comprehensive hierarchical model for the evaluation of alternative designs. Key remanufacturing activities, the associated evaluation criteria and performance measures are identified for the purpose of sustainable remanufacturing. Second, the research advances the use of fuzzy MCDM methods as an effective and realistic modelling approach for evaluating design alternatives from the remanufacturing perspective. Compared to other approaches, e.g. LCA, the integrated fuzzy approach proposed in this paper provides a practical design evaluation solution. While the analysis result is in line with the findings from the study of Schau et al. [11], our approach is simple and less demanding upon the computational power and time needed to make a decision. In addition, the proposed approach is less constraining to rigorous data that is required to conduct a conventional LCA. It is still tractable enough to capture the uncertainty of a product remanufacturing life cycle and provides the efficiency and flexibility to tap the subjectivity and preferences of decision-makers. Third, through the case study, it provides some insights into how the application of the proposed integrated fuzzy approach can support a rational product design selection decision in order to achieve sustainable remanufacturing.

Despite the various advantages outlined in the paper, the presented approach also has its own limitations. For example, decision-makers have to make subjective decisions in the pairwise comparisons in evaluating remanufacturing alternatives. Using reliable data sources instead of subjective decision could lead to more accurate decisions. Therefore, one future research direction is to consider a more objective method such as data envelopment analysis (DEA). Furthermore, the dynamic characteristics and interconnection among the evaluation criteria are not considered in the hierarchy model. The interrelationship between these criteria may generate a different result of the importance weights. Future research may need to tackle this shortcoming by using the decision-making trial and evaluation laboratory (DEMATEL) method or analytic network process (ANP).

Appendix

A hierarchical decision-making matrix $\tilde{D}$ is shown in Table 9.

Table 9 Fuzzy design matrix

			A1			A2			A3	
C_{11}	P_1	2/6	3/6	4/6	2/6	3/6	4/6	2/6	3/6	4/6
	P_2	4/6	5/6	1	2/6	3/6	4/6	0	1/6	2/6
	P_3	2/6	3/6	4/6	2/6	3/6	4/6	2/6	3/6	4/6
	P_4	3/6	4/6	5/6	2/6	3/6	4/6	1/6	2/6	3/6
C_{12}	P_1	2/6	3/6	4/6	2/6	3/6	4/6	2/6	3/6	4/6
	P_2	3/6	4/6	5/6	2/6	3/6	4/6	2/6	3/6	4/6
	P_3	3/6	4/6	5/6	3/6	4/6	5/6	3/6	4/6	5/6
	P_4	3/6	4/6	5/6	2/6	3/6	4/6	1/6	2/6	3/6
C_{13}	P_1	2/6	3/6	4/6	2/6	3/6	4/6	2/6	3/6	4/6
	P_2	2/6	3/6	4/6	2/6	3/6	4/6	2/6	3/6	4/6
	P_3	2/6	3/6	4/6	3/6	4/6	5/6	3/6	4/6	5/6
	P_4	2/6	3/6	4/6	3/6	4/6	5/6	4/6	5/6	1
C_{21}	P_1	2/6	3/6	4/6	2/6	3/6	4/6	2/6	3/6	4/6
	P_2	2/6	3/6	4/6	2/6	3/6	4/6	2/6	3/6	4/6
	P_3	2/6	3/6	4/6	3/6	4/6	5/6	4/6	5/6	1
	P_4	3/6	4/6	5/6	2/6	3/6	4/6	1/6	2/6	3/6
C_{22}	P_1	2/6	3/6	4/6	2/6	3/6	4/6	2/6	3/6	4/6
	P_2	1/6	2/6	3/6	2/6	3/6	4/6	3/6	4/6	5/6
	P_3	1/6	2/6	3/6	2/6	3/6	4/6	3/6	4/6	5/6
	P_4	2/6	3/6	4/6	2/6	3/6	4/6	2/6	3/6	4/6

Table 9 Fuzzy design matrix *(Continued)*

C_{23}	P_1	2/6	3/6	4/6	2/6	3/6	4/6	2/6	3/6	4/6
	P_2	3/6	4/6	5/6	3/6	4/6	5/6	3/6	4/6	5/6
	P_3	3/6	4/6	5/6	2/6	3/6	4/6	3/6	4/6	5/6
	P_4	2/6	3/6	4/6	2/6	3/6	4/6	2/6	3/6	4/6
C_{31}	P_1	1/6	2/6	3/6	2/6	3/6	4/6	3/6	4/6	5/6
	P_2	1/6	2/6	3/6	2/6	3/6	4/6	3/6	4/6	5/6
	P_3	2/6	3/6	4/6	2/6	3/6	4/6	2/6	3/6	4/6
	P_4	3/6	4/6	5/6	1/6	2/6	3/6	0/6	1/6	2/6
C_{32}	P_1	2/6	3/6	4/6	2/6	3/6	4/6	2/6	3/6	4/6
	P_2	3/6	4/6	5/6	2/6	3/6	4/6	1/6	2/6	3/6
	P_3	1/6	2/6	3/6	2/6	3/6	4/6	3/6	4/6	5/6
	P_4	4/6	5/6	1	2/6	3/6	4/6	0/6	1/6	2/6
C_{33}	P_1	2/6	3/6	4/6	2/6	3/6	4/6	2/6	3/6	4/6
	P_2	1/6	2/6	3/6	2/6	3/6	4/6	3/6	4/6	5/6
	P_3	3/6	4/6	5/6	2/6	3/6	4/6	1/6	2/6	3/6
	P_4	3/6	4/6	5/6	2/6	3/6	4/6	1/6	2/6	3/6
C_{34}	P_1	4/6	5/6	1	2/6	3/6	4/6	0/6	1/6	2/6
	P_2	1/6	2/6	3/6	2/6	3/6	4/6	3/6	4/6	5/6
	P_3	2/6	3/6	4/6	2/6	3/6	4/6	2/6	3/6	4/6
	P_4	3/6	4/6	5/6	2/6	3/6	4/6	1/6	2/6	3/6
C_{41}	P_1	1/6	2/6	3/6	3/6	4/6	5/6	2/6	3/6	4/6
	P_2	4/6	5/6	1	2/6	3/6	4/6	0/6	1/6	2/6
	P_3	2/6	3/6	4/6	2/6	3/6	4/6	2/6	3/6	4/6
	P_4	4/6	5/6	1	2/6	3/6	4/6	1/6	2/6	3/6
C_{42}	P_1	3/6	4/6	5/6	2/6	3/6	4/6	1/6	2/6	3/6
	P_2	4/6	5/6	1	2/6	3/6	4/6	0/6	1/6	2/6
	P_3	2/6	3/6	4/6	2/6	3/6	4/6	2/6	3/6	4/6
	P_4	3/6	4/6	5/6	1/6	2/6	3/6	1/6	2/6	3/6
C_{43}	P_1	2/6	3/6	4/6	2/6	3/6	4/6	2/6	3/6	4/6
	P_2	2/6	3/6	4/6	2/6	3/6	4/6	2/6	3/6	4/6
	P_3	2/6	3/6	4/6	2/6	3/6	4/6	2/6	3/6	4/6
	P_4	3/6	4/6	5/6	2/6	3/6	4/6	1/6	2/6	3/6
C_{51}	P_1	2/6	3/6	4/6	2/6	3/6	4/6	2/6	3/6	4/6
	P_2	2/6	3/6	4/6	2/6	3/6	4/6	2/6	3/6	4/6
	P_3	2/6	3/6	4/6	2/6	3/6	4/6	2/6	3/6	4/6
	P_4	3/6	4/6	5/6	2/6	3/6	4/6	1/6	2/6	3/6
C_{52}	P_1	2/6	3/6	4/6	3/6	4/6	5/6	4/6	5/6	1
	P_2	2/6	3/6	4/6	3/6	4/6	5/6	4/6	5/6	1
	P_3	2/6	3/6	4/6	2/6	3/6	4/6	2/6	3/6	4/6
	P_4	4/6	5/6	6/6	2/6	3/6	4/6	0	1/6	2/6
C_{53}	P_1	2/6	3/6	4/6	2/6	3/6	4/6	2/6	3/6	4/6
	P_2	3/6	4/6	5/6	2/6	3/6	4/6	1/6	2/6	3/6
	P_3	2/6	3/6	4/6	2/6	3/6	4/6	2/6	3/6	4/6
	P_4	4/6	5/6	6/6	2/6	3/6	4/6	1/6	2/6	3/6

Abbreviations
AHP: Analytic hierarchy process; ANP: Analytic network process; DEA: Data envelopment analysis; DEMATEL: Decision-making trial and evaluation laboratory; LCA: Life cycle assessment; MCDM: Multi-criteria decision-making; TOPSIS: Technique for Order of Preference by Similarity to Ideal Solution.

Competing interests
The authors declare that they have no competing interests.

Authors' contributions
XW contributed to the design of the study, developed the integrated fuzzy methodology and carried out the case study. HKC participated in the design of the study, carried out the literature review and developed the hierarchical model for the product design evaluation from the remanufacturing perspective. Both authors helped draft the manuscript, read and approved the final manuscript.

Authors' information
Dr. XW is a Senior Lecturer in Management at the School of Economics, Finance and Management, University of Bristol. He obtained his PhD in Management Studies from the University of Liverpool, MSc in e-Business Management from University of Warwick and a bachelor's degree in Computer Science from Zhejiang University, China. His research interests include operations and supply chain management, multi-criteria decision analysis and applied operational research. Dr. HKC is a Senior Lecturer in Operations and Supply Chain Management in the Norwich Business School. He gained his Ph.D., an MSc (with distinction) and a bachelor degree from the University of Hong Kong. He also earned a bachelor's degree in Economics and Management from London School of Economics and Political Science. HKC is a member of the Institution of Engineering and Technology, the Chartered Institute of Marketing, and a senior member of the Institute of Electrical and Electronics Engineers. His main research topics are operations and supply chain management.

Acknowledgements
An early version of this paper was presented at the 6th International Conference on Design and Manufacture for Sustainable Development, Hangzhou, China in April 2013. The authors are grateful to the Editor-in-Chief of the Journal for giving us the opportunity to revise and extend the paper for this submission.

Author details
[1]Department of Management, University of Bristol, Bristol BS8 1TN, UK. [2]Norwich Business School, University of East Anglia, Norwich NR4 7TJ, UK.

References

1. Beamon, BM: Designing the green supply chain. Logist. Inf. Manag. **12**(4), 332–342 (1999)
2. Pigosso, DC, Zanette, ET, Ometto, AR, Rozenfeld, H: Ecodesign methods focused on remanufacturing. J. Clean. Prod. **18**(1), 21–31 (2010)
3. Kaya, O: Incentive and production decisions for remanufacturing operations. Eur. J. Oper. Res. **201**(2), 442–453 (2010)
4. Ilgin, MA, Gupta, SM: Remanufacturing Modelling and Analysis. CRC, Taylor and Francis, Florida (2012)
5. Guide, VDR: Production planning and control for remanufacturing: industry practice and research needs. J. Oper. Manag. **8**(4), 467–483 (2000)
6. Subramoniam, R, Huisingh, D, Chinnam, RB: Remanufacturing for the automotive aftermarket-strategic factors: literature review and future research needs. J. Clean. Prod. **17**(13), 1163–1174 (2009)
7. Savaskan, RC, Bhattacharya, S, Van Wassenhove, LN: Closed-loop supply chain models with product remanufacturing. Manag. Sci. **50**(2), 239–252 (2004)
8. Yung, WKC, Chan, HK, Wong, DC, So, JHT, Choi, ACK, Yue, TM: Eco-design of a personal electronic product subject to energy-using product directive. Int. J. Prod. Res. **50**(5), 1411–1423 (2012)
9. Junnila, S: Life cycle management of energy-consuming products in companies using IO-LCA. Int. J. Life Cycle Assess. **13**(5), 432–439 (2008)
10. Chan, HK, Wang, X, White, G, Nick, Y: An extended fuzzy AHP approach for the evaluation of green product designs. IEEE Trans. Eng. Manag. **60**(2), 327–339 (2013)
11. Schau, EM, Traverson, M, Finkbeiner, M: Life cycle approach to sustainability assessment: a case study of remanufactured alternators. J. Remanuf **2**(5), 1–14 (2012)
12. Saaty, TL: The Analytic Hierarchy Process. McGraw-Hill, New York (1980)
13. Ho, W: Integrated analytic hierarchy process and its applications–a literature review. Eur. J. Oper. Res. **186**(1), 211–228 (2008)
14. Zadeh, LA: Fuzzy sets. Inf. Control. **8**(3), 338–353 (1965)
15. Chang, DY: Applications of the extent analysis method on fuzzy AHP. Eur. J. Oper. Res. **95**(3), 649–655 (1996)
16. Chan, HK, Wang, X: Fuzzy Hierarchical Model for Risk Assessment: Principles, Concepts, and Practical Applications. Springer, London (2013)
17. Kahraman, C, Cebeci, U, Ruan, D: Multi-attribute comparison of catering service companies using FAHP: the case of Turkey. Int. J. Prod. Econ. **87**(2), 171–184 (2004)
18. Celik, M, Er, ID, Ozok, AF: Application of fuzzy extended AHP methodology on shipping registry selection: the case of Turkish maritime industry. Expert Syst. Appl. **36**(1), 190–198 (2009)

19. Wang, X, Li, D, Shi, X: A fuzzy enabled model for aggregative food safety risk assessment in food supply chains. Prod. Plan. Control **23**(5), 377–395 (2012)
20. Bottani, E, Rizzi, A: A fuzzy TOPSIS methodology to support outsourcing of logistics services. Supply Chain Manag. **11**(4), 294–308 (2006)
21. Ertuğrul, İ, Karakaşoğlu, N: Comparison of fuzzy AHP and fuzzy TOPSIS methods for facility location selection. Int. J. Adv. Manuf. Technol. **39**(7–8), 783–795 (2008)
22. Hwang, CL, Yoon, K: Multiple attributes decision making methods and applications. Springer, Berlin (1981)
23. Krohling, RA, Campanharo, VC: Fuzzy TOPSIS for group decision making: a case study for accidents with oil spill in the sea. Expert Syst. Appl. **38**(4), 4190–4197 (2011)
24. Wang, X, Durugbo, C: Analysing network uncertainty for industrial product-service delivery: a hybrid fuzzy approach. Expert Syst. Appl. **40**(11), 4621–4636 (2013)
25. Wang, X, Chan, HK: A hierarchical fuzzy TOPSIS approach to assess improvement area when implementing green supply chain initiatives. Int. J. Prod. Res. **51**(10), 3117–3130 (2013)
26. Biswas, WK, Duong, V, Frey, P, Islam, MN: A comparison of repaired, remanufactured and new compressors used in Western Australian small-and medium-sized enterprises in terms of global warming. J. Remanuf. **3**(1), 1–7 (2013)
27. Kerr, W, Ryan, C: Eco-efficiency gains from remanufacturing: a case study of photocopier remanufacturing at Fuji Xerox Australia. J. Cleaner Prod. **9**(1), 75–81 (2001)
28. Seliger, G, Franke, C, Ciupek, M, Başdere, B: Process and facility planning for mobile phone remanufacturing. CIRP Annals-Manuf. Technol. **53**(1), 9–12 (2004)
29. Sundin, E, Bras, B: Making functional sales environmentally and economically beneficial through product remanufacturing. J. Clean. Prod. **13**(9), 913–925 (2005)
30. Hatcher, GD, Ijomah, WL, Windmill, JF: Design for remanufacturing in China: a case study of electrical and electronic equipment. J. Remanuf. **3**(1), 1–11 (2013)
31. Kim, H-J, Skerlos, S, Severengiz, S, Seliger, G: Characteristics of the automotive remanufacturing enterprise with an economic and environmental evaluation of alternator products. Int. J. Sustainable Manuf. **1**(4), 437–449 (2009)

6

Remanufacturing process for used automotive electronic control components in China

Junjun Wang and Ming Chen*

* Correspondence: mingchen@sjtu.edu.cn
School of Mechanical Engineering, Shanghai Jiao Tong University, Shanghai 200240, People's Republic of China

Abstract

China's recycling roadmap and technology scheme for used automotive electronic control components are investigated. The mathematical analysis model of the remanufacturing process is established on the basis of stochastic network technology, as well as on the graphical evaluation and review technique (GERT). In addition, the calculation method used for estimating single-product remanufacturing time is examined. The objective of this study is to determine the probability of success for the remanufacturing of used automotive electronic control components and remanufacturing time. On the basis of experimental parameters, we simulate the remanufacturing process using the Monte Carlo simulation in Crystal Ball. Compared with the result of the GERT model (8.5114 h), the simulation error rate is 0.225%. This consistency in results indicates that both the stochastic network model and Crystal Ball can accurately simulate the remanufacturing process of used automotive electronic control components, making these techniques feasible approaches for such processes. Aside from numerical experiments on and sensitivity analyses of key processes, the relationship between total remanufacturing time and five influencing factors is identified. Total remanufacturing time can be significantly reduced by optimizing the key processes. The optimization methods are also investigated.

Keywords: Used automotive electronic control components; Recycling; Stochastic network model; Monte Carlo simulation in Crystal Ball; Simulation; China

Introduction

According to the statistics of the China Association of Automobile Manufacturers, the vehicle market of China sets a new sales record of approximately 18,264,700 in 2010, bringing the total number of vehicles in the country to 90,860,000. With the rise in vehicle sales, relevant agencies forecasted that the volume of scrap vehicles will reach 6.44 million in 2015. With the rapid development of electronic information technology, the use of electronic components in vehicles has dramatically increased. Electronic control units (ECUs) control almost all of the functions of a vehicle [1]. According to the statistics of Bayerische Motoren Werke AG, automotive electronic components have exceeded more than 30% of the cost of vehicles. There are 48 microcomputers in some luxury cars, and the electronic components have accounted for more than 50% of their cost. Thus, the volume of used automotive electronic control components in China is foreseen to soar in the next few years. Used automotive electronic control components in the country are currently scrapped together with end-of-life vehicles (ELVs) for material recycling. These vehicles are discarded inappropriately because of lack of technology support and profitable models. This improper disposal results in serious environmental pollution and resource wastage, thereby creating an urgent need for the development of recycling processes for used electronic control components.

In developed countries, ECUs in ELVs can be reused after remanufacturing [2]; they can also be exported after strict bench tests and are marked 'recycled products' [3]. China automotive experts and scholars have realized that recycling and reusing used automotive electronic control components are important issues. Figure 1 shows the recycling technology roadmap based on the current research and development situation of ECUs, as well as on the actual conditions, in China [4]. The three characteristics of used automotive electronic control components are as follows: (1) These components contain heavy metals (such as Pd), which if not properly recycled can seriously endanger the environment as well as the health and survival of human beings and other organisms. (2) These components are high-tech, highly reliable, and high value-added products. (3) The residual life of these components is very long. Thus, the remanufacture and reuse of these components are feasible initiatives. On the basis of the three characteristics mentioned above and the current technical strength of China, we infer that the best recycling scheme for used automotive electronic control components follows the order reuse > remanufacture > material recycle > harmless treatment.

However, a number of uncertainties are associated with the recycling of used automotive electronic control components. These include the estimation of operation and cost-related parameters for the recycling process, as well as the decisions about resolution for product returns and costs of coordination along the recycling system [5]. These uncertainties are related to time, quality, quantity, and variety of

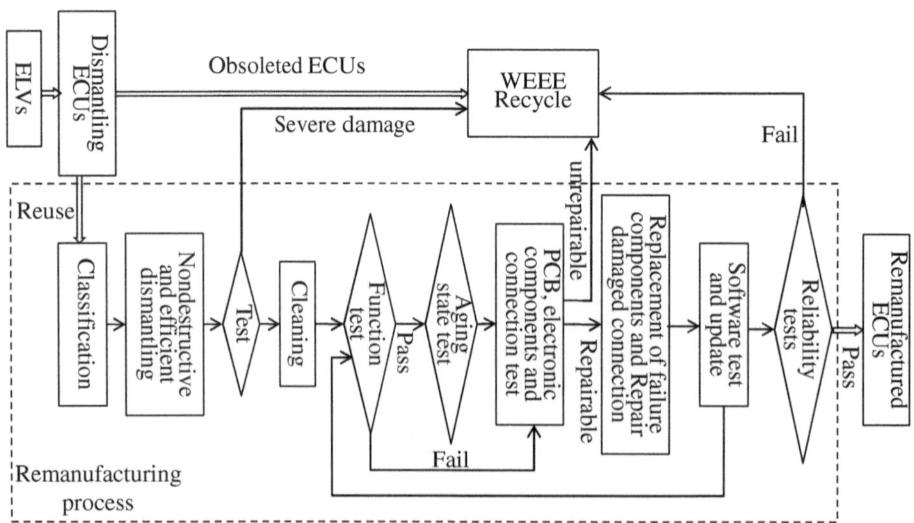

Figure 1 Recycling technology roadmap of used automotive electronic control components in China.

returns; they are also dependent on the influence of regulations and consumer opinion.

The graphical evaluation and review technique (GERT) is an approach used for analyzing a class of networks with the following characteristics: (1) a probability that a branch of the network is part of a realization of the network and (2) an elapsed time or time interval associated with the branch if the branch is part of the realization of the network. Such networks are referred to as stochastic networks and consist of a set of branches and nodes. The GERT research was conceptualized by the RAND Corporation at the Apollo Checkout Study for Headquarters and sponsored by the National Aeronautics and Space Administration (NASA) under contract NASr-21(08); it evolved from a study of the terminal countdown of an Apollo space system in April 1966 [6]. The GERT is a procedure intended for the analysis of a stochastic network composed of EXCLUSIVE-OR, INCLUSIVE-OR, and AND nodes (vertices), as well as multi-parameter branches (transmittances or edges). In the GERT, the branches of the networks are described by two or more parameters: (1) a probability that the branch is traversed and (2) the time (or other attribute) required to traverse the branch if it is taken. A transformation is developed to combine these two parameters into a single parameter. For EXCLUSIVE-OR nodes, a method is derived for the evaluation of networks in terms of the probability of generating an output node. This method is also used to assess the moment-generating function (mgf) of the time spent in generating the output node. The overall concept that comprises the stochastic networks, transformation, and evaluation method is labeled the GERT. The memorandum presents the general concepts and fundamentals of the GERT and introduces its applications for future research.

Samli and Bellas [7] present a discussion of the special characteristics of the GERT and illustrated its capabilities as a tool in the planning and control of marketing research; the authors compare the differences between GERT and program/ project evaluation and review technique (PERT). Randolph and Ringeisen [8] use the GERT to analyze the teaching-learning process and indicated its potential in learning theory. They demonstrate that this approach is very similar to the Markov processes, which have been used extensively to analyze learning models. Sharif and Chin [9] present an analytical technique for technological forecasting. The developed procedure is based on an integrated approach, which combines normative and exploratory approaches to technological forecasting using a stochastic network system. Generally, it can also be applied to technological forecasting. Singh and Shankar [10] apply the GERT to model a two-stage attribute sampling plan based on an exponential failure model. Guo and Qiao [11] establish a hierarchical GERT network to describe the product development process. The activity-time parameters in the process are calculated, and the resource conflicts among parallel activities can be resolved through the readjustment of the activities according to the weighted time-resource priority rule.

Hayashi et al. [12] evaluate the research and development (R&D) of polymer electrolyte fuel cell application technologies using a R&D process model constructed with the GERT. The authors take uncertainties in technology developments into consideration. Abdi et al. [13] model and analyze the mechanization projects of wheat production using the GERT networks. Their results show that the network model can answer any statistical questions concerning the project. Network models, particularly the GERT networks, are increasingly becoming powerful tools for modeling, scheduling, planning, controlling, and analyzing agricultural mechanization projects. Chia-Nan et al. [14] resolve the reentry problem in wafer manufacturing using the GERT and solve the uncertainty problem using a vague set. The algorithm steps for vague GERT are proposed on the basis of the manufacturing procedure of the lithographic area in 300-mm wafer fabrication. A simple decision support system is developed to process the complex calculation procedure and provide more information to managers.

The recycling system for used automotive electronic control components can be described in terms of stochastic networks, and many problems encountered in this system can be solved using the GERT [15].

The present study aims to determine the probability of success for remanufacturing used automotive electronic control components and remanufacturing time, as well as investigate the feasibility of reusing automotive electronic control components. The proposed model is set up on the basis of the recycling technology roadmap of used automotive electronic control components (Figure 1). Calculations are then performed using an analytical method. After the simulation of the remanufacturing process in the Crystal Ball software, the feasibility of the stochastic network model is verified through the comparison and analysis of the numerical calculation and software simulation results.

GERT model formulation and calculation using an analytical method

Components of stochastic networks

Stochastic networks are composed of directed branches (arcs, edges, and transmittances) and logical nodes (vertices). A directed branch is associated with nodes at which it terminates. Two parameters are associated with a branch: (1) the probability that a branch is taken, p_a, given that the node from which it emanates is realized; and (2) if a branch is taken, a time t_a required to accomplish the activity that the branch represents, where t_a can be a random variable.

A node in a stochastic network consists of an input (receiving and contributive) side and an output (emitting and distributive) side. Three logical relations are found on the input side, and two types of relations are found on the output side. The former is shown in Table 1.

Table 1 GERT input side node

Name	Symbol	Characteristics
EXCLUSIVE-OR		The realization of any branch leading into the node causes the node to be realized. However, only one of the branches leading into this node can be realized at a given time
INCLUSIVE-OR		The realization of any branch leading into the node causes the node to be realized. The time of realization is the smallest of the completion times of the activities leading into the INCLUSIVE-OR node
AND		The node is realized only if all the branches leading into the node are realized. Thus, the realization time is the largest of the completion times of the activities leading into the AND node

Table 2 GERT output side node

Name	Symbol	Characteristics
DETERMINISTIC		All branches emanating from the node are taken if the node is realized (i.e., all branches emanating from this node have a p parameter equal to 1)
PROBABILISTIC	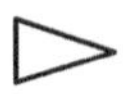	Exactly one branch emanating from the node is taken if the node is realized

GERT network calculation

The steps involved in applying the GERT are as follows:

1. A qualitative description of a system or problem is converted into a model in network form.
2. The necessary data are collected to describe the branches of the network.
3. An equivalent one-branch function between two nodes of the network is obtained.
4. The equivalent function is converted into the following two performance measures of the network:
   a. The probability that a specific node is realized.
   b. The mgf of the time associated with an equivalent network.
5. Inferences concerning the system under study are made on the basis of the information obtained in step 4.

The two relations on the output side are shown in Table 2.

For notational convenience, the input and output symbols are combined to show that six types of nodes are possible (Table 3).

Moment-generating function and *w*-function

A function w_k is defined for each arc k. If p_k is the probability that arc k is chosen and if $M_K(s)$ is the mgf of random variable X_k, which represents the length of the kth arc, then w_k is given by

$$w_k(s) = p_k M_k(s) = p_k \int_{-\infty}^{+\infty} e^{sx_k} dF_{X_k}(x_k), \quad (1)$$

where F_{X_k} is the distribution function of X_k and the integral is the familiar Riemann-Stieltjes integral.

A closing arc is added to apply a topology equation from flow graph theory because this equation pertains only to closed networks. Any network can be depicted by a black box, in which the closing arc creates a single-loop network, where w_c represents the w-function of the closing arc, and w_E is the w-function of the entire network (see Figure 2).

Table 3 GERT node symbols

Output	Input		
	EXCLUSIVE-OR	INCLUSIVE-OR	AND
DETERMINISTIC	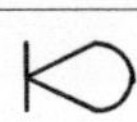	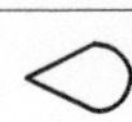	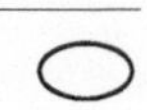
PROBABILISTIC			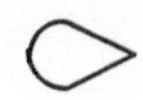

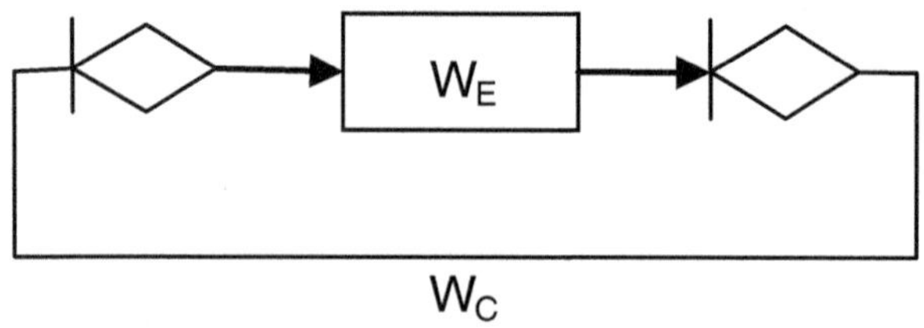

Figure 2 *w*-function of a network.

We define $w_E(s)$ by $w_E(s) = p_E m_E(s)$, where p_E is the probability that the black box is entered and $M_E(s)$ is the mgf of the entire network. Given that $M_E(0) = 1$,

$$p_E = w_E(0). \tag{2}$$

Thus, for any value of s, function $w_E(s)$ yields the mgf from the equation [8]:

$$M_E(s) = \frac{w_E(s)}{w_E(0)}. \tag{3}$$

The mgf of common distributions are shown in Table 4.

Equivalent network

An equivalent network is defined as a reduction of a multi-branched network into a one-branch network, where the parameters of the latter are derived from the parameters of the multi-branched network branches.

Table 5 illustrates the equivalent one-branch network for series, parallel, and self-loop networks.

To obtain the expected value of X, the total network characteristic should identify the first moment from the mgf:

$$\mu_E = \frac{dM_E(s)}{ds}|s = 0. \tag{4}$$

Formulation and calculation of the model for used automotive electronic control components

A number of uncertainties are associated with the recycling process of used automotive electronic control components. The recycling process is of stochastic type. The relationship between the individual remanufacturing processes and the working time of each process is random. Because of the different work environments, the damage

Table 4 Moment-generating function of some distributions

Name	Distribution law (density function)	Mean	Variance	mgf
Uniform	$f(x) = \begin{cases} 1/(b-a), (a < x < b) \\ 0, (\text{else}) \end{cases}$	$(a+b)/2$	$(b-a)^2/12$	$(e^{bs} - e^{as})^2/s(b-a)$
Normal	$f(x) = \exp\left[-(x-\mu)^2/2\sigma^2\right]/\sqrt{2\pi}\sigma$ $(-\infty < \mu < +\infty, \sigma > 0)$	μ	σ^2	$\exp\,(\mu s + \sigma^2 s^2/2)$

Table 5 Network reduction employing the topological equation

Network type	Graphical representation	Equivalent network
Series		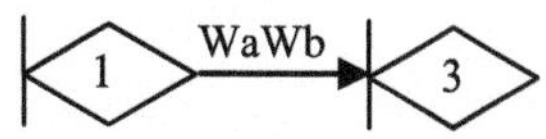
Parallel	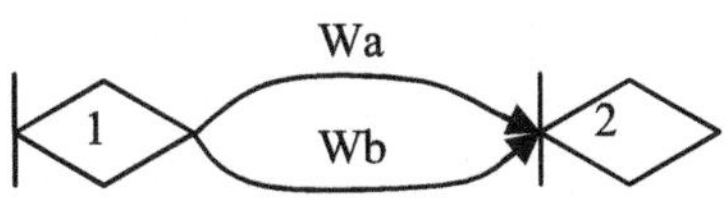	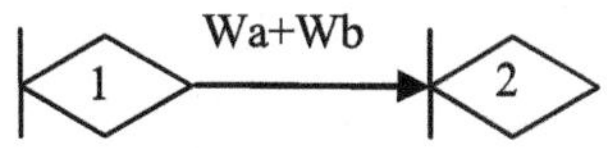
Self-loop(1)	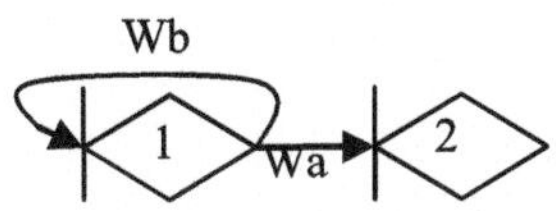	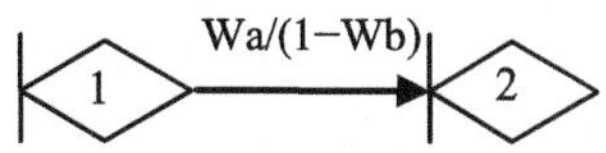
Self-loop(2)	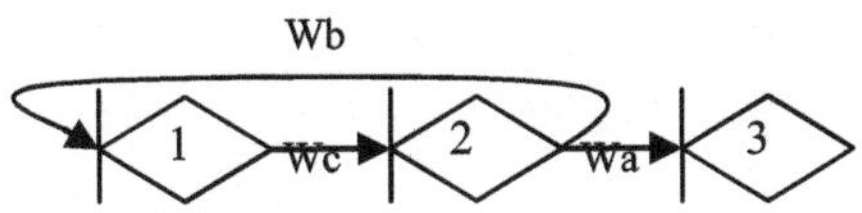	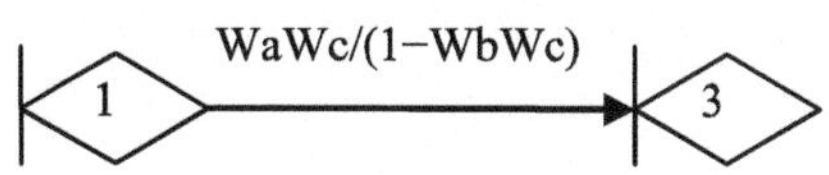

imposed on the ECU is not constant, and its disassembly, cleaning, test, and remanufacturing processes are random. According to the recycling technology roadmap of used automotive electronic control components in China (Figure 1), the remanufacturing process is a circular loop. Thus, modeling the recycling process using the GERT is a feasible approach. The GERT model is shown in Figure 3.

In accordance with the experimental data and statistical methods, we depict the manner by which probability is derived, and the time probability distribution for each activity is performed in the GERT network model graph (Figure 3) in Table 6.

The GERT network model graph (see Figure 3) can be broken down into two parts, as shown in Figures 4 and 5.

The branches connected to node 2 of the first step simplification of the GERT network model graph (I, Figure 4) represent the activity 'WEEE recycle,' which is not the focus of this study.

Figure 5 shows the remanufacturing process of used automotive electronic control components. After being remanufactured, these components can be reused and labeled 'remanufactured products.' This recycling scheme can help minimize environmental pollution and facilitate ECU technology breakthroughs in China. In addition, it is the most economical and high value-added recycling mode. This study focuses on this recycling scheme.

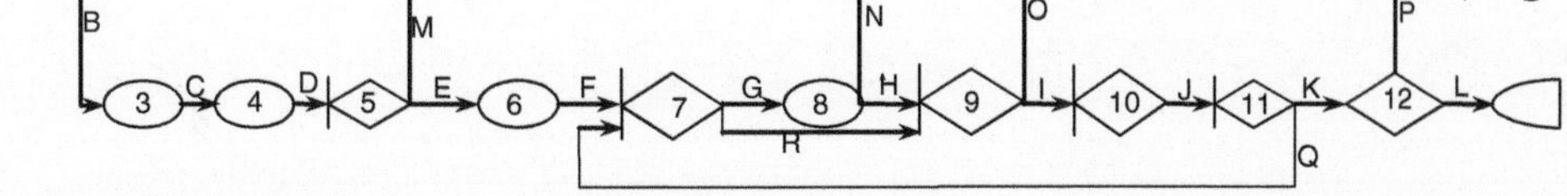

Figure 3 GERT network model graph of the recycling process of used automotive electronic control components in China.

Table 6 Activity code and description

No.	Activity	Branch	Probability of success	Time distribution (h)	mgf	Equivalent function
A	WEEE recycle	1-2	0.55	Constant (0.5)	exp(0.5 s)	0.55exp(0.5 s)
B	Classification	1-3	0.45	Normal $\mu = 1, \sigma = 1$	exp(s + 0.5 s^2)	0.45exp(s + 0.5 s^2)
C	Dismantling	3-4	1	Normal $\mu = 1, \sigma = 1$	exp(s + 0.5 s^2)	exp(s + 0.5 s^2)
D	Test	4-5	1	Normal $\mu = 1, \sigma = 1$	exp(s + 0.5 s^2)	exp(s + 0.5 s^2)
E	Cleaning	5-6	0.96	Constant (0.025)	exp(0.025 s)	0.96exp(0.025 s)
F	Function test	6-7	1	Constant (0.3)	exp(0.3 s)	exp(0.3 s)
G	Aging state test	7-8	0.95	Constant (0.5)	exp(0.5 s)	0.95exp(0.5 s)
H	Hardware test	8-9	0.94	Constant (0.5)	exp(0.5 s)	0.94exp(0.5 s)
I	Hardware repair	9-10	0.96	Normal $\mu = 1, \sigma = 1$	exp(s + 0.5 s^2)	0.96exp(s + 0.5 s^2)
J	Software test and update	10-11	1	Constant (0.2)	exp(0.2 s)	exp(0.2 s)
K	Reliability tests	11-12	0.98	Constant (2)	exp(2 s)	0.98exp(2 s)
L	Reuse ECUs	12-13	0.99	Normal $\mu = 1, \sigma = 1$	exp(s + 0.5 s^2)	0.99exp(s + 0.5 s^2)
M	WEEE recycle	5-2	0.04	Constant (0.5)	exp(0.5 s)	0.04exp(0.5 s)
N	WEEE recycle	8-2	0.06	Constant (0.5)	exp(0.5 s)	0.06exp(0.5 s)
O	WEEE recycle	9-2	0.04	Constant (0.5)	exp(0.5 s)	0.04exp(0.5 s)
P	WEEE recycle	12-2	0.01	Constant (0.5)	exp(0.5 s)	0.01exp(0.5 s)
Q	Function test	11-7	0.02	Constant (0.025)	exp(0.025 s)	0.02exp(0.025 s)
R	Hardware test	7-9	0.05	Constant (0.5)	exp(0.5 s)	0.05exp(0.5 s)

According to the equivalence calculations of basic networks, the branches from nodes 7 to 9 (Figure 5) can be simplified into a single branch (Figure 6). The branches from nodes 7 to 12 (Figure 5) can also be simplified into a single branch (Figure 7).

Finally, the GERT network model graph (see Figure 2) can be simplified, as shown in Figure 8.

The equivalent function is as follows:

$$w_E(s) = w_B w_C w_D w_E w_F w_V w_L = \frac{w_B w_C w_D w_E w_F w_L w_U w_K}{1 - w_Q w_U} = \frac{w_B w_C w_D w_E w_F (w_G w_H + w_R) w_I w_J w_K w_L}{1 - w_Q (w_G w_H + w_R) w_I w_J}. \tag{5}$$

After being remanufactured, the probability that a used automotive electronic control component can be safely reused is expressed as follows:

$$p_E = w_E(0) = 38.64\% \tag{6}$$

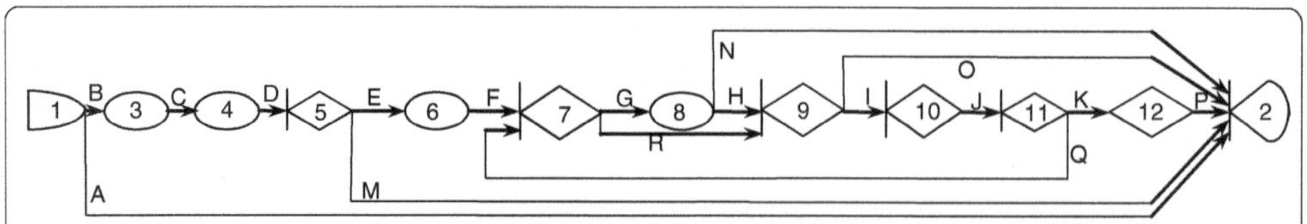

Figure 4 GERT network model graph for waste electrical and electronic equipment (WEEE) recycling (I).

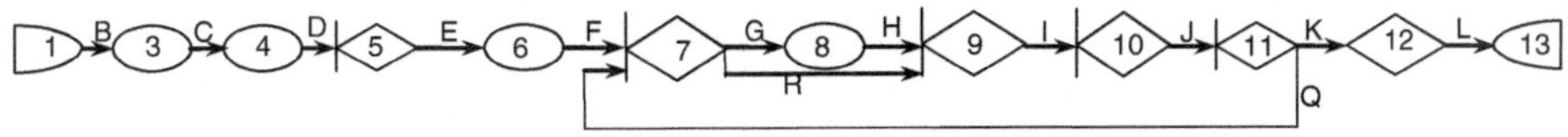

Figure 5 GERT network model graph for remanufacturing (II).

Similarly, after being remanufactured, the probability that a used ECU can be safely reused is 38.64%.

$$T = \frac{dM_E(s)}{ds}|s = 8.5114\,\mathrm{h} \tag{7}$$

The remanufacturing process of used automotive electronic control components takes about 8.5 h to complete.

At present, an automotive ECU is designed for models of automotives. After one type of automobile is released in the market, its ECU can no longer be used. In addition, 55% of used automotive electronic control components are recycled as WEEE (branches 1 and 2). If the ECU design is standardized (i.e., has standardized interfaces, pin assignments, function modules, etc.), the reuse probability can be improved to 95%. Then,

$$p_E = w_E(0) = 81.57\%. \tag{8}$$

Using Monte Carlo simulation in Crystal Ball to analyze the recycling process

Monte Carlo simulation in Crystal Ball

Crystal Ball [16] is a tool that helps executives, analysts, and others analyze risk by performing simulations on spreadsheet models. The forecasts that result from these simulations help quantify areas of risk so that decision-makers can have as much information as possible to enable sound decision-making.

A type of spreadsheet simulation is the Monte Carlo simulation, which repeatedly generates random values for uncertain variables to simulate a model [15]. Monte Carlo simulation is implemented in Crystal Ball [17] to generate a range of values for assumptions and to overcome the limitations encountered in traditional spreadsheet analyses.

A range of possible values can be described for each uncertain cell in the spreadsheet. With the Monte Carlo simulation, Crystal Ball displays results in a forecast chart that shows the entire range of possible outcomes and the likelihood of achieving each outcome. In addition, it keeps track of the results of each scenario.

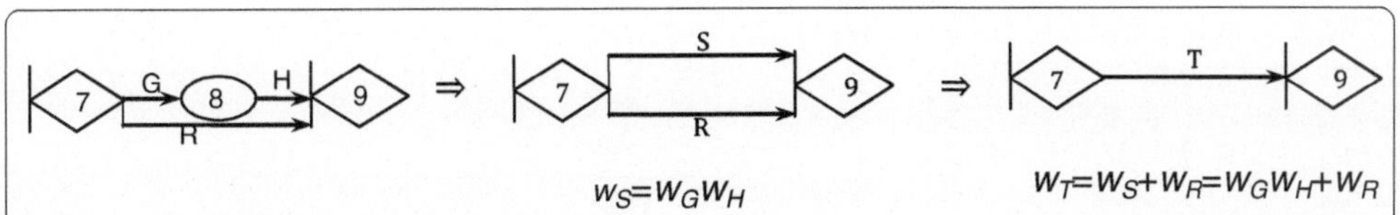

Figure 6 Simplification of branches from nodes 7 to 9.

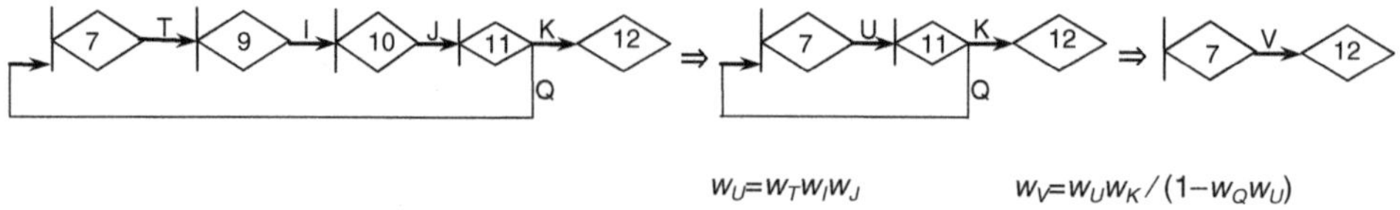

Figure 7 Simplification of branches from nodes 7 to 12.

Steps in using Crystal Ball

The general steps in creating and interpreting simulations with Crystal Ball are as follows:

1. A spreadsheet model is created in Microsoft Excel format with data and formula cells that represent the scenario.
2. Crystal Ball is initiated.
3. The spreadsheet model is loaded.
4. The assumptions and forecasts are then defined using Crystal Ball.
5. Run preferences are set.
6. The simulation is run.
7. The results are analyzed, beginning from viewing the forecast charts. First, overlay charts, sensitivity charts, and trend charts are created and used. Then, reports may be created or data may be extracted for use in other applications.

Monte Carlo simulation in Crystal Ball for the recycling process of used automotive electronic control components

Simulation model and calculation

The remanufacturing process can be obtained (Figures 9 and 10) according to the Monte Carlo simulation and remanufacturing process of used automotive electronic control components.

Figure 9 shows the direct reuse of used automotive electronic control components. The Crystal Ball spreadsheet model is subsequently established (Figure 11). In this model, the used automotive electronic control components are successfully tested and recovered. No element is damaged and replaced. The assumption parameters of the Crystal Ball model (see Figure 11, define assumptions) are taken from Table 6. Figure 10 illustrates the reuse of used automotive electronic control components after these are remanufactured; this process is similar to that of the direct reuse Crystal Ball simulation. In this paper, only the direct reuse of used automotive electronic control components (Figure 9) is simulated in Crystal Ball.

Figure 8 Simplified GERT network model graph.

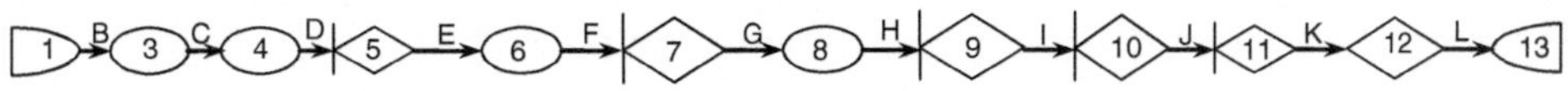

Figure 9 Direct reuse of used automotive electronic control components.

The simulation is performed by calculating the total remanufacturing time (Figure 12, define forecasts). After 10,000 trials (Figure 13), the simulation results are shown in the 'result reports' (Figure 14).

Analysis of results

According to the result reports, the median of the total remanufacturing time is 8.50594 h (8.5114 h in the GERT model) and the minimum total time is 4.42469 h, while the maximum total time is 12.64542 h.

The sensitivity charts (Figure 15) show that the key factors that affect the total remanufacturing time are dismantling (20.6%) > reuse ECUs (20.3%) > test (20.0%) > hardware repair (19.9%) > classification (19.2%).

Reducing the total time using these factors means

- *Highly efficient dismantling without damages.* Dismantling is the most critical factor that affects the total remanufacturing time. Special dismantling technologies and equipment must be developed and investigated because these can efficiently dismantle ECUs without damaging them. The dismantling time is reduced to 0.025 h, as shown in Figure 16.
- *Reuse ECUs (assembly, packaging, and storage).* This process includes reassembly, repackaging, and re-storage. Given that the ECUs are difficult to dismantle and some damages may occur during dismantling, more time is need for reassembly, repackaging, and re-storage. Thus, special reassembly, repackaging, and re-storage technologies and equipment must also be developed and studied. The design for dismantling should be considered when designing ECUs. The process time for ECU reuse is reduced to 0.025 h.
- *High-speed test technology.* Such innovations are important means for reducing total remanufacturing time, which in turn, enables the development of special test technology and equipment. The test process time drops to 0.025 h.
- *Fast hardware repair technology.* ECU hardware repair is a precise and complex process. Thus, repair knowledge and experience should be accumulated to significantly reduce repair process time. The hardware repair process time is reduced to 0.025 h.
- *Fast classification.* ECUs should be selected and stored by category during ELV dismantling. These tasks can save classification time. The classification process time decreases to 0.025 h.

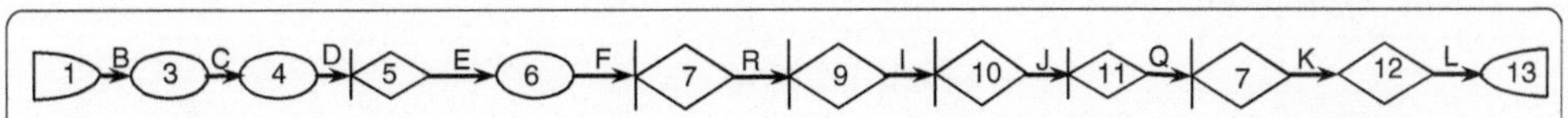

Figure 10 Reuse of used automotive electronic control components after remanufacturing.

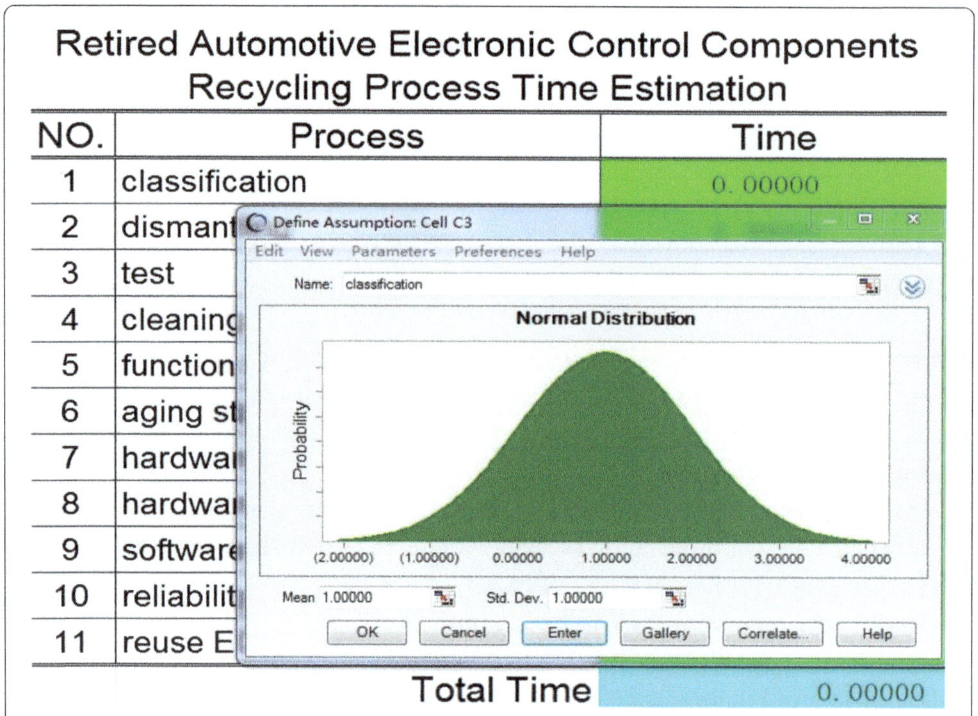

Figure 11 Direct reuse of the Crystal Ball model and define assumptions.

After 10,000 trials, the simulation results are shown in the 'result reports after the optimization of key processes' (Figure 17).

As shown in Figure 17, the median of the total remanufacturing time decreases by 63.85% from 8.50594 h before optimization to 3.07487 h after the optimization.

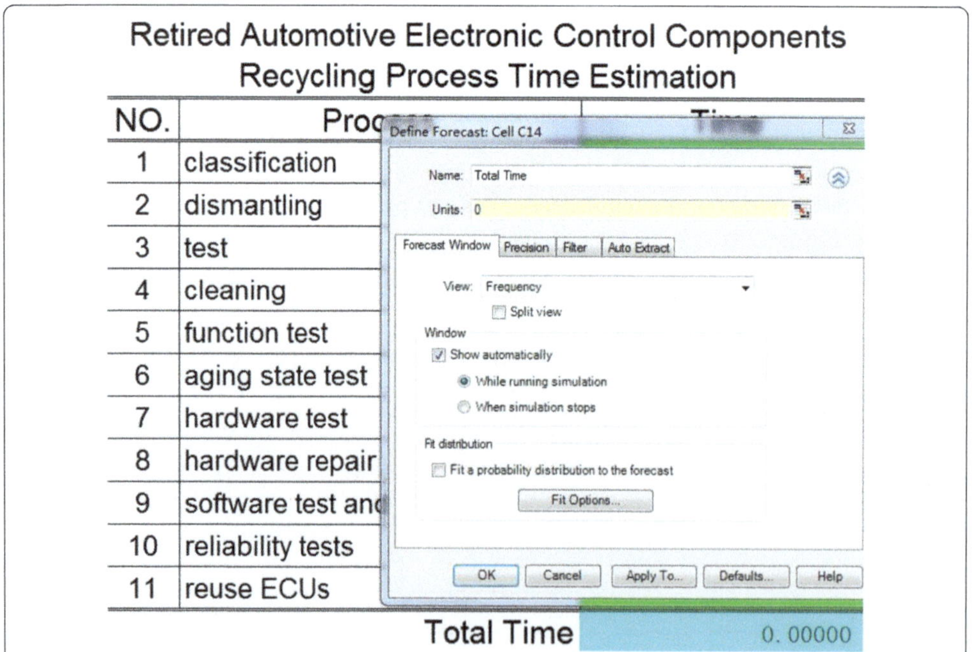

Figure 12 Define forecasts.

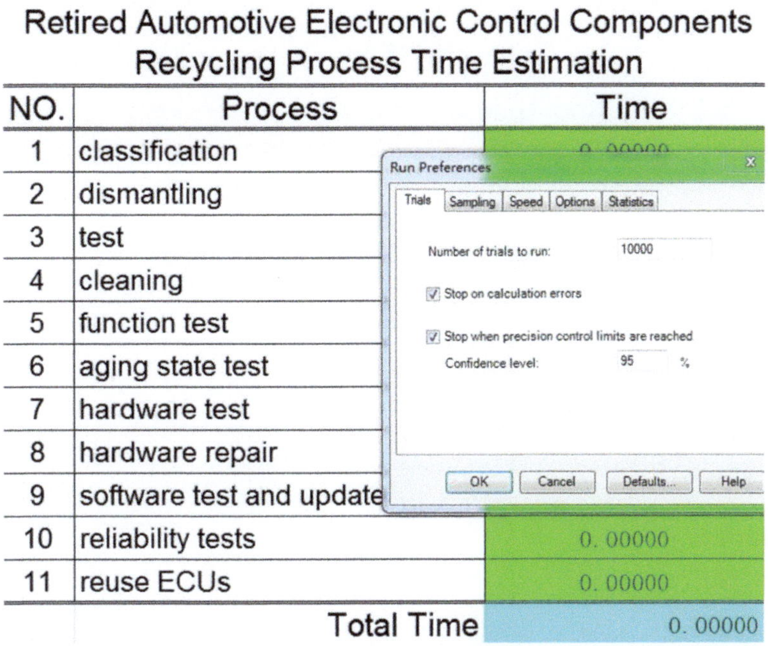

Figure 13 Set run preferences.

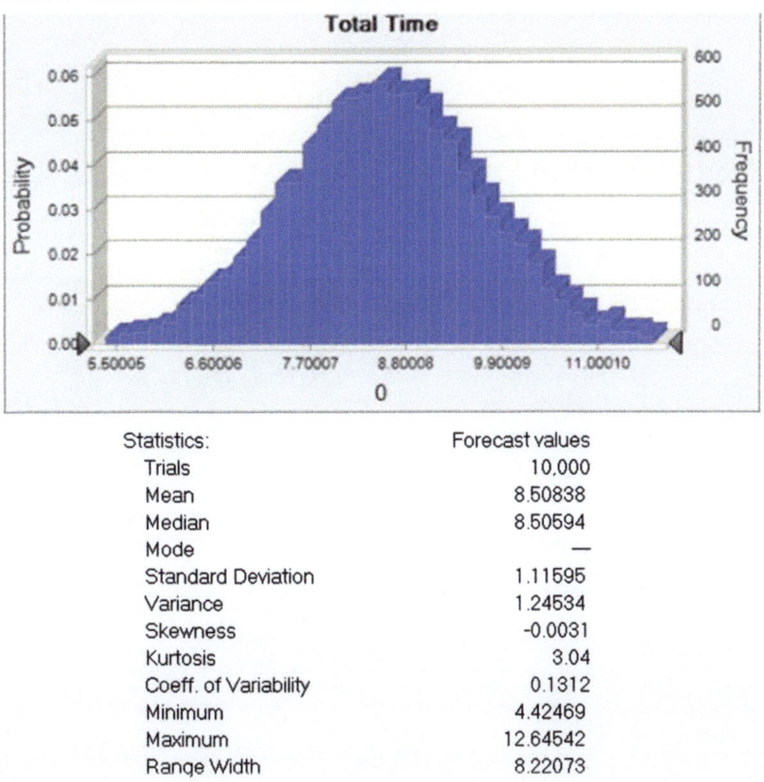

Figure 14 Result reports.

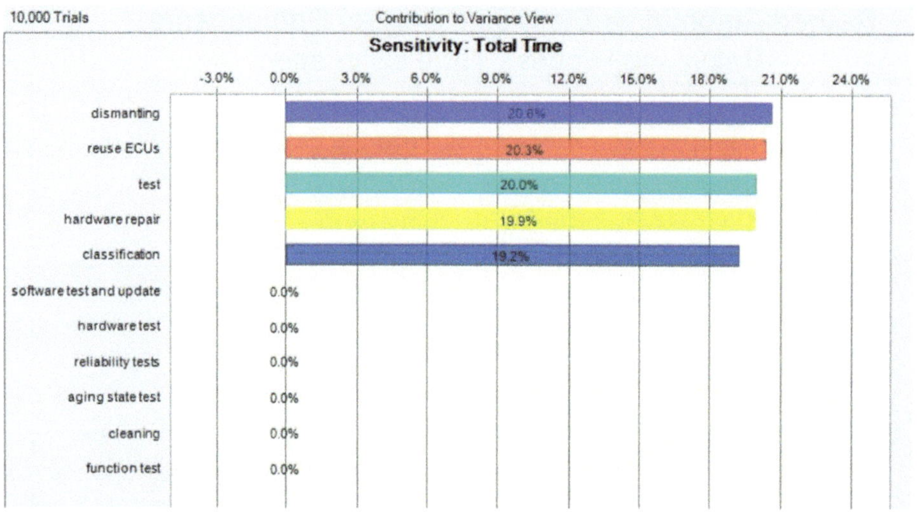

Figure 15 Sensitivity charts.

Numerical experiments

The number of trials is set to 100, 1,000, 10,000, 100,000, and 1,000,000. The simulation results are shown in Table 7. The average total remanufacturing time for all five simulations is as follows:

$$\frac{8.53620 + 8.55494 + 8.50594 + 8.52929 + 8.52645}{5} = 8.530564 \text{ h}. \tag{9}$$

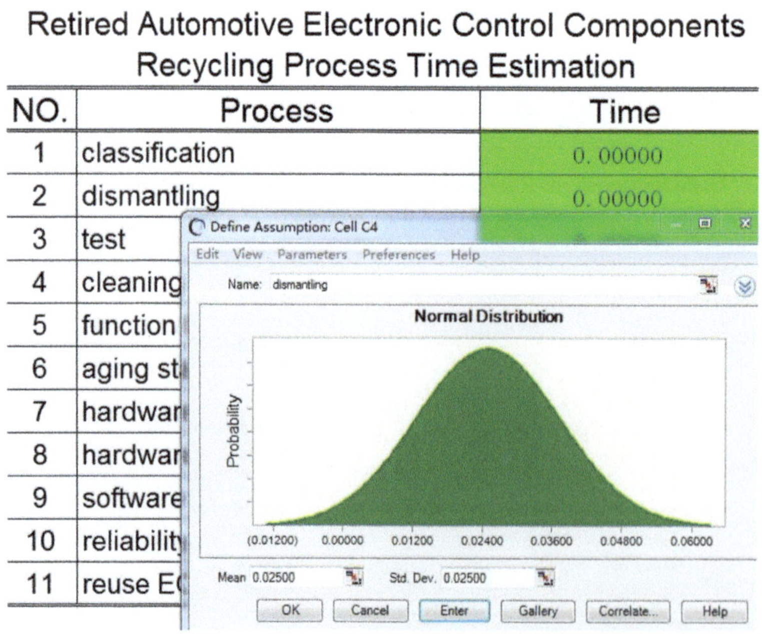

Figure 16 Highly efficient dismantling without damages: dismantling process time.

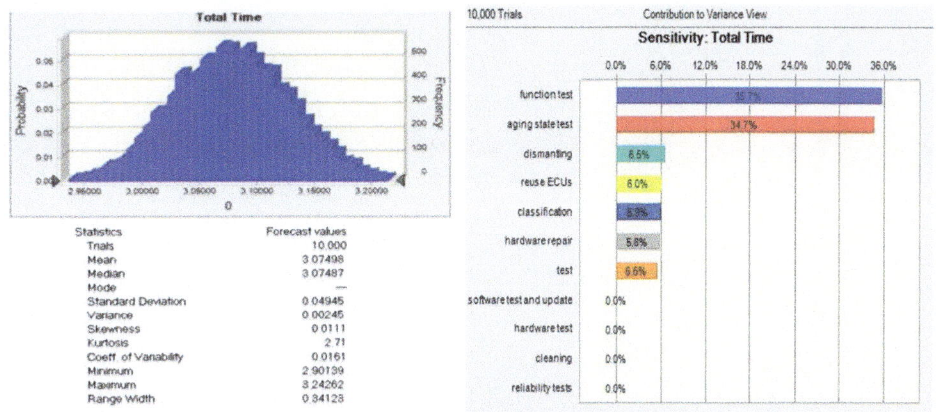

Figure 17 Result reports after the optimization of key processes.

Comparison and analysis of the results of the GERT calculation and Crystal Ball simulation

Compared with the result of the GERT model (8.5114 h), the simulation error rate is

$$\frac{8.530564-8.5114}{8.5114} = 0.225\%. \tag{10}$$

The low error rate validates the suitability of both methods (GERT calculation and Crystal Ball simulation) for modeling and simulating the used automotive electronic control component recycling process. As shown in Figure 18, the results of the GERT and Crystal Ball are almost the same.

The consistency in results indicates that the stochastic network model is feasible for the remanufacturing process of used automotive electronic control components. In addition, Crystal Ball can accurately simulate the said process.

Discussion and conclusion

In this study, China's recycling roadmap and technology scheme for used automotive electronic control components are investigated. The mathematical analysis model for the remanufacturing process of used automotive electronic control components is established on the basis of stochastic network technology and the GERT. These models can calculate the probability of success for the remanufacturing process of used automotive electronic control components and total remanufacturing time. The remanufacturing process of used automotive electronic control components is simulated by Monte Carlo simulation in Crystal Ball. The error rate of the results is

Table 7 Comparison and analysis of the GERT calculation and Crystal Ball simulation results

Trials	100		1,000		10,000		100,000		1,000,000	
Median time (h)	8.53620		8.55494		8.50594		8.52929		8.52645	
Min. and max. time (h)	5.29	10.94	4.73	13.06	4.44	13.06	3.65	13.84	3.11	13.84
GERT model result (h)					8.5114					

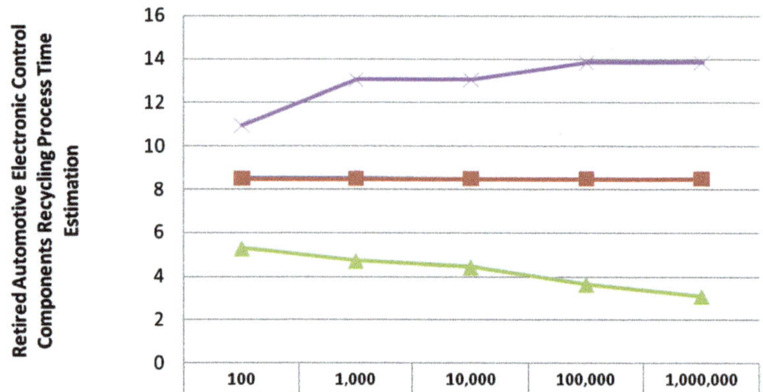

	100	1,000	10,000	100,000	1,000,000
Median time(h)	8.5362	8.55494	8.50594	8.52929	8.52645
GERT model result(h)	8.5114	8.5114	8.5114	8.5114	8.5114
Mini time(h)	5.29	4.73	4.44	3.65	3.11
Maxi time(h)	10.94	13.06	13.06	13.84	13.84

Figure 18 Comparison and analysis of the results of the GERT calculation and Crystal Ball simulation. Blue line, Crystal Ball simulation result; red line, GERT result. These two lines overlap.

only 0.225%, which demonstrates that both methods (GERT calculation and Crystal Ball simulation) are suitable for modeling and simulating the recycling process of used automotive electronic control components. In addition, the numerical experiments for the Crystal Ball simulation are conducted. The total time can be significantly reduced by analyzing and optimizing the sensitivity of the key processes. The optimization methods are also investigated.

The following aspects will be the focus of future investigations: First, the GERT modeling and calculation should be investigated further. Second, remanufacturing key technologies, such as test technology, fast hardware repair technology, and high-efficiency dismantling without damages, should be continuously examined. Finally, corresponding key equipment should be developed. Last year, the Industry-University-Research Center for Auto Recycling Industrial Technology Innovation was built in Wuxi City of Jiangsu Province, where these key technologies and equipment are currently being developed and studied.

Abbreviations

ECU: Electronic control unit; ELV: End-of-life vehicle; GERT: Graphical evaluation and review technique; mgf: Moment-generating function; NASA: National Aeronautics and Space Administration; PERT: Program/project evaluation and review technique; R&D: Research and development; WEEE: Waste electrical and electronic equipment.

Competing interests

The authors declare that they have no competing interests.

Authors' contributions

JW carried out the used ECU remanufacturing technology studies, participated in the model simulation, and drafted the manuscript. MC carried out the parameter definition. JW participated in the design of the study and performed the statistical analysis. Both authors read and approved the final manuscript.

Acknowledgements

The authors express sincere thanks to the Ministry of Science and Technology, People's Republic of China for financing this research within the project 'The Key Technique Investigation & Equipment Development of ELV Green Dismantling & Recycling' under the label 2011BAF11B05, and sincere thanks to the National Natural Science Foundation of China for financing this research within the project 'Fundamental Research on Automobile Shredder Residues (ASR) Pyrolysis/Gasification Mechanism and Its Recovery in China' under the label 51175342.

References

1. Wohlgemuth, F, Hofmann, P: Automotive system design: today and tomorrow. In: Digital Avionics Systems Conference, 2nd edn, pp. I12/1–I12/7. Proceedings, 17th DASC: the AIAA/IEEE/SAE, Bellevue, WA, USA (1998)
2. Sawyer-Beaulieu, SS, Tam, EKL: Applying life cycle assessment (LCA) to North American end-of-life vehicle (ELV) management processes, SAE paper 2005, 01st edn, p. 0846. SAE (2005). doi: 10.4271/2005-01-0846
3. Xia, XF, Xi, BD: Dissembling and Utilization of ELV, p. 78. National Defense Industry Press, Beijing (2008)
4. Wang, JJ, Chen, M: Recycling of electronic control units from end-of-life vehicles in China. JOM. **63**, 42–47 (2011)
5. Wang, J, Zhao, J, Wang, X: Optimum policy in hybrid manufacturing/remanufacturing system. Comput. Ind. Eng. **60**, 411–419 (2011)
6. Pritsker, AAB: GERT: Graphical Evaluation and Review Technique, report, p. RM-4973. NASA, USA (1966)
7. Samli, AC, Bellas, C: Use of GERT in planning and control of marketing research. JMR. **8**, 335–339 (1971)
8. Randolph, PH, Ringeisen, RD: A network learning model with GERT analysis. J. Math. Psychol. **11**, 59–70 (1974)
9. Sharif, MN, Chin, WH: Stochastic network technique for technological forecasting. Technol. Forecast Soc. Change. **10**, 259–276 (1977)
10. Singh, HR, Shankar, G: GERT analysis of two-stage attribute sampling plans based on an exponential failure model. Microelectron. Reliab. **28**, 319–321 (1988)
11. Guo, JF, Qiao, LH: Modeling of product development process with graphical evaluation and review technique. Comput. Integr. Manuf. Syst. **10**, 758–763 (2004)
12. Hayashi, A, Kosugi, T, Yoshida, H: Evaluation of polymer electrolyte fuel cell application technology R&Ds by GERT analysis. Int. J. Hydrogen Energy. **30**, 931–941 (2005)
13. Abdi, R, Ghasemzadeh, HR, Abdollahpour, S, Sabzeparvar, M, Nasab, ADM: Modeling and analysis of mechanization projects of wheat production by GERT networks. Agric. Sci. China. **9**, 1078–1083 (2010)
14. Wang, CN, Yang, GK, Hung, KC, Chang, KH, Chu, P: Evaluating the manufacturing capability of a lithographic area by using a novel vague GERT. Expert. Syst. Appl. **38**, 923–932 (2011)
15. Kurihara, K, Nishiuchi, N: Efficient Monte Carlo simulation method of GERT-type network for project management. Comput. Ind. Eng. **42**, 521–531 (2002)
16. Ragsdale, CT: Spreadsheet Modeling and Decision Analysis, pp. 474–492. Publishing House of Electronics Industry, Beijing (2006)
17. Shakhsi-Niaei, M, Torabi, SA, Iranmanesh, SH: A comprehensive framework for project selection problem under uncertainty and real-world constraints. Comput. Ind. Eng. **61**, 226–237 (2011)

Offering branded remanufactured/recycled products: at what price?

Leila Hamzaoui-Essoussi[1] and Jonathan D Linton[2*]

* Correspondence: Linton@UOttawa.ca
[2]Power Corporation for the Management of Technological Enterprises, University of Ottawa, 55 Laurier E, Ottawa K1N 6 N5, Ontario, Canada
Full list of author information is available at the end of the article

Abstract

This paper investigates the impact of product category, perceived risk, and brand name on consumers' willingness to pay (WTP) for greener (recycled/remanufactured) products. Results provide an understanding on how consumers differentiate between types of products when stating their WTP. The findings suggest that WTP for greener versus branded greener or new products varies with product category. For paper, toner cartridges, and cell phones, brand effects are apparent. However, for cameras and printers brand does not appear to counterbalance perceived risk. As the importance of brand is a function of product category and is significantly related to WTP, original equipment manufacturers (OEMs) should carefully consider this relationship for their products before making decisions on the use of brand as a part of remanufacturing/recycling strategy.

Keywords: Brand name; Willingness to pay; Remanufacturing

Introduction

The relationship between brand and willingness to pay (WTP) for products that are 'greener', because they are made of recycled content and/or utilize parts from post-consumer products (i.e., remanufactured products), is considered. Remanufacturing is more environmentally friendly and inherently greener as it recaptures the value of a material that was already engineered for product use. Remanufacturing not only captures the value of the engineered materials but also the physical form of the product through the use of parts and assemblies that are either in or can be upgraded to a like-new or better than new condition. While environmental marketing has been considered for decades, there is limited research on recycled and remanufactured products in the marketing literature. This paper helps close this gap.

Production by industry, consumption patterns, and behavior of the consumers are causing environmental concerns [1]. Marketers recognize that responding to environmental degradation can be done through the introduction of green products [2]. Although greener products remain a niche market [3], firms are developing products and processes that are both economically and environmentally attractive [4-6]. Simultaneously, legislation and regulations requiring the reuse of material and end of product life take-back have been put in place [7-10], and advances in operations and manufacturing management have enabled the reuse of material and part contents for existing products [11]. Moreover, consumers' growing concerns for environmental issues are leading to new market opportunities, justifying an increased interest in product remanufacturing [12].

In this context, the manufacturing and operational issues associated with the incorporation of reused and recycled materials are increasingly addressed by both academe and practice [13-18]. While the topic of green marketing is well researched [19,20] and shows that consumers express strong concerns for the environment and for recycling [21,22], the influence of consumers' attitudes on their consumption practices with recycled and remanufactured products has yet to be explicitly studied. Recycled and remanufactured products offer an under-explored new subfield to the new product development and marketing literature.

Consumers often perceive recycled and remanufactured products to be inferior to new conventional products [23,24]. This research shows that a lack of familiarity with the use of recycled/used materials in some products makes consumers less willing to pay premium prices for products like refurbished tires. This highlights how the perceived functional risk associated to recycled or remanufactured products is considered due to its influence on the WTP for a product [25]. Whether this risk can be addressed by using risk reduction strategies such as brand names - to positively influence consumer attitude - towards WTP for recycled or remanufactured products still needs further study. Brands reduce perceived risk by acting as a proxy for product quality [26,27] and can command price premiums [28]. However, the role of brands has not yet been considered for recycled and remanufactured products. This is important as it helps determine if consumer assessment of quality and risk can be counterbalanced by the presence of a brand when the consumer considers greener (recycled and remanufactured) products. Therefore, the focus is on the impact of original equipment manufacturer (OEM) brand on the relative attractiveness of greener products - i.e., consumers' WTP for these products.

This study examines consumers' WTP price premiums for branded recycled/remanufactured products versus conventional products. Two specific objectives are addressed: (1) the impact of product category and its related perceived risk on WTP for recycled/remanufactured products and (2) the impact of the brand name on WTP for these products.

Background

Sustainable development requires 'sustainable marketing' as marketing efforts must focus not only on competition and product survival but also to protect the environment for future generations [19,29]. In response to the concern over environmental degradation, marketers understand the importance of introducing products that involve recycling, packaging reduction, re-consumption, dematerialization, use of sustainable raw materials, increased product durability, improving repairability, post-consumer use, compostable products and packaging, and lowering environmental impact during use [19,29-31]. Success requires understanding tradeoffs between product quality and functionality with green attributes as well as how much consumers are willing to pay for 'green' products.

Greener products: recycled and remanufactured products

Sustainability initiatives, more specifically remanufacturing, are becoming more prevalent [12,14,32,33]. Public and government pressure motivates firms to adopt remanufacturing and to maximize profits from product returns [34]. For example, ReCellular Inc. processes more than 400,000 returned cell phones every month, of which nearly 70% are remanufactured and resold [35]. Products such as personal computers, fridges, mobile phones, photography equipment, toner cartridges, and copiers are already remanufactured by OEMs [12].

Remanufactured products undergo 'the process of disassembling, cleaning, inspecting, repairing, replacing, and reassembling the components of a part or product to like-new conditions' ([36], p.35). Because products are restored to a like-new condition, remanufacturing is characteristically distinct from repair or reuse activities - remanufactured components keep their original function. Recycling does not capture the value associated with the shape of a part or product but only the residual value of the material. For a material like aluminum, there is no difference in quality between a virgin and recycled material. However, for other materials such as paper and plastic, recycling processes substantially reduce material quality. Consequently, products made of recycled materials may be inferior to products made of virgin materials and thus offer lower product performance. Moreover, consumers may lack knowledge about remanufacturing or even confuse remanufactured products with pre-owned/used products [21,22], further affecting perceptions of product quality.

Mobley et al. [23] found that recycled products are considered eco-friendly and are positively evaluated by consumers. An increased demand for these products and green marketing is driven by rising consumer awareness and consciousness about the environment [37] and by growing environmental crises. Since the 1970s, the environment is an increasingly important construct within the marketing literature. Sensitivity to environmental issues shifts consumer behavior towards the growth and diffusion of green marketing and other ecologically conscious behavior - including preference for greener firms, increased demand for greener products, and greater acceptance of recycling [20,23]. While environmental awareness and concern have increased, an attitude-behavior gap still exists [20]. While there is a general tendency of consumers to be prepared to pay more for environmentally friendlier products [38], there is still skepticism about paying premiums as consumers have residual concerns regarding product functionality and performance [21,22]. Not much is known about consumers' attitudes and WTP for recycled and remanufactured products. Consequently, the link between perceived quality, perceived risk, and consumers' willingness to pay for recycled/remanufactured product is considered.

Perceived quality, perceived risk, and willingness to pay

Perceived product performance is a significant barrier to the selection for consumption of environmentally sustainable products [39], which in turn influences consumers' WTP for these products [38,40-42].

It is often assumed that one's WTP, the maximum amount of money that a consumer will pay for a good [42], corresponds with perception of relative quality level. A lower perceived quality of remanufactured products is associated with a reduced WTP for remanufactured products [43]. Past research suggests that consumers often do not fully understand and trust the processes and procedures involved in remanufacturing - that is, remanufactured products are seen as inferior in quality and have a lower value than new products [24,44]. The actual level of quality in remanufactured products may even exceed that of new products [21,22,45]. The WTP for products with remanufactured or recycled components and materials appears linked to perceptions of quality and the risk that reflect uncertainty of the actual levels of quality. WTP also relates to the perceived potential outcomes of product use.

Consumers differ with respect to the risk level they are willing to incur in a given situation [46]. The concept of risk consists of (1) uncertainty of outcome, (2) importance of the negative consequences associated with the outcome [47], and (3) perceived severity of negative consequences of the outcome [48]. Both perceived uncertainty of outcomes and the level of negative consequences may differ from situation to situation [49]. Moreover, many studies show that consumers perceive risk differently, and their risk-reducing behavior depends very much on the product class [50-52]. Consumers' perceived quality and WTP for recycled/remanufactured products are influenced by the perceived ambiguity inherent in the recycling or remanufacturing process [23,44]. If the evaluation of a recycled/remanufactured product is influenced by the type of product and by the perceived risk associated to it, consumers' WTP will vary with the product category. Therefore:

H1: Consumers' WTP a premium price for recycled/remanufactured products differs between product categories.

Brand name and willingness to pay

Product and brand management studies show the effect of risk aversion on consumer decision-making and willingness to purchase [51,53]. Zeithaml [54] and Bauer [55] suggest that consumers use extrinsic cues to form perceptions of product quality (or benefits) and perceptions of monetary sacrifice (or cost) to determine product value [52]. The most commonly studied extrinsic cues are price and brand [51].

A brand is 'a name, term, sign, symbol, or design, or a combination of them, that is intended to identify the goods and services of one seller or a group of sellers and to differentiate them from those of competitors' ([56], p.443). Researchers view a brand name as a summary construct [57,58] for quality. That is, consumers often infer product quality based on brand and pay premium prices for branded products, because brands are perceived to offer quality and reduced risk of failure [51]. Brands play multiple roles in consumer decision-making [59]. Brands help alleviate consumers' feelings of threat and ambiguity of novel situations that make them reluctant to try new products. Consumer perception of risk is based on untried products being uncertain and unknown compared to established products and brands [60]. Quality uncertainty, especially in the presence of risk aversion, results in consumers searching for additional information regarding product quality prior to making a purchase [61,62].

Aaker [26] suggests that, all else being equal, strong brands are associated with high quality. Pickett-Baker and Ozaki [39] show that consumers associate well-known brands with effective products. Furthermore, the perceived risk of whether a product will perform as expected is lower for a well-known and respected brand than a less well-known brand [63]. Brands are also seen as credible signals as they embody the cumulative effect of prior marketing-mix strategies and activities (e.g., advertising, product development, and pricing). The notion that credibility is the sum of prior behaviors is referred to as 'reputation' in the information economics literature [64]. The literature also considers the importance of brand credibility under consumer uncertainty [65-67]. Brand credibility is the believability of the product information contained in a brand. This requires that consumers perceive a brand to have the ability (expertise) and willingness (trustworthiness) to deliver as promised [65]. This aspect of a brand is relevant as using the brand name of established product manufacturers to provide new conventional products on the market, not brand names of specific remanufacturing companies, is being considered. Moreover, Michaud and Llenena [12] suggest that consumers are more confident with a remanufactured product offered by an OEM well known for conventional products rather than by unknown third parties.

According to Subramanian and Subramanyam [68], remanufacturer identity significantly explains price differentials for remanufactured electronic products: consumers pay higher prices for products remanufactured by OEMs than for those remanufactured by third parties. Consumers show purchase preferences for green products offered by established brands [69]. The cost of brand development leads some companies to use existing brands for the introduction of new products, including green products. Aaker and Keller [70] (1990) indicate that there are risks associated with extending a brand name. A lower perceived quality for recycled/remanufactured products may have a negative impact on a brand. This is critical to note as OEMs rarely remanufacture products [71].

Branding can signal product position when there is consumer uncertainty about a product. If consumers perceive the brand as having the ability and willingness to deliver what is expected, then expected quality level will increase and perceived risk will decrease, leading to an overall increase in expected utility [27]. Consequently, it is worth examining the relevance of brand to WTP for recycled/remanufactured products as lower perceived quality of recycled/remanufactured products is assumed to lead to reduced WTP for these products [43]. But as strong brands command great price premiums [28,72,73], brands are expected to influence perceived quality and perceived risk related to greener (recycled/remanufactured) products, and thus consumers' WTP for these products.

H2: Consumers' WTP a premium price for branded recycled/remanufactured products will be higher than for unbranded recycled/remanufactured products.

H3: Consumers' WTP a premium price for unbranded recycled/remanufactured products will be lower than for branded new products.

H4: Consumers' WTP a premium price for branded new products will be higher than for branded recycled/remanufactured products.

Methods

Five product categories were chosen: personal electronics (cell phone, single-use camera), home or business electronics (toner cartridge, printer), and paper. The selection criteria were based on the different degrees of perceived risk associated to these categories and on the varying degrees of hedonic and functional dimensions related to product consumption (personal electronics entail more hedonic than functional dimensions than home office electronics or paper). The choice of product categories was guided by current products' availability in a remanufactured form. These products are commonly used by younger adults that are likely to have knowledge of remanufactured products and high levels of environmental consciousness. Each subject was exposed to information on all five products.

For each category, the most respected brand names were utilized. This selection was based on interviews with 14 students regarding brand awareness and brand knowledge regarding brands available on the Canadian market. Brand names of OEMs offering conventional products were chosen in order to test the effect of known brand names on perceived quality and WTP for recycled/remanufactured products. The selected brands were as follows: Domtar - paper, LG - cell phones, NEC - printers, Konica - single-use cameras, and Epson - toner cartridges.

A survey on WTP for recycled and remanufactured products was developed (see Additional file 1). The first section considered constructs relating to familiarity [74] and involvement with recycling [75], perceived product quality, environmental knowledge [76], and perceived risk [77] related to recycled/remanufactured products (in general and not for each product category). The second section was based on [15,25] and considered willingness to pay for recycled/remanufactured products, branded recycled/remanufactured products, and branded new products. For each product category, respondents were given the average price of the conventional version of the product and then asked to state their willingness to pay more, the same, or less than that regular amount. The respondents indicated the increase or decrease in their WTP. Finally, a measure of environmental consciousness [76] and demographic information is collected.

A two-phase pilot test was conducted to assess the survey. Four students and three academicians completed the survey, reviewing it for content appropriateness and clarity. Only minor changes to the format of the instrument were made after this pilot test. Next, the survey was deployed to targeted respondents - undergraduate management students. Hazen et al. [44] indicate that university students often have ample experience with recycled/remanufactured products and thus represent potential consumers of these products. Others suggest that younger adults are more knowledgeable about remanufactured items than older adults [78,79]. Furthermore, the product categories considered are commonly used by university students. Consequently, they are a suitable target sample.

The survey was offered in both English and French to 550 students at a large Canadian university (in 2nd and 3rd year undergraduate classes); 359 completed surveys were completed. Three-hundred and twenty-two surveys were usable. SPSS was utilized for descriptive statistics and hypotheses testing.

The final sample ($n = 322$) was 52%, and 88.9% of the respondents were between 18 and 24 years old. Annual family income was under $49,000 (35.8%), in the range of $50,000 to $99,000 (28.3%), and more than $100,000 (35.8%) for respondents.

Results

Familiarity with recycling, environmental knowledge, involvement with recycled products, environmental consciousness, and perceived risk (five-point Likert scales) were considered.

The respondents are moderately familiar with recycling and have a good general environmental knowledge (Table 1). The respondents consider themselves as involved with recycling and having a high level environmental consciousness. The perceived risk linked to recycled/remanufactured products suggests that recycled/remanufactured products are considered to be somewhat risky.

Table 1 Means for environmental variables and perceived risk

Factors	Mean	Standard deviation	Sample size
Familiarity with recycled products	2.58	0.92	312
Environmental knowledge	3.36	0.79	318
Involvement with recycling	3.58	0.70	317
Environmental consciousness	3.99	0.65	318
Perceived risk	2.52	0.85	317

Recycled paper is considered as a product of good quality, followed by remanufactured toner cartridge, remanufactured printers and cell phones, and with reused single-use cameras following behind (Table 2). Overall, these greener versions of products are considered to be of above average quality, although differences exist between product categories.

Analysis of consumers' willingness to pay was conducted based on the comparison of (1) branded recycled products versus unbranded recycled products, (2) recycled products versus branded products, and (3) branded products versus branded recycled products, for each product category.

H1: Consumers' WTP a premium price for recycled/remanufactured products differs between product categories - supported.

Table 3 summarizes the differences in consumers' WTP between the product categories. The first notable difference is between recycled paper and the other product categories: paper has the most people willing to pay a price premium and the smallest number of consumers expecting a price discount. Cell phones have the most people willing to pay a discounted price. Interestingly, printers and toner cartridges present the second most people willing to pay a discounted price; and toner cartridges the least people willing to pay a premium price. Consumers' WTP a discounted or a premium price varies with product category.

H2: Consumers' WTP a premium price for branded recycled/remanufactured products will be higher than for unbranded recycled/remanufactured products - partially supported.

Three categories (Table 4) show a statistically significant difference in WTP.

Table 4 shows how branded recycled paper is described as being more valuable than recycled paper. Similar results apply for cell phones and the toner cartridges. In some cases, it is worth branding recycled products, but not in all cases. The difference for single-use cameras, printers, and tires is insignificant - i.e., consumers will not always pay extra for branded products. As only one brand is considered, comments cannot be made relating to the importance of specific brands.

H3: Consumers' WTP a premium price for unbranded recycled/remanufactured products will be lower than for branded new products - partially supported.

Table 2 Perceived quality of recycled/remanufactured products

Perceived quality of recycled/remanufactured	Mean	Standard deviation	Sample size
Paper	3.70	0.9	319
Toner cartridges	3.61	1.02	320
Printers	3.42	1.08	320
Cell phones	3.41	1.12	320
Single-use cameras	3.12	1.08	320

Table 3 Summary of differences in willingness to pay

Willingness to pay	Recycled paper	Reused camera	Remanufactured cell phone	Remanufactured printer	Remanufactured toner
Decreased WTP					
0.10% to 0.50%	12 (2.8%) 5.4	19 (6.0%) 8.1	43 (13%) 19.5	30 (9.4%) 14.6	28 (8.7%) 12.7
0.60% to 0.80%	32 (10%) 23.9	41 (13%) 30.7	65 (20%) 47.1	56 (18%) 41.4	64 (20%) 47.2
0.81% to 0.99%	45 (14%) 41.1	70 (22%) 63.5	71 (22%) 62.4	79 (25%) 70.0	72 (22%) 63.5
Same WTP					
1.00	144 (45%) 144	140 (44%) 140	99 (31%) 99	113 (35%) 113	123 (38%) 123
Increased WTP					
1.01% to 1.20%	84 (26%) 91.81	46 (14%) 50.8	36 (11%) 39.7	37 (12%) 40.4	33 (10%) 36.2
1.21% to 1.50%	3 (0.9%) 3.8	3 (0.9%) 3.8	6 (1.9%) 7.9	4 (1.2%) 5.4	1 (0.3%) 1.25
Total respondents	320	319	320	319	321

Number of respondents at each level, percentage, and sum for each reported level.

Table 5 indicates that the WTP for paper is not a function of newness. It seems that 'saving a tree' matters to many people and is considered when purchasing paper. Recycled paper is so commonplace that it is seen as holding no disadvantage in relation to virgin pulp. In fact, recycled paper is often preferred. Branding does not seem to be important for commodities like paper. For other product categories, used material or components result in a discount. The greatest discount is for cell phones: presumably due to a mixture of personal and functional risk.

H4: Consumers' WTP a premium price for branded new products will be higher than for branded recycled/remanufactured products - partially supported.

Table 6 indicates that people are often willing to pay less for branded recycled/remanufactured products than for branded new products.

Table 6 shows that consumers will pay more for a branded new product than for a branded greener product - except in the case of paper. It shows that, for a product that has been on the market for a long time and closely related to a benefit such as 'saving trees and forests', people will buy recycled paper rather than conventional paper and are indifferent to brand identity. In some cases - like paper - the buying choice is between green and new, and brand may not matter.

Tables 2, 3, 4, 5 and 6 show that respondents differ in their behavior for purchasing paper in terms of WTP for branded greener products, unbranded greener products, and branded new products. Additionally, for cell phones and automobile tires, respondents behave the same as for paper in terms of comparing branded greener products to unbranded greener products.

Table 4 Recycled/remanufactured products versus branded recycled/remanufactured products

Product category	Sample size	*T*-statistics	Statistical significance
Paper	313	−1.793	0.037*
Camera	316	−0.802	0.21
Cell phone	321	−1.91	0.028*
Printer	315	−0.3	0.38
Toner	319	−2.548	0.006*

Test of whether the difference between respondent values for each price is statistically and significantly different from zero.
*Statistically significant at the 5% level.

Table 5 Recycled/remanufactured products versus branded new products

Product category	Sample size	*T*-statistics	Statistical significance
Paper	315	−1.57	0.059
Camera	316	−5.2	0.001*
Cell phone	320	−8.9	0.001*
Printer	315	−7.7	0.001*
Toner	317	−6.9	0.001*

Test of whether the difference between respondent values for each price is statistically and significantly different from zero.
*Statistically significant at the 5% level.

After comparing consumers' WTP for branded greener products, unbranded greener products, and branded new products, the relationship between perceived quality of branded recycled/remanufactured products and WTP was tested using regression analysis for each product (see Table 7).

There is a significant relation between *perceived quality* and WTP for branded recycled/remanufactured products. The effects of low strength, might be explained by other variables - such as perceived risk.

Finally, *perceived risk* was considered in relation with the branding's impact on WTP. Based on the measure *perceived risk for recycled/remanufactured products*, we ran means comparisons between WTP for recycled/remanufactured products and WTP for branded recycled/remanufactured products for those who perceive the risk of buying recycled/remanufactured products to be high (perceived risk >2.5) and those who perceived it to be low (perceived risk <2.5). While respondents are unwilling to pay more for a recycled/remanufactured product than for a conventional product, low-risk respondents are willing to pay a slightly higher price than their counterparts that consider the risk of these purchases to be higher. The difference in means for WTP increases slightly for high-risk respondents if brand is included ($t = -3.55$ at the 1% level). However, this is not the case for respondents that perceive risk to be low.

Discussion

Insight into the impact of the product category, related functional risk, and brand name on consumers' WTP for greener (recycled and remanufactured) products is offered. The results indicate differences between product categories with respect to consumers' WTP for recycled/remanufactured versus new products. The effect of brand name on consumers' WTP for recycled/remanufactured versus new products varies in magnitude from product to product. Inclusion of brand name either has a significantly positive impact or has no difference in WTP.

Table 6 Branded new products versus branded recycled/remanufactured products

Product category	Sample size	*T*-statistics	Statistical significance
Paper	312	1.09	0.138
Camera	315	4.37	0.001*
Cell phone	320	9.11	0.001*
Printer	313	8.64	0.001*
Toner cartridge	316	6.41	0.001*

Test of whether the difference between respondent values for each price is statistically and significantly different from zero.
*Statistically significant at the 5% level.

Table 7 Regression results of WTP for branded recycled/remanufactured products on their perceived quality

Product	Perceived quality on WTP		
	Coefficient	**R2**	***F*value**
Paper	0.197***	0.039	10.918
Toner	0.206***	0.042	12.783
Cell phone	0.161**	0.026	8.000
Camera	0.177**	0.031	8.999
Printer	n.s	-	-

Statistically significant at the (**) 5% level and (***) 1% level.

Contribution to theory

The consumers' WTP premium prices for recycled/remanufactured products is clearly product specific. Recycled/remanufactured products do not have the same value as 'new' products. It is evident that some skepticism exists regarding many recycled/remanufactured products. This attitude seems to be based on a perception that greener versions of *some* product categories are of lesser quality. Moreover, our results for the perceived quality of recycled/remanufactured products indicate that consumers expect some recycled/remanufactured product categories to be more acceptable than others: paper, toner cartridges, printers, cell phones, and then single-use cameras in terms of relative quality between new and remanufactured versions. This study supports earlier assumptions that a consumer's perceptions of quality of recycled/remanufactured products influences consumer WTP for these products.

Perceived functional risk of a product category influences consumers' WTP for new products versus products containing reused or recycled materials. Ferguson and Toktay [43] suggest that a reduced WTP for recycled/remanufactured products is influenced by a lower perceived quality of these products. Although levels of quality found in some greener products may exceed that of new products, the risk of a lack of quality or uncertain levels of quality is still present. However, one must consider each product category individually, as there are examples such as paper that are clearly not the case.

The respondents displayed a moderate level of familiarity with recycled products (mean = 2.58); this seems insufficient to enable them to assess the uncertainty and perceived risk associated with the recycled/remanufactured products' quality and performance. Again, recycled paper is the exception - a commonplace, low risk, commodity product. Elliot and Yannopoulou [53] suggest that when consumers are faced with a potential purchase involving a low level of perceived risk, familiarity will suffice for purchase - this is the case for paper. But for other product categories considered here, there is a higher level of perceived risk and familiarity. Consumers are facing more 'unknowable' recycled/remanufactured products. In such cases, confidence and/or trust are needed to counterbalance the perceived risk. Consequently, a brand name can reduce the risk by offering consumers assurance of quality and security. As perceived risk increases and price levels are high, brand names may offer comfort to consumers.

Tables 3 and 5 allow comparison of the attractiveness of greener (recycled and remanufactured), branded greener, and branded new products. Consumers' WTP differs depending on context. For example, the WTP for recycled and remanufactured products is influenced by brand name for paper, toner cartridges, and cell phones. With these categories, a brand offers some value or warranty increasing WTP. For cameras, printers, and tires, the brands have insufficient impact to significantly influence consumers' WTP for the recycled and remanufactured products. Although a strong brand acts as a proxy for quality [26], enabling consumers to better understand and accept uncertainty and risk [53], the brand did not impact consumers' WTP for cameras, printers, and tires. For product categories perceived as being of lower quality due to the inclusion of recycled or reused parts (thus higher perceived risk), the brand may not counterbalance the uncertainty. In summary, products with higher perceived risk are less influenced by the inclusion of a brand name - consumers are unwilling to pay more in these situations.

A lack of familiarity with specific brands (means range from 1.81 to 2.15 on a five-point Likert) may explain the lack of impact of brand for these product categories. Indeed, the use of different brand names - with different brand familiarity levels and brand equity - could provide an additional insight. In order to affect WTP for remanufactured/recycled products, a strong brand with sufficient credibility is required. This relationship should be investigated further in future research.

Paper is a notable exception. Consumers express WTP more for recycled paper than for either branded recycled or even branded new paper. The value of recycling is greater than brand value for paper. While respondents were unfamiliar with the paper brand (mean = 2.09), the product category is familiar, predictable, and credible leading to a high WTP. Recycled paper is thus an interesting exception as respondents show a higher WTP for the recycled version than for branded and/or new paper.

Implications to practice

Marketers need to ensure and emphasize that the quality of recycled/remanufactured products is perceived as equal to traditional products - that is, decrease perceptions of high risk and low performance that may exist in the consumer's mind. This relation is directly linked to consumer familiarity with product categories. As recycled/remanufactured products become a more commonplace (like recycled paper), in the future, the associated ambiguity will decrease.

Information campaigns promoting recycling/remanufacturing could assist in establishing an 'as new' quality image for these products, which in turn would positively impact consumers' WTP. Besides quality, the environmental characteristics of recycled and remanufactured products need to be communicated. As in the absence of environmental information, consumers are willing to pay less for the remanufactured products than for its substitutes [21,22]. That is, environmentally conscious consumers aware of the environmental impacts of a conventional and a remanufactured product will select the remanufactured option offered by the OEM producing both new and remanufactured products, thereby benefiting from brand and reputation.

Next, companies should assess the potential impact of their brand on recycled/remanufactured product versions. Brand may act as a warranty - overcoming perceived risk associated with a product containing recycled/remanufactured materials and components. This impact may or may not be greater than the difference in value between new and un-new products. In some cases - such as recycled paper - the brand has no apparent value. Therefore, companies should assess the impact on WTP of linking a brand name to recycled/remanufactured products. Companies should also consider launching a 'brand extension' - a recycled/remanufactured version of current product offerings. Two issues arise: (1) does a brand extension negatively or positively affect brand value and (2) will the extension negatively impact total profit through cannibalization of existing sales [15]. Both issues should be considered prior to launching a branded recycled/remanufactured product. Companies might offer a remanufactured version of their products in order to reach consumers who are at a lower price point but are looking for product functionality and brand name performance. For these consumers, the benefits related to the brand name may include the offer of extended or full warranty to support the remanufactured version of the product. As with conventional products, an extended or full product warranty clearly signals the confidence the company has in their products' quality. This is important for OEMs if there is consumer uncertainty regarding their remanufactured products' quality and performance.

Finally, investigating the role of branding for recycled/remanufactured products is of critical concern for the OEM in terms of both customer retention and profit maximization. Failure to offer a remanufactured product creates an opportunity in the marketplace that is often filled by a third party. The third party then benefits from the new markets that remanufactured products create and also captures the function-oriented part of the existing market currently satisfied by the new product [80]. If brand adds value to remanufactured products and the OEM is first to market, it is unlikely that third parties will enter the market as the need is fulfilled, and the OEM has a competitive advantage. If a third party is already in the market, the value offered by the brand may provide sufficient advantage for the OEM to enter and compete/reclaim a part of this market. Profit maximization is often supported by remanufacturing as production costs are lower. This is especially the case if price discount for remanufactured products is lowered through tactics such as branding.

Limitations and future research

A brand's importance differs depending on a product category. Sometimes this is very important in purchasing decisions related to recycled/remanufactured products. *What exactly determines the importance of brand* is an important question. Perceived risk and quality of the product appear to be important. Consumer assessment of quality and risk may or may not be counterbalanced by branding. A lack of brand value can be explained in part by the equity of a specific brand. Additional consideration of comparing the effects of brands having different levels of equity and their impact on consumers' WTP for branded greener (recycled/remanufactured) products is needed.

There is a tension between the 'lower' value of recycled/remanufactured parts of a product and the 'higher quality/value' association of a brand name. The balancing point of these two drivers is worth determining. In summary, for a recycled/remanufactured product, *what does the consumer consider first: greeness or brand? Does* a consumer first look for a greener product and then select an available brand, or does the consumer select a brand and then decide on a new or greener product? While Guide and Li [21] suggest that companies with well-established brand names should offer remanufactured versions of their products, it is unclear whether this is a call to compete for market share for greener products or to offer a low price point version of their existing products.

Finally, the broader question of *under what circumstances do brands offer substantial/little value* is an issue critical to firm marketing strategy. Additional brands with varying brand equity for a broader range of products should be considered in future studies. In-depth and analytical research is needed to assess the impact of brand/corporate identity on the WTP of products with reused or recycled content.

Subjects may declare a higher WTP to be seen as socially responsible than they would reveal by actual behavior. Consequently, it is likely that stated willingness to pay is higher than actual behavior. Future research should provide a better understanding of how stated preference and behavior differ and can be better calibrated.

Conclusions

The effect of branding on the WTP for greener (recycled/remanufactured) products is considered. This is important as failure to introduce recycled/remanufactured versions of a product creates the opportunity for third parties to compete against an OEM's existing market through an environmental differentiation strategy. This has occurred in several markets - including single-use cameras and toner cartridges - resulting in a loss of OEM market share. OEMs that sell recycled/remanufactured versions of their products increase corporate social responsibility (CSR) while profiting and protecting markets from the emergence of new competitors. Considering the impact of brand name on willingness to pay offers insights on the competitive advantage that OEMs have over third parties that compete via a 'greener' recycled/remanufactured version of the OEM's product. Brand tends to raise the value of recycled/remanufactured products according to stated willingness to pay. However, willingness to pay varies tremendously as a function of brand name and product category. Consequently, OEMs must consider the value of their brand for the products they offer. The approach used in this study can be utilized for this purpose. For theoreticians, the outstanding question is *what type of risks do brands alleviate?* For policy makers, *what is the potential of education programs to reduce the uncertainty associated with products that have recycled/remanufactured content - and nullify the need for branding?*

Additional file

Additional file 1: Survey utilized.

Competing interests

The authors declare that they have no competing interests.

Authors' contributions

LH and JL and developed the survey and were involved in data collection. Entry of data was conducted by a student. Statistical analysis was conducted by JL and LH. Writing of paper and revising of paper was conducted by LH and JL. All authors read and approved the final manuscript.

Acknowledgements

The assistance of Hervé Sibomana Bouchard in entering the surveys into a database is greatly appreciated. Thanks also goes to the Telfer School of Management for support through a SMRF grant, Canadian Foundation for Innovation for support for the Emerging Technology and Innovation Management Laboratory and the Power Corporation through support provided by the Power Corporation Professorship for the Management of Technological Enterprises.

Author details

[1]Telfer School of Management, University of Ottawa, 55 Laurier E, Ottawa K1N 6 N5, Ontario, Canada. [2]Power Corporation for the Management of Technological Enterprises, University of Ottawa, 55 Laurier E, Ottawa K1N 6 N5, Ontario, Canada.

References

1. Kilbourne, W, Pickett, G: How materialism affects environmental beliefs, concern, and environmentally responsible behavior. J Business Res **61**(9), 885–893 (2007)
2. Prakash, A: Greening the Firm – The Politics of Corporate Environmentalism. Cambridge University Press, Cambridge (2000)
3. Schaefer, A, Crane, A: Addressing sustainability and consumption. J Macromarketing **25**(1), 76–92 (2005)
4. Ewadinger, M, Mouw, S: Recycling creates jobs and boosts economy. Biocycle **46**(10), 41–43 (2005)
5. Farrell Tucker, M: Becoming the world's coolest recycler. In Business **28**(5), 14 (2006)
6. Jung, KS, Hwang, H: Competition and cooperation in a remanufacturing system with take-back requirement. J Intell Manuf **22**(3), 427–433 (2011)
7. Shih, LH: Reverse logistics system planning for recycling electrical appliances and computers in Taiwan. Resour Conserv Recycling **32**, 55–72 (2001)
8. European Union (EU 2003a): Directive 2002/96/EC of the European Parliament and of the Council of 27 January 2003 on waste electrical and electronic equipment (WEEE), Official Journal, L 037, 13/02/2003, pp. 24–39. European Union, Brussels (2003)
9. European Union (EU 2003b): Directive 2002/95/EC of the European Parliament and of the Council of 27 January 2003 on the restriction of the use of certain hazardous substances in electrical and electronic equipment, pp. 19–23. European Union, Brussels (2003). Official Journal, L 037, 13/02/2003
10. State of California: Electronic Waste Recycling Act of 2003 (SB 20) SHER. Statutes of 2003, California (2003)
11. Linton, JD, Klassen, R, Jayaraman, V: Sustainable supply chains: an introduction. J Oper Manage **25**, 1075–1082 (2007)
12. Michaud, C, Llenena, D: Green consumer behavior: an exploratory analysis of willingness to pay for remanufactured products. Business Strategy Environ **20**, 408–420 (2011)
13. Arndt, M: Cat sink its claws into services. Bus Week **3692**(December 5), 56–59 (2005)
14. Linton, JD, Johnston, DA: A decision support system for the planning of remanufacturing at Nortel. Interfaces **30**, 17–31 (2000)
15. Linton, JD: Assessing the economic rationality of remanufacturing products. J Product Innovation Manage **25**, 287–302 (2008)
16. Finisterra Do Paco, A, Barata Raposo, ML, Filho, WL: Identifying the green consumer: a segmentation study. J Target Meas Anal Mark **17**(1), 17–25 (2009)
17. Barquet, AP, Rozenfled, H, Forcellini, FA: An integrated approach to remanufacturing: model of a remanufacturing system. J Remanufacturing **3**(1), 1–11 (2013). doi:10.1186/2210-4690-3-1
18. Xu, Y, Feng, W: Develop a cost model to evaluate the economic benefit of remanufacturing based on specific technique. J Remanufacturing **4**(4), 1–12 (2014). doi:10.1186/2210-4690-4-4
19. Kinoti, M: Green marketing intervention strategies and sustainable development: a conceptual paper. Int J Business Soc Sci **2**(23), 263–273 (2011)
20. Tsen, C, Phang, G, Hasan, H, Buncha, MR: Going green: a study of consumers' willingness to pay for green products in Kota Kinabalu. Int J Business Soc **7**(2), 40–54 (2006)
21. Guide, VD, Li, J: The potential for cannibalization of new products sales by remanufactured products. Decision Sci **41**(3), 547 (2010)
22. Hanss, D, Böhm, G: Sustainability seen from the perspective of consumers. Int J Consum Stud **36**, 678–687 (2012)
23. Mobley, AS, Painter, TS, Untch, EM, Rao Unnava, H: Consumer evaluation of recycled products. Psychol Mark **12**, 165–176 (1995)
24. Smith, VM, Keoleian, GA: The value of remanufactured engines: life-cycle environmental and economic perspectives. J Ind Ecol **8**, 193–221 (2004)
25. Hamzaoui-Essoussi, L, Linton, JD: New or recycled products: how much are consumers willing to pay? J Consum Mark **27**(5), 458–468 (2010)
26. Aaker, DA: Managing Brand Equity. The Free Press, New York (1991)
27. Erdem, T, Swait, J: Brand equity as a signalling phenomenon. J Consum Psychol **7**(April), 131–157 (1998)
28. Agrawal, D: Effects of brand loyalty on advertising and trade promotions: a game theoretic analysis with empirical evidence. Mark Sci **15**, 80–108 (1996)
29. Polonsky, M, Mintu, A: The future of environmental marketing: food for thought. In: Polonsky, M, Mintu, A (eds.) Environmental Marketing Strategies Practices, Theory and Research, pp. 389–391. Routledge, Oxford, UK (1997)
30. Ottman, J: What sustainability means to consumer product marketers. In: The Ottman Report on Environmental Marketing. (1998). available at: www.greemnarketing.com/articles/Ottman-Report.html
31. Charter, M, Peattie, D, Ottman, J, Polonsky, MJ: Marketing and Sustainability. The Centre for Business Relationships, Accountability, Sustainability and Society (BRASS), Cardiff (2002)
32. Barros, AI, Dekker, R, Scholten, V: A two-level network for recycling sand: a case study. European J Oper Res **110**, 199–214 (1998)
33. Krikke, HR, Van Harlen, A, Schuur, PC: Business case OCE: reverse logistics network redesign for copiers. OR Spectrum **21**, 381–409 (1999)
34. Zhang, T, Chu, J, Wang, X, Liu, X, Cui, P: Development pattern and enhancing systemof automotive components remanufacturing industry in China. Resour Conserv Recycling **55**(6), 613–622 (2011)
35. Nasr, N: Real-world remanufacturing. Ind Engineer **43**(6), 24 (2011)
36. Thorn, BK, Rogerson, P: Take it back. IIIE Solutions **34**(4), 34–40 (2002)
37. Basgoze, P, Tektas, O: Ethical perceptions and green buying behavior of consumers: a cross-national exploratory study. J Econ Behav Stud **4**(8), 477–488 (2012)
38. Laroche, M, Bergeron, J, Barbaro-Forleo, G: Targeting consumers who are willing to pay more for environmentally friendly products. J Consum Mark **18**(6), 503–520 (2001)
39. Pickett-Baker, J, Ozaki, R: Pro-environmental products: marketing influence on consumer purchase decision. J Consum Mark **25**(5), 281–293 (2008)
40. Cason, TN, Gangadharan, L: Environmental labelling and incomplete consumer information in laboratory markets. J Environ Econ Manage **43**, 113–114 (2001)

41. Kirchhoff, S: Green business and blue angels: a model of voluntary overcompliance with asymmetric information. Environ Resour Econ **15**(4), 403–420 (2000)
42. Miyake, M: Convergence theorems of willingness-to-pay and willingness-to-accept for non market goods. Soc Choice Welfare **34**(4), 549–570 (2010)
43. Ferguson, ME, Toktay, LB: The effect of competition on recovery strategies. Prod Oper Manage **15**(3), 351–368 (2006)
44. Hazen, BT, Overstreet, RE, Jones-Farmer, LA, Field, HS: The role of ambiguity tolerance in consumer perception of remanufactured products. Int J Prod Econ **135**, 781–790 (2012)
45. Lund, R: Remanufacturing. Technol Rev **87**, 18–23 (1984)
46. Mandrik, CA, Bao, Y: Exploring the concept and measurement of general risk aversion. In: Menon, G, Rao, AR (eds.) Advances in Consumer Research, 32nd edn, pp. 531–539. (2005)
47. Rousseau, DM, Sitkin, SB, Butt, RS, Camerer, C: Not so different after all: a cross-discipline view of trust. Acad Manage Rev **23**(3), 393–404 (1998)
48. Mitchell, VW: Consumer perceived risk: conceptualisations and models. European J Mark **33**(1/2), 163–195 (1999)
49. Matzler, K, Grabner-Kräuter, S, Bidmon, S: Risk aversion and brand loyalty: the mediating role of brand trust and brand affect. J Product Brand Manage **17**(3), 154–162 (2008)
50. Germunden, HG: Perceived risk and information search: a systematic meta-analysis of the empirical evidence. Int J Res Mark **2**, 79–100 (1985)
51. Agarwal, S, Teas, RK: Perceived value: mediating role of perceived risk. J Mark Theory Pract **9**(4), 1–14 (2001)
52. Dodds, WB, Monroe, KB, Grewal, D: Effects of price, brand, and store information on buyers' product evaluations. J Mark Res **28**(3), 307–319 (1991)
53. Elliot, R, Yannopoulou, N: The nature of trust in brands: a psychosocial model. European J Mark **41**(9/10), 988–998 (2007)
54. Zeithaml, VA: Consumer perceptions of price, quality and value: a means-end model and synthesis of evidence. J Mar Sci **52**(3), 2–22 (1988)
55. Bauer, RA: Consumer behavior as risk taking. In: Hancock, RS (ed.) Dynamic Marketing for a Changing World, Proceedings of the 43rd Conference of the American Marketing Association, pp. 389–398. (1960)
56. Kotler, P: Marketing Management, 9th edn. Prentice Hall, New Jersey (1997)
57. Han, CM: Country image: halo or summary construct? J Mark Res **26**(May), 222–229 (1989)
58. Johansson, JK: Determinants and effects of the use of the 'Made In' labels. Int Mark Rev **6**(January), 47–58 (1989)
59. Swait, J, Erdem, T: Brand effects on choice and choice set formation under uncertainty. Marketing Science **26**(5), 679–697 (2007)
60. Steenkamp, J-BEM, Hofstede, F, Wedel, M: A cross-national investigation into the individual and national cultural antecedents of consumer innovativeness. J Mar Sci **63**, 55–59 (1999)
61. Money, BR, Gilly, MC, Graham, JL: Explorations of national culture and word-of-mouth referral behaviour in the purchase of industrial services in the United States and Japan. J Mar Sci **62**(October), 76–78 (1998)
62. Bearden, WO, Shimp, TA: The use of extrinsic cues to facilitate product adoption. J Mark Res **19**, 229–239 (1982)
63. Fowler, GA: Green sales pitch isn't moving many products. Wall Street J **6**, March (2002). B1 and B4
64. Herbig, P, Milewicz, J: The relationship of reputation and credibility to brand success. J Consum Mark **12**, 5–10 (1995)
65. Erdem, T, Swait, J: Brand credibility and its role in brand choice and consideration. J Consum Res **31**(1), 191–199 (2004)
66. Erdem, T, Keane, MP: Decision-making under uncertainty: capturing dynamic brand choice processes in turbulent consumer goods markets. Mark Sci **15**(1), 1–2 (1996)
67. Eckert, C, Louviere, JJ, Islam, T: Seeing the forest despite the trees: brand effects on choice uncertainty. Int J Res Mark **29**(3), 256–264 (2012)
68. Subramanian, R, Subramanyam, R: Key factors in the market for remanufactured products. Manuf Serv Oper Manage **14**(2), 315–326 (2012)
69. Mooth, R: Winning a green innovation: room for growth in untapped markets. (2009). available at http://www.nielsen.com/us/en/insights/news/2009/winning-at-green-innovation-room-for-growth-in-untapped-markets.html - accessed July 25, 2014
70. Aaker, DA, Keller, KL: Interpreting cross-cultural replications of brand extension research. Int J of Res in Marketing **10**(1), 55–59 (1993)
71. El Korchi, A, Millet, D: Conditions of emergence of OEM's reverse supply chains. J Remanufacturing **4**(3), 1–17 (2014). doi:10.1186/2210-4690-4-3
72. Park, C, Srinivasan, V: A survey-based method for measuring and understanding brand equity and its extendibility. J Mark Res **31**(2), 271–288 (1994)
73. Sethuraman, R: A model of how discounting high-priced brands affects the sales of low-priced brands. J Mark Res **33**, 399–409 (1996)
74. Tepper, K: The role of labeling processes in elderly consumers' responses to age segmentation cues. J Consum Res **20**(March), 503–519 (1994)
75. Cho, CH, Lee, JG, Tharp, M: Different forced-exposure levels to banner advertisements. J Advert Res **41**, 45–56 (2001)
76. Bohlen, G, Schlegelmilch, BB, Diamantopoulos, A: Measuring ecological concern: a multi-construct perspective. J Mark Manage **9**(4), 415–430 (1993)
77. Cox, D, Cox, A: Communicating the consequences of early detection: the role of evidence and framing. J Mar Sci **65**(July), 91–103 (2001)
78. Ko, Y, Hwang, H: Efficient operation policy in a closed-loop tire manufacturing system with EPR industry. Engineering Manage Syst **8**(3), 162–170 (2009)
79. Linnanen, L: An Insider's Experiences with Environmental Entrepreneurship Making Ecopreneurs: Developing Sustainable Entrepreneurship, 2nd edn. Gowen Publishing Ltd, Farnham, England (2010)
80. Atasu, A, Guide Jr, VDR, Van Wassenhove, LN: So what if remanufacturing cannibalizes my new product sales? Calif Manage Rev **52**(2), 56–76 (2010)

8

Develop a cost model to evaluate the economic benefit of remanufacturing based on specific technique

Yuchun Xu* and Wei Feng

* Correspondence:
yuchun.xu@cranfield.ac.uk
School of Applied Sciences,
Cranfield University, Cranfield,
Bedford MK43 0AL, UK

Abstract

Remanufacturing is a process of recovering used products to a like-new condition. It can potentially achieve considerable economic, environmental and social benefits in many applications. However, its economic benefit varies for different products and remanufacturing processes. This research aims to develop a framework and cost model to quantitatively evaluate the benefits of remanufacturing techniques to assist the decision making on end-of-life strategies. Additive manufacturing-based remanufacturing process has been modelled first, then cost breakdown structure for the process has been created, and the cost model has been developed. Validation of the cost model has been conducted based on expert judgement, and a case study has been carried out by using the developed cost model to compare the benefit of remanufacturing a specified component or making a new one.

Keywords: End of life; Remanufacturing; Product recovery; Cost estimation; Cost engineering; Economic benefit analysis

1 Introduction

Remanufacturing is a process of recovering the used products to a like-new condition [1]. It enables the customers to have products in lower cost and also enables the manufacturers to save raw materials and other production costs. Remanufacturing can potentially achieve considerable economic, environmental and social benefits.

In some high-tech industrial sectors, e.g. aerospace, remanufacturing is an appropriate strategy to be adopted because (1) high-value parts are well worth to keep if possible and still meet safety requirements; (2) Product Service System (PSS) business model is widely adopted by some manufacturers, such as the aero engine manufacturers Rolls-Royce Plc. and General Electric (GE), which makes remanufacturing possible because the used products are owned by the original manufacturers in the PSS model, and the logistic channel and customer relationships are relatively mature and (3) new techniques provide potential ways to remanufacture high-value parts.

Despite the potential benefits, remanufactured products still account a relatively small portion compared to new manufactured products in general. One of the reasons is the economic benefit of remanufacturing is not clear when considering variety of products to be remanufactured and variety of remanufacturing processes that can be used.

Some previous research have developed methods to assess remanufacturing and other end-of-life options in economic, environmental and social aspects. For example, the decision making approach has been used by a number of researchers. Dunmade introduced a sustainability-based approach for evaluating remanufacturing and other product recovery processes, which named 'product lifecycle extension techniques selection (PLETS)' model. This model designed by fuzzy logic and multi-attribute decision can score each product recovery process against the selected criteria [2]. Bufardi et al. proposed a multi-criteria decision aid (MCDA) approach to help manufacturers select suitable products for end-of-life alternatives (remanufacturing or other product recovery processes) [3]. Jun et al. developed a multi-objective algorithm to evaluate the end-of-life alternatives including remanufacturing [4]. Fernández et al. developed a fuzzy algorithm to select product recovery policies. In that algorithm, product characteristics such as product value, volume, useful life and customer characteristics (such as customer location) are considered as the variable factors [5]. Iakovou et al. designed the 'Multicriteria Matrix' methodology for decision making in end-of-life alternatives. Multi-criteria analysis was performed based on the residual value, environment impact, weight, quantity and ease of disassembly [6]. Ghazalli and Murata developed an analytical hierarchy process (AHP) with case-based reasoning (CBR) approach to support end-of-life strategy selection. The economic cost and environmental cost were selected as the evaluation criteria [7]. Jiang Z et al. [8] proposed a multi-criteria decision making (MCDM) model for selecting remanufacturing technology, the pair-wise comparison approach of the analytic hierarchy process (AHP) was employed for the remanufacturing technology portfolio selection.

Research has also been found particularly for assessing the economic benefit of remanufacturing and other end-of-life (EoL) options. Chen and Chang researched the economics of a closed-loop supply chain (CLSC) incorporating remanufacturing based on a newsvendor framework bearing case. The economic analysis was carried out at high level for the hybrid system (including both manufacturing and CLSC and remanufacturing) to support decision making at remanufacturing system level [9]. Sutherland et al. presented a cost model at system level, and the model is used to optimise the remanufacturing facility size. The model did not consider the variation of remanufacturing methods, it was not able to optimise the selection of EoL strategies [10]. Azadivar and Ordoobadi estimated the cost of remanufacturing at high level by taking into account the assembly cost, disassembly cost, inventory cost and disposal cost. The cost estimation was more focused on how to model the lot size of remanufacturing and build the impact of lot size in the cost estimation, and it did not consider the difference between different remanufacturing technologies and processes [11].

From the above-reviewed research, most of evaluation and cost estimation of remanufacturing and EoL options are conducted at high level (EoL strategy level); however, research on evaluating remanufacturing based on specific techniques and processes remains limited. This paper presents a study on how reliable cost estimation for remanufacturing can be realised by cost modelling at detailed techniques and process levels.

2 Remanufacturing process modelling

In this research, the investigated remanufacturing process is based on additive manufacturing technique, which delivers a wire to the working area and the wire is heated and melted and eventually deposited into the working area. This remanufacturing process consists of four phases, which are reverse engineering, data processing, depositing and grinding, as can be seen in Figure 1.

In Figure 1, the reverse engineering phase is to identify what areas/features to remanufacture. It consists of the following steps:

1) Returned products are measured by high-resolution digital cameras or any other contact or non-contact approach to build up the digital geometry of used part.
2) The digital geometry of the used part is compared with a reference model to identify the difference in geometry.

The data processing phase includes four activities: generating reconstruction specification, generating machining specification, generating tool path in the remanufacturing machine tool and simulating tool path. The reconstruction specification is transformed into machining specification, and then the tool path is generated. Simulation is conducted to pre-check the remanufacturing process at the beginning. The data processing step is normally carried out semi-automatically.

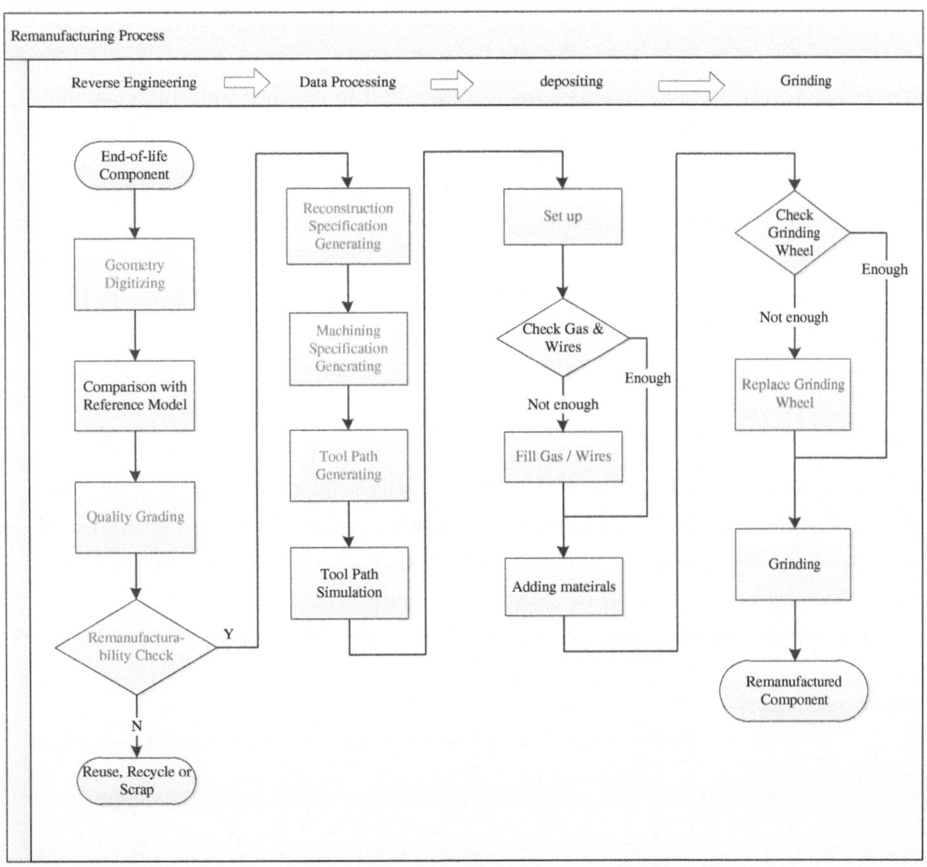

Figure 1 Additive manufacturing-based remanufacturing process model.

The machining process consists of two phases, which are depositing phase and grinding phase. The depositing phase includes setup, gas/wire check and adding materials; the grinding phase includes grinding wheel check and grinding process itself. Except the setup, other machining processes are operated automatically.

3 Cost model development

The cost model in this study was developed by following the cost estimation process suggested by the NASA Cost Estimating Handbook [12]. The key steps are explained in the following sections.

3.1 Cost estimation methodology selection

An appropriate cost estimating methodology needs to be selected first. The available cost estimating methodologies, such as parametric estimation [13-17], engineering build up estimation [18-22,16,23,17], analogy-based estimation and vendor quotes estimation [13,16], are compared in terms of required time, required resource, required data and scope of applications, as shown in Table 1. Letter H, M and L represent high, medium and low level for each term, respectively.

For the remanufacturing process based on additive manufacturing technique, there is no ready available history cost data about this process, so it is impossible to use parametric cost estimation method. Also, there is no similar valid remanufacturing cost estimation cases based on additive manufacturing technique, thus analogy-based estimation and vendor quotes estimation are not suitable in this study. Considering the data and resource available, engineering build-up cost estimation method has been selected for developing the cost estimation model in this study.

3.2 Cost breakdown structure development

The cost element associated with the remanufacturing process has been identified according to the process modelled. This identification process complies with the mutually exclusive and collectively exhaustive (MECE) principle, which means elements in the cost breakdown structure (CBS) should have 'no overlaps' and 'no gaps' [24]. The whole CBS is like a solid 'pyramid' without any omission and duplication.

According to this guideline below, the process is followed for CBS establishment. The first step is to define the first level of the CBS, this level covers all the major elements of the total remanufacturing cost, and also have no dependency between each other. After that, the lower levels of the CBS are developed based on the same principle until the level where cost value can be collected directly or calculated through the detailed remanufacturing process information. The CBS developed in this study is shown

Table 1 Features of cost estimating methodologies

	Time needed	Resource needed	Data needed	Scope of the method
Parametric estimation	M	M	M	M
Engineering build up estimation	H	H	H	H
Analogous estimation	L	M	H	L
Vendor quotes estimation	L	L	L	L

in Figure 2. As can been seen from it, the remanufacturing process includes three cost elements in the first level: pre-production cost, production cost and overhead cost.

1) Pre-production cost: Pre-production cost includes reverse engineering cost and data processing cost; both are labour-intensive costs and can be calculated by labour hourly rate and required working time.
2) Production cost: Production cost has been divided into three cost elements: setup cost, depositing cost and grinding cost. Setup cost is the labour cost of inputting the manufacturing specification, specifying the tool path and clamping the components onto the machine table. Depositing cost and grinding cost are machine-related costs, they can be divided into material cost, consumable cost (such as shielding gas cost, wheel replacement cost, cooling liquid cost, etc.),

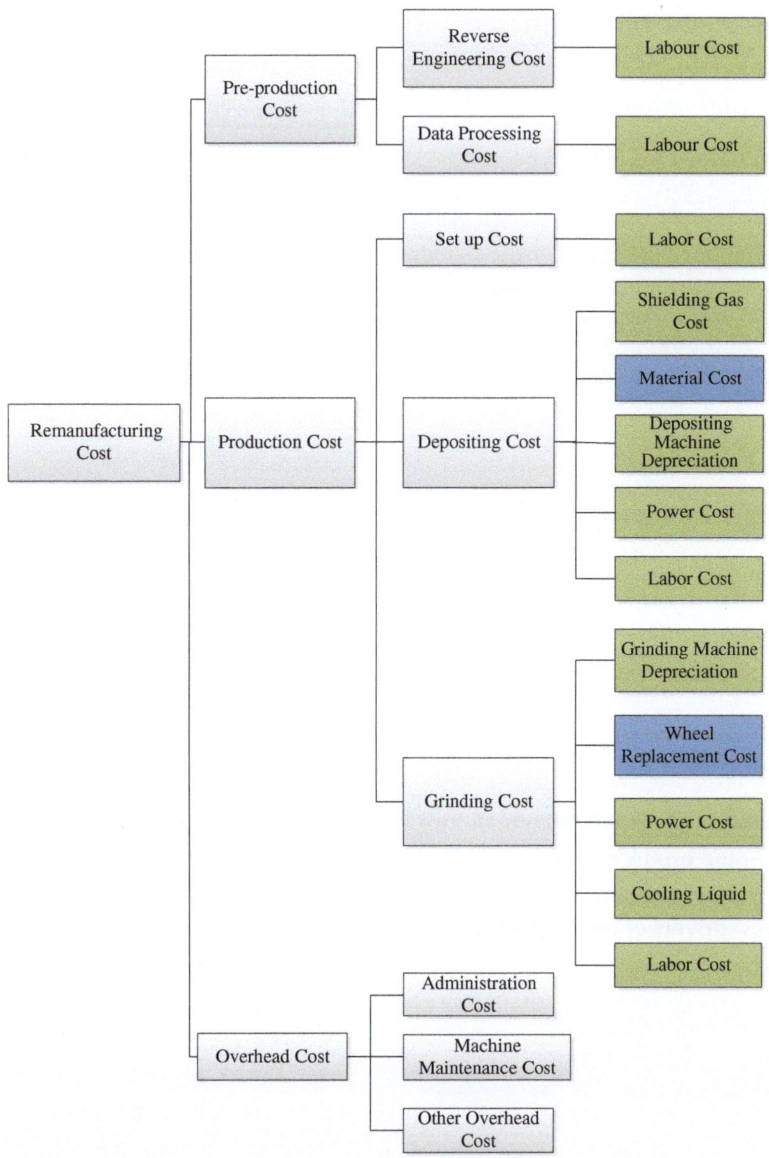

Figure 2 **Cost breakdown structure for remanufacturing process.**

machine depreciation cost, power cost and labour cost. Most of them (except material cost) are functions of operation time.

3) Overhead cost: The last cost element in the first level of the CBS is overhead cost. It includes administration cost, machine maintenance cost and other overhead cost, such as lighting, insurance cost, etc. As overhead cost is indirect cost, it varies by cases.

3.3 Cost driver identification

Cost drivers are those attributes of remanufacturing process that affect the cost of the different activities in remanufacturing process. The cost drivers have been mainly identified based on the information gathered from the author's knowledge, reviewed literature and interviewed field experts. The main cost drivers for each phase of remanufacturing process are listed in Table 2.

3.4 Cost estimation relationships development

Cost estimation relationships (CERs) need to be developed for each cost element in the CBS using the identified cost drivers. It means that each cost element is expressed as a function of the cost drivers. For example the labor cost of depositing and grinding process is calculated as below:

$$C_l = R_l \times \left(T_w + T_g\right) \tag{1}$$

where C_l is the labour cost (£), R_l is the labour hourly rate (£/h) and T_w and T_g are the times for depositing process and grinding process, respectively.

$$\mathrm{T}_w = \frac{T_d}{1 - i_w} = \frac{M_d}{D \times (1 - i_w)} \tag{2}$$

$$\mathrm{T}_g = \frac{T_{\mathrm{rm}}}{1 - i_g} = \frac{M_g}{R \times \left(1 - i_g\right)} \tag{3}$$

where T_d and T_{rm} are the deposition time in depositing process and material removal time in grinding process (h), i_w and i_g are the idle rate of depositing and grinding processes (%), D is the deposition rate of depositing machine (kg/h), R is the removal rate of grinding machine (kg/h).

In summary, the total cost of remanufacturing is calculated by aggregating all the cost elements:

$$C_{\mathrm{total}} = C_{\mathrm{pre}} + C_{\mathrm{set}} + C_w + C_s + C_g + C_d + C_l + C_p + C_o \tag{4}$$

where C_{total} is the total cost of remanufacturing, C_{pre} is the pre-production cost, C_{set} is the setup cost in the production stage, C_w, C_s and C_g are the cost of depositing material,

Table 2 Main cost drivers of remanufacturing process

Remanufacturing processes	Cost drivers
Reverse engineering	Complexity of the shape to remanufacture
Data processing	Complexity of the shape to remanufacture
Depositing	Volume of the part to remanufacture
	Complexity of the shape to remanufacture
Grinding	Volume of the part allowance to remove
	Complexity of the shape to remanufacture

shielding gas cost and grinding wheel consumption cost in the production stage, C_d, C_l, C_p and C_o are the machine depreciation cost, labour cost, power cost in the production stage and overhead cost, respectively.

3.5 Process parameters identification

The parameters of the remanufacturing process depend on the input parameters for remanufacturing and are determined by a knowledge base, which stores the relationship between these two sets of parameters. This knowledge base is designed previously based on the optimization results of experiments and the experts' experiences; this concept is shown in Figure 3.

The process parameters include deposition current and voltage, wire feed speed (WFS), travel speed (TS) and gas flow rate. Below is an example of knowledge-based rules used in Figure 3:

If material = aluminium (specification: 4,043 A),
and if geometry complexity = straight wall,
and if wall width = 5 mm,
then process current = 75 A, process voltage = 12.5 V, WFS = 4 m/min, TS = 0.4 m/min, gas flow rate = 16 L/min.

3.6 Cost model validation and implementation

The cost model has been validated conceptually on the cost estimation methodology, cost breakdown structure (CBS), cost estimating relationships (CERs) and cost estimation process by three experienced experts:

- Expert 1: A project manager of additive manufacturing project with more than three years of field experience has been used to validate the remanufacturing process model and process data.
- Expert 2: A cost engineering expert with over 10 years of experience in cost estimating has been used to validate the cost estimation methodology and the CBS.
- Expert 3: Another cost engineering expert with over 4 years of experience in cost estimating has been used to validate the cost estimation methodology.

In addition, the data collected in the cost estimation model also has been initially validated. The validation has confirmed that the data used in the cost model are within the

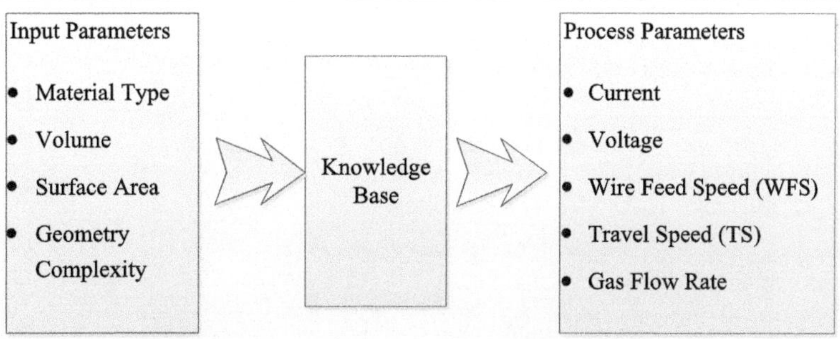

Figure 3 **Relationship between input parameters and process parameters.**

reasonable ranges; the cost model is developed logically and it is valid to meet the purpose of evaluating the economic benefit of remanufacturing based on specific techniques. The improvements made to the cost estimation model through the validation are as follows:

- In the remanufacturing process modelling, two scenarios of depositing equipment are considered in the depositing cost estimation. One is a robot arm assembled in a depositing machine, while the other one is a gantry combined with a depositing machine. Operators can choose a suitable setup for their needs in remanufacturing application.
- In the cost modelling, definition of an input parameter, geometry complexity, is revised for clarity so that the model users can input value correctly for their cases.

The cost model developed has been implemented in MS Excel® platform. Related depositing and grinding specifications are embedded into the spreadsheet. Figure 4 shows the model interface including input parameters and normalised outputs.

4 Case study

4.1 Remanufacturing of slat track

An end-of-life slat track has been selected to demonstrate the application of the developed cost model, i.e. to evaluate if this slat track is economically viable to remanufacture in comparison with making a new one. Slat track is a component linked to slats in aircraft wing, it conducts the motion trails of the slats. The slat track used in this study (as shown in Figure 5) is designed by Thomas Falvey in his MSc thesis at Cranfield

Inputs		
	Value	Unit
Part Volume	10000	mm3
Wall Width	3	mm
Material Type	ti	al/ti/st
Complexity	1	(1-5)
Part Quantity	1	

Output		
	Normalised Cost	Unit
Variable Cost	24.11	£
Setup Cost	12	£
Depositing Cost	11.28	£
Grinding Cost	0.83	£
One-off Cost	32.56	£
Data Processing Cost	10	£
Test Run Cost	22.56	£
Total Cost	56.67	£

Figure 4 Interface of MS Excel® based cost model.

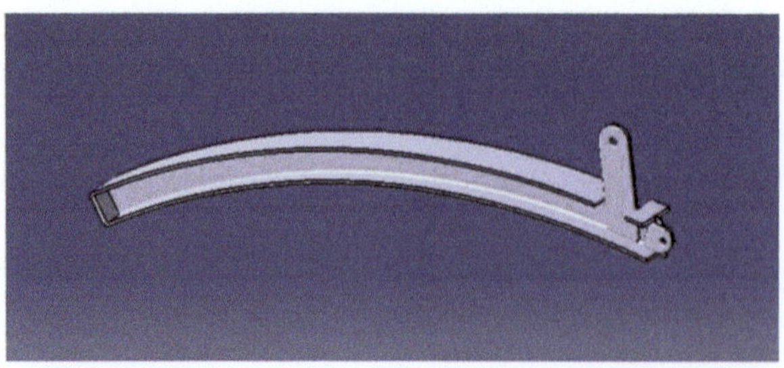

Figure 5 CAD model of the slat track.

Table 3 Geometry and material of the slat track

	Unit	Value
Volume	mm^3	1,013,700
Weight	kg	4.4887
Length (*Y*)	mm	910
Width (*X*)	mm	85
Height (*Z*)	mm	205

Material: Ti6Al4V.

University [25]. The related information about the slat track's geometry and material is shown in Table 3.

As part of the control mechanism, the slat track bears intensive cyclic stress, it tends to crack after a certain period in operation. In this study, an end-of-life slat track with damaged lugs is researched. The assumed geometry of the used slat track is shown in Figure 6. This slat track can be remanufactured by using additive manufacturing process to its original like-new shape shown in Figure 5.

The end-of-life slat track with damaged lugs has been analysed in reverse engineering stage. The reconstructed geometry is as shown in Figure 7. The reconstruction specification of the lug is shown in Table 4.

4.2 Cost distribution analysis

Cost distribution analysis is used to identify the main cost contributor, which accounts for the largest share of the total cost. A pie chart in Figure 8 shows the cost distribution in the remanufacturing process obtained from the developed cost model.

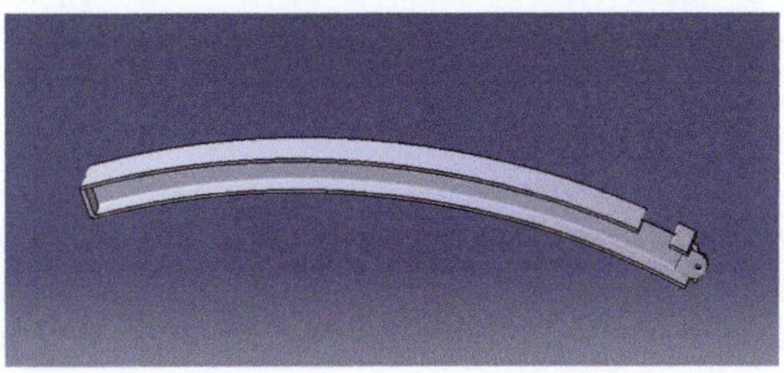

Figure 6 CAD model of the end-of-life slat track.

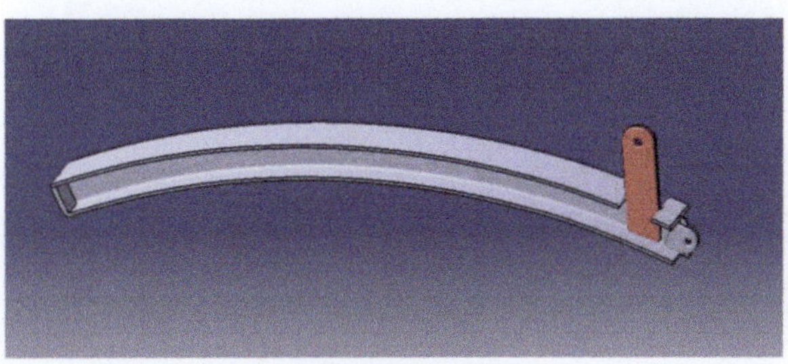

Figure 7 Reconstructed geometry of the end-of-life slat track.

Table 4 Reconstruction specification of the lug

	Unit	Value
Volume	mm^3	35,860
Weight	kg	0.3176
Length (*Y*)	mm	48
Width (*X*)	mm	7
Height (*Z*)	mm	178
Quantity		2

Material: Ti6Al4V.

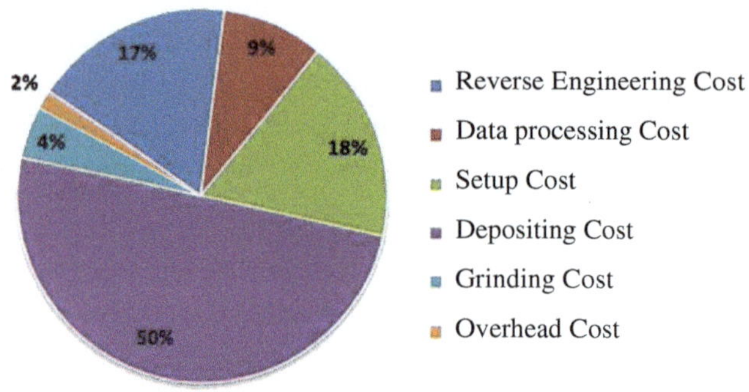

Figure 8 Cost distribution of the total remanufacturing cost.

The depositing cost has the largest share, which is about 50% of the total cost. The following big contributors are setup cost, reverse engineering cost and data processing cost, which are all pre-production activities. If these three cost elements are considered as an integrated one, the pre-production activities will account for around 44% of the total cost, which is the second biggest cost contributor. Because these activities are all manual or semi-automatic, the efficiency can be improved by using more experienced staff or by using more automatic process.

4.3 Comparison with traditional manufacturing

To assess the economic benefits of additive manufacturing-based remanufacturing option, the traditional production cost of the slat track is also calculated.

The raw material for manufacturing the slat track is a titanium block as shown in Figure 9. The traditional production process for this slat track is a combination of rough milling and finish milling, which has been modelled in Figure 10.

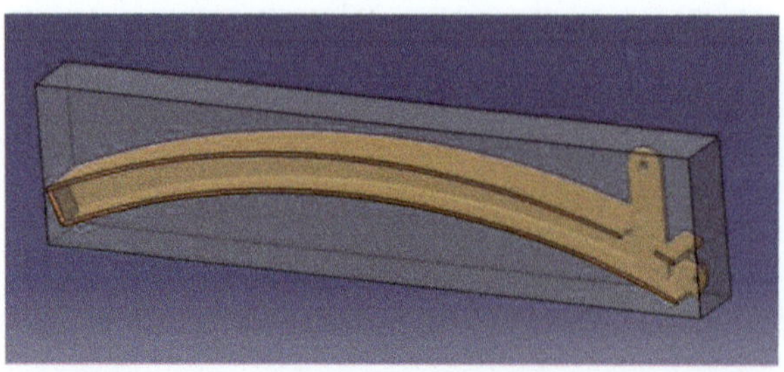

Figure 9 Raw material billet for manufacturing the slat track.

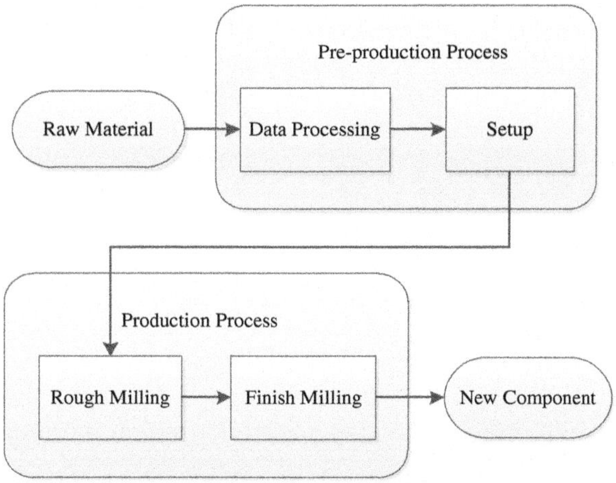

Figure 10 Process model of traditional production of the slat track.

For cost estimation of the new built slat track, SEER-MFG software has been used. The main input parameters used in this estimation are the same as used in the remanufacturing cost estimation in this study. The cost rates, like hourly labour rate and material unit cost are also the same as used in the remanufacturing cost estimation.

The result shows that by using additive manufacturing technique to remanufacture the selected slat track in this study, over 90% of the cost saving can be achieved in comparison with making a new slat track. This significant cost saving is achieved because of the raw material saving in remanufacturing, particularly in this case study the buy-to-fly ratio, reduces from 18 in making a new slat track to 1.12 in remanufacturing it.

5 Discussion and conclusion

In the analysis of economic benefit of remanufacturing as an EoL option, engineering build-up methodology for cost estimation can account for the details of specific remanufacturing techniques and processes. It was found that material type is a main cost driver in additive manufacturing-based remanufacturing process.

This cost model aims to evaluate the economic benefit of remanufacturing process based on specific techniques, so only the remanufacturing cost incurred in the remanufacturing premises is considered in this work. However for supporting decision making on end-of-life strategy (inc. remanufacturing) at high level, cost estimation of the whole remanufacturing system (including the cost of reverse logistics, inventory, etc.) need to be conducted.

Through this research, a cost estimation model based on engineering build-up methodology has been developed, the model has been initially validated by the experts and implemented in MS Excel® platform. The developed cost model has been used in a case study of remanufacturing a slat track in aircraft wing, and it is able to evaluate the economic benefit of remanufacturing process based on specific techniques. In this particular case study, the application of the developed cost model indicates that great cost saving can be made in remanufacturing the slat track than making a new slat track.

Authors' contributions

WF carried out literature review, developed the cost model and carried out case study. YX supervised the work and drafted the manuscript. All authors read and approved the final manuscript.

Acknowledgments
The authors acknowledge Dr. Jorn Mehnen, Dr. Fude Wang, Ms Jialuo Ding for their suport in this study.

References

1. Lund, RT: Remanufacturing: the experience of the United States and implications for developing countries. World Bank, Washington, D.C (1984)
2. Dunmade, I: PLETS model: a sustainability-concept-based approach to product end-of-life management. Proc. SPIE 5583 Environ. Conscious. Manufact. IV **118** (2004). doi:10.1117/12.569629
3. Bufardi, A, Sakara, D, Gheorghe, R, Kiritsis, D, Xirouchakis, P: Multiple criteria decision aid for selecting the best product end of life scenario. Int. J. Comp. Integr. Manufact. 16 **7**(8), 526–534 (2003)
4. Jun, HB, Cusin, M, Kiritsis, D, Xirouchakis, P: A multi-objective evolutionary algorithm for EOL product recovery optimization: turbocharger case study. Int. J. Prod. Res. **45**(18–19), 4573–4594 (2007)
5. Fernández, I, Puente, J, García, N, Gómez, A: A decision-making support system on a products recovery management framework: a fuzzy approach. Concurr. Eng. **16**(2), 129–138 (2008)
6. Iakovou, E, Moussiopoulos, N, Xanthopoulos, A, Achillas, C, Michailidis, N, Chatzipanagioti, M, Koroneos, C, Bouzakis, K, Kikis, V: A methodological framework for end-of-life management of electronic products. Resour. Conserv. Recycl. **53**(6), 329–339 (2009)
7. Ghazalli, Z, Murata, A: Development of an AHP-CBR evaluation system for remanufacturing: end-of-life selection strategy. Int. J. Sustain. Eng. **4**(1), 2–15 (2011)
8. Jiang, Z, Zhang, H, Sutherland, JW: Development of multi-criteria decision making model for remanufacturing technology portfolio selection. J. Clean. Prod. **19**(17), 1939–1945 (2011)
9. Chen, J-M, Chang, CI: The economics of a closed-loop supply chain with remanufacturing. J. Operat. Res. Soc. **63**(10), 1323–1335 (2011)
10. Sutherland, JW, Jenkins, TL, Haapala, KR: Development of a cost model and its application in determining optimal size of a diesel engine remanufacturing facility. CIRP. Annals-Manufact. Technol. **59**(1), 49–5 (2010)
11. Azadivar, F, Ordoobadi, S: A simulation model to justify remanufacturing policies, pp. 1592–1600. Proceedings of the IEEE Simulation Conference 2010 Winter, Baltimore, MD. doi:10.1109/WSC.2010.5678909
12. National Aeronautics and Space Administration (NASA): NASA Cost Estimating Handbook. NASA Headquarters, Cost Analysis Division, Washington, DC (2008)
13. Niazi, A, Dai, J, Seneviratne, L, Balabani, S: Product cost estimation: technique classification and methodology review. J. Manuf. Sci. Eng. **128**(2), 563–575 (2006)
14. Hajare, AD: Parametric Costing—Steel Wire Mill, pp. 172–178. Proceedings of the Annual Convention of the Wire Association International, Cleveland, Ohio, USA (1998)
15. Cavalieri, S, Maccarrone, P, Pinto, R: Parametric vs. neural network models for the estimation of production costs: a case study in the automotive industry. Int. J. Prod. Econ. **91**(2), 165–177 (2004)
16. Curran, R, Raghunathan, S, Price, M: Review of aerospace engineering cost modeling: the genetic causal approach. Prog. Aerosp. Sci. **40**(8), 487–534 (2004)
17. Langmaak, S, Wiseall, S, Bru, C, Adkins, R, Scanlan, J, Sóbester, A: An activity-based-parametric hybrid cost model to estimate the unit cost of a novel gas turbine component. Int. J. Prod. Econ. **142**(1), 74–88 (2013)
18. Jung, JY: Manufacturing cost estimation for machined parts based on manufacturing features. J. Intell. Manuf. **13**(4), 227–238 (2002)
19. Kiritsis, D, Neuendorf, KP, Xirouchakis, P: Petri Net techniques for process planning cost estimation. Adv. Eng. Software. **30**, 375–387 (1999)
20. Xu, Y, Elgh, F, Erkoyuncu, JA, Bankole, O, Goh, Y, Cheung, W, Baguley, P, Wang, Q, Arundachawat, P, Shehab, E, Newnes, L, Roy, R: Cost engineering for manufacturing: current and future research. Int. J. Comput. Integr. Manuf. **25**(4-5), 300–314 (2011). iFirst article
21. Xu, Y, Roy, R, Cassaro, G, Ramsden, J: Development of a cost estimating framework for nanotechnology based products, pp. 193–201. Proceedings of the 16th ISPE International Conference on Concurrent Engineering, Taibei, Taiwan (2009)
22. Xu, Y, Wang, J, Tan, X, Curran, R, Raghunathan, S, Doherty, J, Gore, D: Manufacturing cost modeling for aircraft wing, pp. 817–824. Proceedings of Sixth International Conference on Manufacturing Research (ICMR08), Brunel University
23. Wei, Y, Egbelu, P: A framework for estimating manufacturing cost from geometric design data. Int. J. Comput. Integr. Manuf. **13**(1), 50–63 (2000)
24. What is MECE, and is it MECE? http://timvangelder.com/2010/06/04/what-is-mece-and-is-it-mece/. Accessed 12th August 2011
25. Falvey, T: A9 Dragonfly: composite leading edge slats designer and Cg, mass and inertia manager. Cranfield University, MSc thesis (2010)

9

Exploring inter-organizational relationships in automotive component remanufacturing

Sebastian Lind, David Olsson and Erik Sundin*

* Correspondence: erik.sundin@liu.se
Division of Manufacturing Engineering, Department of Management and Engineering, Linköping University, Linköping SE-58183, Sweden

Abstract

One of the industry sectors with the longest history in remanufacturing is the automotive industry. Remanufactured parts include brake calipers, engines, servo pumps and alternators. A big challenge for automotive component remanufacturers is to achieve a steady flow of cores (parts that are used for remanufacturing). This flow could be secured by making agreements with core suppliers, such as an original equipment manufacturer (OEM), a core broker or another actor in the market. The remanufacturer can also choose to collect the cores without closer collaboration with the core suppliers. One crucial aspect in choosing how to collect the cores is that it has to be lucrative.

The aim of this paper is to explore how remanufacturers manage their inter-organizational relationships in the closed-loop supply chain. A case study was conducted within the European research project 'CAN-REMAN', and empirical data was collected from six participating companies within the project, all European small and medium-sized (SME) remanufacturers of automotive components. These companies were investigated, and their relationships, defined in earlier research with core suppliers, were evaluated.

A key finding of the research is that the most problematic parameter with supplier relationships is to receive the ordered quantity of cores from the supplier. This parameter is continually ranked as one of the most important, and the participating companies also claim to have problems with it. A successful relationship and take-back system was pointed out by one of the companies to never be the owner of the actual cores, and only perform the remanufacturing activity (service) for an OEM. This new relationship, called *reman-contract,* is where the OEM owns the core and the remanufacturer just performs remanufacturing including some sorting and storing. It was found that with this kind of relationship, the ordered quantity of cores was fulfilled to a higher degree, and thus the challenge of achieving a steady flow of cores was met.

Keywords: Reverse logistics; Reverse supply chain; Remanufacturing; Automotive; SME; Inter-organizational; CAN-REMAN

Background

Remanufacturing is an industrial process whereby used/broken-down products (or components) - referred to as 'cores' - are restored to useful life [1]. Remanufacturing means that a product is reprocessed or upgraded in an industrial process. During this process, the core passes through a number of remanufacturing operations, e.g. *inspection, cleaning, disassembly, part reprocessing, reassembly* and *testing,* to ensure it meets the desired product standards [2].

The remanufacturing business in the automotive aftermarket has existed for a long time, but in recent years has become more interesting due to higher consciousness of environmental issues and directives from the European Union such as the end-of-life vehicles (ELV) directive. The ELV directive was proposed to decrease the amount of waste in the automotive industry by increasing the percentage of a car's weight that has to be reused; in the year 2000, 75% of the car's weight was reused, while in 2015, the aim is to reuse 95% [3].

A fairly new and interesting area within remanufacturing in the automotive industry concerns mechatronics (e.g. power steering systems, central locking systems and anti-lock braking systems) and electronic systems (e.g. engine control units and distance control units) communicating *via* a controller area network (CAN) bus. These kinds of CAN bus components were also the focus in the research project called 'CAN-REMAN,' of which this research is a part. The purpose of CAN-REMAN was to develop innovative diagnosis methods and technologies for automotive mechatronic and electronic remanufacturing and investigate the closed-loop supply chain; in other words, how the collection systems of cores are formed in the participating companies today [4]. The project, funded by the European Union, was conducted by Bayreuth University (Germany), Linköping University (Sweden), the University of Applied Sciences Coburg (Germany), the Fraunhofer Project Group Process Innovation (Germany) and eight European small and medium-sized enterprises (SMEs).

When the work on this study was initiated, interviews had already been completed and data collected at five of the six remanufacturers participating in the CAN-REMAN project. These interviews are reported in Dunbäck and Sundin [5,6]. Based on these interviews, four different types of core suppliers were identified:

- Original equipment manufacturer (OEM),
- Independent car dealerships (ICDS),
- Core brokers, and
- Scrap yards.

These four kinds of core suppliers and their relationships to the remanufacturing companies will be further investigated in this study.

The aim of this paper is to explore how remanufacturers of automotive components in Europe manage their inter-organizational relationships in the reverse supply chain, and to identify which parameters are important to determine the efficiency of the relationships with their suppliers.

Previous research on relationships in a closed-loop supply chain

There are three main activities for a remanufacturer: the collection of cores, the remanufacturing process and the redistribution of remanufactured products [7]. As this paper addresses the relationships between remanufacturer and core supplier, it is the collection of cores that will be in focus. One of the biggest challenges when collecting cores is to manage the quantity, timing and quality of returns [8]. This theory is supported by Fleischmann et al. [9], who state that returns are dependent on the former user's requirements, and not the remanufacturer's. Geyer and Jackson [10] argue that there is limited access to end-of-life products in the closed-loop supply chain, and therefore also for remanufacturers. This, together with Wit and Meyer's [11] argument that no firm can be autarchic and must collaborate with other firms, makes it interesting to study how remanufacturers, being a firm, collaborate with their suppliers of cores.

According to Östlin et al. [12], there are seven different types of common relationships between a remanufacturer and a customer or supplier. Östlin et al. [12] have identified the following relationships within a closed-loop supply chain: *ownership-based, service-contract, direct-order, deposit-based, credit-based, buy-back* and *voluntary-based.* The level of relationship focus and core control varies between the relationships, as shown in Table 1.

Table 1 Relationships identified by Östlin et al. [12]

Relationship	Description
1. Ownership-based	The manufacturer owns the product and leases it to a customer; the manufacturer often provides service for the product, including remanufacturing.
2. Service-contract	The customer owns the product but the manufacturer performs service on it.
3. Direct-order	One core is sent to a remanufacturer; after it has been remanufactured, the exact same core is returned to the customer.
4. Deposit-based	The customer must return a similar core to be allowed to purchase a remanufactured one.
5. Credit-based	The customer receives a credit when returning a core that can be used for future purchases.
6. Buy-back	The remanufacturer simply buys the cores needed.
7. Voluntary-based	Cores are given to the remanufacturer.

Table 2 Overview of the case companies

Company	Country	Empl.	Annual turnover [M€]	Type of company	Main products
A	Germany	100	14	Independent remanufacturer	Cylinder heads, engine parts
B	Germany	24	1.6	Independent remanufacturer	Alternators, starters
C	Germany	80	4	Contracted/independent remanufacturer	Gear boxes, engines
D	Germany	80	13	Contracted/independent remanufacturer	Engines, powertrains
E	Sweden	130	7.4	Contracted/independent remanufacturer	Brake calipers, diesel particulate filters
F	Spain	64	-	Independent remanufacturer	Diesel pumps, injectors and gearboxes

This study will focus on the seven relationships that remanufacturers have with the four kinds of suppliers further described in the 'Description of core suppliers' section: OEMs, independent car dealerships (ICDS), scrap yards and core brokers.

Previous interviews made in the CAN-REMAN project were used to complement the information. The employees contacted were plant and logistics managers with several years of experiences at the companies. An overview of the participating companies in the CAN-REMAN project can be seen in Table 2 and in Figure 1, where information about size and products is described. The company labels A to E are the same as in Dunbäck and Sundin [5,6], and refer to the same specific companies.

In Table 3, the survey status is presented, as well as which sources of data were used for each case company. The total amount of data in the study was considered to be sufficient to facilitate the exploring nature of this paper.

Figure 1 Examples of products that the companies are remanufacturing. From left to right: a brake caliper, a gearbox and an alternator (from HowStuffWorks© [13]).

Table 3 Survey status

Company	Questionnaire	Interviewed by authors	Interviewed in 'CAN REMAN'
A	Yes	No	Yes
B	No	Yes	Yes
C	Yes	No	Yes
D	No	No	Yes
E	Yes	Yes	Yes
F	Yes	Yes	No

In addition, these case companies were collaborated with during the CAN-REMAN project from a more reverse engineering perspective, as described by Freiberger et al. [4].

Closed-loop supply chain relationships for a remanufacturer

A number of relationships have been identified in the closed-loop supply chain in the case study, as shown in Figure 2. Here, relationship denotes *an exchange of cores between two actors.*

Starting with the end customer in the upper left corner of Figure 2, a core could travel several ways back into the closed-loop supply chain. One way could be that the end customer chooses to go to a licensed or independent repair shop, and from there, the core is delivered to the remanufacturer. Another way could be from scrapping, where the core could be sent directly from the scrap yard to the remanufacturer, or pass through a core broker.

In our study, the Original Equipment Service (OES) is the service organization associated with the OEM. The OES and OEM are highly interconnected, and are therefore linked in the dotted square in Figure 2. The OES in the automotive industry consists of

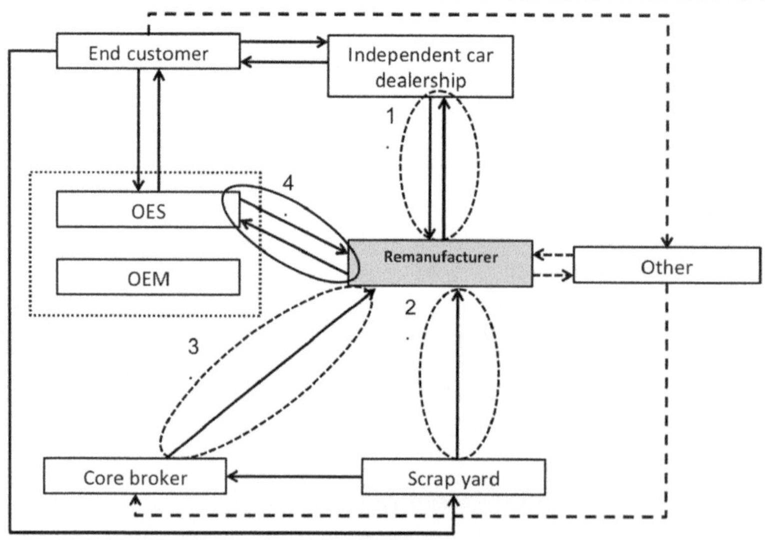

Figure 2 Relationships identified between actors in a closed-loop supply chain.

the dealership network and licensed mechanics, and thus makes up a part of the aftermarket. *Other* could be another kind of actor that has not been identified in this study.

The circles numbered 1 to 3 in Figure 2 contain only relationships defined by Östlin et al. [12]; meanwhile, circle 4 also includes a relationship that does not fit directly into their definitions, which will be further developed in the following chapter.

Description of core suppliers

Each case company was asked to list what relationship they generally *have* with each of the four types of suppliers, what kind of relationship they *would like to have*, and thereafter to evaluate a few specific relationships. The types of core suppliers were (numbers corresponding to the circles indicated with the same numbers in Figure 2):

1. *Independent car dealership (ICDS).* It was shown that with this type of supplier, remanufacturers generally have a deposit-based take-back system, and sometimes also a credit-based or buy-back system. What they would like to have is a deposit-based or direct-order take-back system. Direct-order relationships have been evaluated specifically, and there are no reported difficulties with variation in lead-time or problems with the geographical location, and the purchasing prices are generally lower than from other suppliers. The delivery performance is not satisfactory, which means the remanufacturer does not receive what is ordered.
2. *Scrap yard.* A common relationship that a remanufacturer has to scrap yards is buy-back systems; what they would also like to have are buy-back and take-back systems.
3. *Core broker.* The most common relationship that remanufacturers generally have with their core broker suppliers is of the buy-back type, and this is the relationship they also wish to have with this kind of supplier. From the specifically evaluated relationships with core brokers, the biggest problem seems to be the delivery performance, with some indications that the quantity delivered may vary more than 50% from what was actually ordered. There are no clear results as to whether the purchasing prices from core brokers generally are lower or higher than other suppliers.
4. *OEM/OES.* The main partner for a remanufacturer in a relationship with this actor is the OES organization, and it is from them that cores are sent and received. An interesting relationship identified between a remanufacturer and the OEM/OES is the case where the OEM owns the cores at all times, and the remanufacturer executes the remanufacturing activity. Furthermore, the remanufacturer has to sort out remanufacturable cores from what is delivered, and store remanufactured as well as non-remanufactured cores in the inventory. The OEM, however, performs all other activities such as collection, transportation and to some degree sorting. It is reminiscent of the ownership-based relationship, but a better name would be *reman-contract.* Another common relationship is the direct-order. From the specific supplier evaluations, it can be concluded that the least important parameter is the geographical position. It seems like most OEMs supply the remanufacturer with the right quantity of cores; the supply is regular without variation. Many of the case companies also report that the average purchasing price is lower than from other suppliers.

An interesting aspect of the *reman-contract* relationship investigated in this study is that the remanufacturer itself has come up with the idea and unique knowledge. The idea was presented to several OEMs, and thereafter developed what could be called an industry standard in the automotive industry with this specific remanufacturing process. The advantage for the remanufacturer is that it is not necessary to tie up capital in owning the cores; a disadvantage is that cores may not arrive sorted, and it is unsure how long the remanufactured cores have to be stored before the OEMs demand them back. It is shown that this relationship is successful and that the remanufacturer is content to work like this. In this *reman-contract* relationship, the remanufacturer provides the service of remanufacturing the OEMs' used/broken-down products (cores).

Further, it is shown that in 75% of the evaluated relationships with core brokers, the correct quantity is not delivered, while in relationships with the OEM, 50% is delivered correctly. A reason for this was e.g. that the suppliers could not provide the remanufacturers with more cores. Hence, the remanufacturer needed to negotiate with its customer or buy in new spare parts to fulfil the customer order. This reduces the profits of the remanufacturers.

Advantages and disadvantages of the relationships in CAN-REMAN

This section intends to summarize the most important advantages and disadvantages with the different kinds of relationships that a remanufacturer can have with its suppliers. Since it has been shown that the most common relationships for the companies within CAN-REMAN have been *reman-contract, buy-back* and *deposit-based*, these three relationships will be discussed. There will also be a discussion of whether or not an independent remanufacturer should collaborate with an OEM.

Reman-contract relationship

As stated earlier, a reman-contract relationship with an OEM has many advantages. Since a major problem and difficulty for a remanufacturer is to obtain a steady flow of cores, it may be a good idea to collaborate with OEMs in a reman-contract relationship, since the OEM provides the remanufacturer with the desired cores and the remanufacturer does not have to look for cores. It has been shown that this does not remove the problem entirely, since the delivery performance of the OEM, within the CAN-REMAN companies, is just as bad as from other suppliers. Another disadvantage is that some OEMs do not sort the cores before sending them to the remanufacturer's facility, which means that the remanufacturer has to sort out the cores that can be remanufactured from those that cannot. Having a reman-contract relationship could be very comfortable, since many otherwise necessary activities could be omitted, but may also be very resource-intensive, since, as Daugherty et al. [14] state, a lot of commitment needs to be achieved to set up the relationship. Also, keeping the remanufacturing lead-times in the contract could sometimes be hard due to the condition of the cores and their individual parts. If well-considered and carefully thought out before setting up the relationship, long-term success could be created as Daugherty et al. [14] claim, and the risk that a partner acts in a way that causes negative consequences could be avoided [15].

Buy-back relationship

In this investigation, it has been shown that Company F, a company that does all of its purchasing in buy-back relationships, is satisfied having this kind of relationship; it does not experience problems obtaining a steady flow of cores. However, some other remanufacturers that use the buy-back relationship only for complementary reasons, such as Company C and Company E, have experienced problems getting a steady flow. Company F has a better relation and trusts its suppliers and therefore gets a steadier flow of cores. The remanufacturers that purchase in a buy-back relationship only for complementary reasons do not commit themselves to their suppliers. These remanufacturers use what Spekman et al. [15] describe as competition-driven purchasing; this could lead to diminishing delivery performance, which also has been shown in this study, and could be one of the reasons that the delivery performance is so poor. On the other hand, it is more flexible to only buy cores when needed since there are no further commitments when the cores are paid for and delivered. A disadvantage found with the buy-back system in this investigation is that the independent remanufacturers are dominated in this kind of relationship, i.e. both of the suppliers (often core brokers and scrap yards) and the customers. This is because the demand for cores is greater than the supply. According to Cox [16], a buyer should only commit itself and trust a supplier if the buyer dominates that supplier. In this case, it is the other way around, and it can be hard for the remanufacturer to trust and commit to the suppliers. In a relationship that lacks both trust and commitment, it is hard to collaborate long-term and is better suited for the short-term. This is however contradicted by the fact that Company F is satisfied with working this way, and it can collaborate with its suppliers on the long-term, even though it is dominated.

Deposit-based relationship

The deposit-based relationship has been a traditional way to work for independent remanufacturers with some types of suppliers. An advantage of using this take-back system is that it is possible to balance supply and demand since the customer needs to return a similar product when purchasing a product, as stated by Östlin [7]. Östlin [7] also discusses how sometimes a core that is returned cannot be remanufactured, and therefore it is not always certain that there is a match between supply and demand, which has been verified by the companies in this study. It has been confirmed that companies that use this relationship often need complementary purchases from other suppliers in order to deliver remanufactured products to the customer. Another disadvantage with this relationship is that it will tie up a lot of capital in stock keeping, due to a core being returned when selling a product.

Close collaboration with the OEM

An advantage when collaborating with an OEM is that the remanufacturer can collect cores from cars with warranties. Company F stated that in its case, it is impossible to remanufacture components from cars with warranties since those end customers go to licensed repair shops, and since Company F does not collaborate with the OEM, these cores will never be available for them. Thus, if independent remanufacturers choose not to collaborate with an OEM, they will miss out on a potential market of cores, or radically increase their inventory costs by purchasing these cores and storing them until the demand for cores outside the warranty period grows sufficiently large enough to be able to sell the stored cores, making the remanufacturer subject to lead-time lag. Company F confirms that this is an issue in its business, and it sometimes stores cores awaiting the demand. A disadvantage for a remanufacturer when choosing to collaborate with an OEM is the risk of becoming too dependent. Company E has changed much of its business, and today is focused on remanufacturing diesel particulate filters. This could be a dangerous choice in the long-term, however, since it is not certain that cars will use diesel fuel in a number of years.

In Table 4, a summary of the advantages and disadvantages with relationships and collaboration with an OEM is compiled.

Importance of supply chain parameters

In a questionnaire sent out to the participating companies, they were asked to rank some parameters and the importance of these parameters for them, depending on what type of supplier it is. The parameters are *on-time delivery, purchasing price, geographical position* and *delivery performance* (ordered quantity corresponds to delivered quantity).

Figure 3 shows results from the general part of the survey. The result shows that the three most important parameters (for all kinds of suppliers) are the following:

- *On-time delivery,*
- *Purchasing price,* and
- *Ordered quantity matches delivered quantity.*

The *geographical position* was found to be not as important.

When comparing one parameter for the four different kinds of suppliers, it was shown that the *purchasing price* is more important when buying from a core broker than from any of the other three suppliers. An explanation for this is that a remanufacturer normally purchases from core brokers for complementary reasons, i.e. when there is not enough supply from the other kinds of suppliers.

The same parameters have been evaluated in the supplier-specific part of the quantitative survey, where specific individual relationships have been evaluated. Notable is that many companies have stated that they do not get the *ordered quantity*; in fact, in 66% of the investigated relationships, the remanufacturers do not get the correct quantity, something which they rank as very important. The parameter *on-time delivery* was not considered to be a problem for the evaluated supplier relationships. Furthermore,

Table 4 Advantages and disadvantages of relationships and collaboration with an OEM

Relation	Advantages	Disadvantages
Reman-contract	Dominating position	Risk of dependency
	Do not need to source cores	Sorting costs
	Access to more cores	
Buy-back	Flexible	Weak power position
	Good for complementary purchases	Hard to fully commit
Deposit-based	Balance supply and demand	Hard to get a steady flow
Close collaboration with an OEM	Access to more cores	Risk of dependency
	Possibility of lower inventory costs	

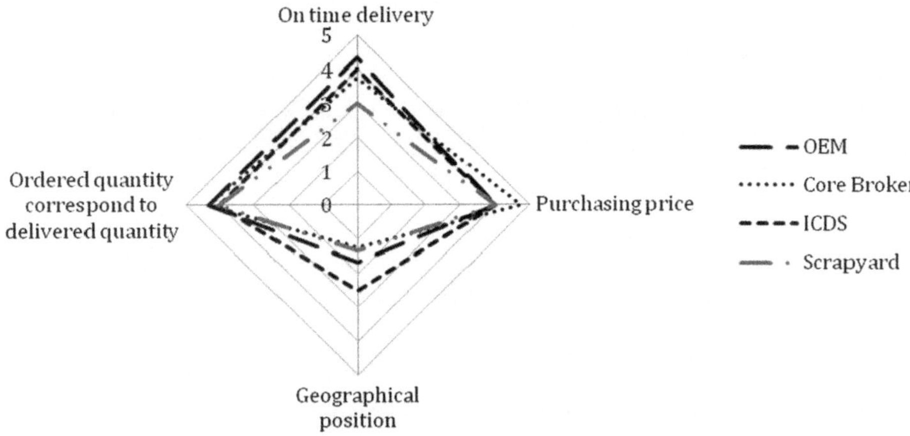

Figure 3 Importance of parameters between a remanufacturer and four different kinds of suppliers. 1 = not important and 5 = very important.

the *geographical position* was not considered to be an important parameter either, nor was it a problem for the participating companies.

On-time delivery

In the questionnaire that was sent to the participating remanufacturers, they were asked to rank how important it is for them to get their orders on time. The ranking ranges from 1 to 5, with 5 meaning very important. As seen in Figure 3, *on-time delivery* was shown to be an important parameter. The kind of supplier that stands out a bit is the scrap yard; *on-time delivery* for it is only ranked at 3. A reason for this is that purchases occur very infrequently from scrap yards; therefore, the need to get the order on time decreases. For the rest of the suppliers, *on-time delivery* is ranked close to 4, and is ranked highest for OEM/OES at 4.33.

On-time delivery is in general considered to be important; however, the remanufacturers do not consider *on-time delivery* to be a problem. The supplier-specific questionnaire showed that only one of nine suppliers (11%) normally has a problem with *on-time delivery*. In that case, the lead-time normally varies between 1 to 3 weeks. Since *on-time delivery* is not seen as a problematic parameter but *delivery performance* is seen as a very problematic parameter, it means that the suppliers of cores prefer to send a few cores on time rather than wait until they have the ordered amount. This is supported by the core broker interviewed in the CAN-REMAN project, who stated that they divide the 'good cores' so that each customer gets at least some good cores; they then fill out the order with cores of lesser quality. In this way, they are able to deliver cores to the remanufacturer on time and provide at least some cores with very good quality.

Delivery performance

Company E says that *delivery performance* in the business from all kinds of suppliers is generally poor, even from the OES. The other remanufacturers share the same opinion, as many do not receive the number of cores ordered. They state that the reason for the poor *delivery performance* is that the availability of cores is limited, with demand greater than supply. The remanufacturers were asked to rank the importance of the *delivery performance* of the supplier, i.e. that the ordered quantity corresponded to the delivered quantity. It was shown that this is a very important parameter for remanufacturers; for all four kinds of suppliers, it ranked as 4 or higher.

Delivery performance is, in general, poor in the remanufacturing of automotive components business. This is because there is not enough supply to cover the demand. Many of the interviewed remanufacturers say that the availability of cores is one of their biggest problems, and due to the shortage of cores, suppliers cannot deliver what they order. This opinion is shared by the core broker, who says that there are not enough cores in the market to satisfy demand. In the supplier-specific evaluation, it was shown that four of the six studied remanufacturers (66%) do not receive the quantity of cores ordered. In addition, at two out of five core suppliers (40%) where the ordered amount was not received, the amount varied by more than 50%. This shows that the remanufacturers have a problem with *delivery performance.*

Figure 3 shows that the remanufacturers rank the importance of *delivery performance* highly; the ranking is 4.33 out of 5 that the ordered amount corresponds to the delivered amount. *Delivery performance* plays a large role when choosing a supplier; if a supplier can deliver an amount that is close to or exactly the same as ordered, that supplier would be very lucrative. However, it has proven that there are no suppliers that can deliver the ordered amount, and even large suppliers such as OEMs cannot provide the exact amount. A reason for this is the lack of cores on the market; a reason for the lack of cores, in turn, is that the quality is unknown and may have to be scrapped by the remanufacturer. This means that in order to satisfy the demand for spare parts, components must be newly manufactured.

Another reason for the lack of cores is the lack of awareness of the value of the cores. Interviews with both remanufacturers and a core broker have shown that scrap yards do not see the value of a single core, and may for that reason mishandle the core and treat it incautiously, or just scrap the whole car instead of cannibalize it into components. It may not be profitable to cannibalize a single car to get a single core and sell it; therefore, scrap yards will not do it for one or many cars, not realizing that it could be profitable to cannibalize a great quantity of cars into components and sell them. This can be compared with what Ellegaard et al. [17] describe as transaction cost economics (TCE), which aims to maximize the total profit. If a TCE approach is implemented, it is meant to maximize the seller's total income in an exchange relationship. This will lead to a minimization of the total work while the revenue and profit are maximized. Every seller, OEM, ICDS and scrap yard needs to assess whether it is lucrative to disassemble parts and sell them to remanufacturers.

Today, there is little awareness among suppliers regarding the value of cores, something that remanufacturers need to increase. If each of the remanufacturers' suppliers does an assessment and comes to the conclusion that the total profit will increase when disassembling cars to get components, more cores will be available and the gap between supply and demand will decrease. A further advantage that remanufacturers will gain if the awareness of the value of cores is raised is that the handling of cores will be more careful. This could lead to the remanufacture of more cores. If assessments are executed and they show that the total profit will increase when cars are disassembled in order to get components, this will lead to a win-win situation where the profit will increase for the scrap yards and more cores will be available for the remanufacturers, so that they in turn can increase their profit.

Purchasing price

As seen in Figure 3, the *purchasing price* is regarded as very important for an independent remanufacturer when choosing supplier, with a ranking of 4. The independent remanufacturers rank this parameter even higher when buying from core brokers, with an average rank of 5. Company E does not consider the *purchase price* when obtaining cores from an OEM/OES since it never owns the cores and therefore does not need to buy them. For the other remanufacturers that do not work in this manner, the purchase price is still a very important parameter, as seen in Figure 3. Also of note is that the remanufacturers consider the *purchasing price* as even more important when buying from core brokers, where it is ranked at 4.75. A reason for this may be that many of the remanufacturers in the CAN-REMAN project procure from core brokers for complementary reasons; they try to satisfy demand by purchasing from suppliers that they normally do not use. These remanufacturers will choose the core broker that can offer the lowest prices. Therefore, this parameter will become even more important when buying from core brokers. This investigation has shown that some remanufacturers never actually purchase the cores; they are only performing a service for an OEM by remanufacturing products. To these remanufacturers, the purchasing price is not a problem and a parameter that they do not take into consideration; the interesting parameter will instead be the selling price for providing the service for remanufacturing one core.

Geographical position

The *geographical position* is ranked as not so important when choosing supplier, as seen earlier in Figure 3. For one kind of supplier, however, it was ranked somewhat more importantly: ICDS ranked at 2.5 compared to the others with an average of 1.4 out of 5. A reason for this is that the size of ICDS ranges from very large, such as e.g. large retailers of car parts, to very small organizations. It is unlikely that the independent remanufacturers would want to collaborate with a very small ICDS that provides few cores and that is situated very far away. In this case, the purchase price per unit will be higher since no economy of scale can be applied for the transport; in that case, the margin would disappear. It can be argued if the same reasoning could be applied to scrap yards and core brokers, which also could be small organizations, but in those cases, the suppliers often can supply several cores. The *geographical position* is not a very important parameter when choosing a core supplier. Since there is a shortage of cores and the remanufacturers normally can sell everything they remanufacture and they are doing everything in their power to procure cores. Therefore, they are not likely to take into consideration the *geographical position* when choosing a supplier.

Quality

An interesting aspect regarding quality of cores is that there are always a percentage of cores that is not possible to remanufacture. The remanufacturers in CAN-REMAN state that they normally cannot remanufacture 20% to 40% of the cores that they purchase, but they do not consider this to be a major problem. In many other businesses, it would be considered a significant problem if 20% to 40% of the products they purchased would be useless. Some of this can be explained by the nature of the remanufacturing business, where remanufacturers take back cores and the quality is unknown. However, it is debatable whether suppliers to the remanufacturers effectively sort out the bad cores from the good. For the supplier, sorting is a resource-intensive activity due to the excessive time and cost involved; therefore, they would want to leave that activity with the remanufacturer. The remanufacturer would not want to do the sorting itself since, as stated earlier, it is an extra cost. However, the demand is much greater than the supply and it can be hard for the remanufacturer to request that the cores arrive sorted. Therefore, the remanufacturer must surrender to the suppliers' wishes, and therefore, there are no incentives for the supplier to sort out bad cores before supplying the remanufacturers. A further positive aspect for the supplier to send the cores unsorted is that it is possible that the remanufacturer lets a bad core 'go through' the sorting process, and the supplier will therefore get paid for a useless core. The suppliers to the remanufacturers are thus in a favourable situation, where they can avoid the sorting cost and also earn more money by selling bad cores.

Conclusions

This paper has identified the most common relationships that independent remanufacturers of automotive components in Europe have with four different kinds of suppliers. While most of the identified relationships have already been described by Östlin et al. [12], one new relationship was found: the *reman-contract* relationship where the OEM owns the core and the remanufacturer just performs remanufacturing including some sorting and storing.

This paper has also ranked parameters considered to be important in a relationship between remanufacturers and their suppliers. The most problematic parameter is to get the ordered quantity to correspond to the delivered quantity, i.e. the *delivery performance*. This was ranked to be the most important parameter in a relationship with a supplier of cores, but in 66% of the cases, the remanufacturers do not receive the quantity that was ordered, and the delivered quantity varied widely. This is mainly due to the lack of cores and knowledge about the core values.

In order to verify and specify the attributes of the *reman-contract* relationship, further investigations and discussions with contracted remanufacturers and OEMs should be conducted. Since using the *reman-contract* relationship with suppliers seems to be a very lucrative strategy for remanufacturers, especially if the remanufacturing process is unique, this would be a very interesting area to continue investigating. It can be argued whether it is sufficient to investigate just six remanufacturers, as in the CAN-REMAN project. It would be an interesting future investigation to include more remanufacturers to verify the result from this paper. In addition, other types of remanufacturers e.g. OEM and contracted remanufacturers' situations would be interesting to explore. This research has been focusing on the European automotive remanufacturing market. It would be interesting to also study other industry sectors and parts of the world where other core suppliers could be available, e.g. insurance companies.

Methods

In this investigation, there were four sources of data and theory collection: 1. Literature study; 2. Questionnaire designed by the authors; 3. Interviews conducted by the authors; 4. Previous interviews made in the CAN-REMAN project.

2. What kind of contract do you or would you prefer to have with a...

a) OEM

Ownership-based	Service contract	Direct order	Deposit-based	Credit-based	Buy-back	Voluntary-based

Figure 4 Snapshot of the questionnaire (general part).

The questionnaire was designed and distributed to the participating companies, and was answered by a person on a managerial level with a good overview of the company's suppliers. The questionnaire was divided into two parts: a general part (Figure 4) and a supplier-specific part (Figure 5). In the general part, the receiver was asked to rank the following parameters:

- On-time delivery
- Purchasing price
- Geographical position
- Ordered quantity corresponding to delivered quantity

The ranking of the parameters is from 1 (not important) to 5 (very important). The parameters were ranked for each type of supplier described earlier. The supplier-specific part was formed to investigate and evaluate certain specific supplier relationships.

The authors also conducted interviews with some of the companies, either by visits in person, telephone or e-mail. These interviews were conducted in order to get a general overview of the business and to identify interesting patterns between the remanufacturers and their suppliers.

10a. Do you normally get the exact quantity ordered from this supplier?

Yes	No

10b. If no, by how much does the quantity vary?

< 10%	10-25%	25-50%	>50%

Figure 5 Snapshot of the questionnaire (supplier-specific part).

Abbreviations

CAN: controller area network; ELV: end-of-life vehicle; ICDS: independent car dealership; TCE: transaction cost economics; OEM: original equipment manufacturer; OES: original equipment service; SME: small- and medium-sized enterprises.

Competing interests

The authors declare that they have no competing interests.

Authors' contributions

All authors have collaborated in writing on the manuscripts of this paper. ES supervised and planned the M.Sc. project. SL and DO planned and performed the survey and interviews within their M.Sc. project, which was the basis for this paper. SL and DO summarized the survey and interviews and wrote the first draft of the manuscript. A previous version of this paper was presented by SL and DO at the 1st International Conference on Remanufacturing in Glasgow last summer. All authors read and approved the final manuscript.

Authors' information

ES (M.Sc., Ph.D.) is an Associate Professor in Sustainable Manufacturing since 2008 at the Division of Manufacturing Engineering, Department of Management and Engineering, Linköping University, Sweden. His main research interests are Remanufacturing, Design for Remanufacturing, Product/Service Systems, EcoDesign. His PhD title is 'Product and Process Design for Successful Remanufacturing.' SL earned his M.Sc. in Industrial Engineering and Management 2011, Linköping University, Sweden. Since 2011, he has been working in consulting within Asset Management, strategic maintenance of production facilities and maintenance in life cycle perspective of such facilities. His M.Sc. thesis title is 'Improving inter-organizational efficiency of reverse supply chain within remanufacturing of automotive components in Europe.' DO earned his M.Sc. in Industrial Engineering and Management 2011, Linköping University, Sweden. Since 2011, he has been working in the automotive industry, implementing lean manufacturing. His M.Sc. thesis title is 'Improving inter-organizational efficiency of reverse supply chain within remanufacturing of automotive components in Europe.'

Acknowledgements

The authors would like to thank the participating companies for answering the questionnaires and questions from the authors. Furthermore, a thank you should be given to the 'CAN-REMAN' colleagues for all interview data widely used in this study, and also to the Swedish Governmental Agency for Innovation Systems (VINNOVA) for its financial support.

References

1. Östlin, J, Sundin, E, Björkman, M: Product life-cycle implications for remanufacturing strategies. J. Clean. Prod. **17**(11), 999–1009 (2009)
2. Sundin, E, Bras, B: Making functional sales environmentally and economically beneficial through product remanufacturing. J. Clean. Prod. **13**(9), 913–925 (2005)
3. Union, E: Directive 2000/53/EC of the European Parliament and of the Council of 18 September 2000 on end-of-life vehicles. Off. J. Eur. Communities **53**, 34–42 (2000)
4. Freiberger, S, Albrecht, M, Käufl, J: Reverse engineering technologies for remanufacturing of automotive systems communicating *via* CAN bus. J. Remanufact. **1**, 6 (2011)
5. Dunbäck, O, Sundin, E: Reverse logistic challenges within remanufacturing of automotive components. International Conference on Remanufacturing, Glasgow (2011)
6. Sundin, E, Dunbäck, O: Reverse logistic challenges in remanufacturing of automotive mechatronics and electronic systems. J Remanufact **3**, 2 (2013)
7. Ostlin, J: On Remanufacturing Systems - Analysing and Managing Material Flows and Process Organisation. PhD thesis. Linköping University, Department of Management and Engineering, Linköping, Sweden (2008)
8. Seitz, MA, Peattie, K: Meeting the closed-loop supply challenge. Calif. Manag. Rev. **46**(2), 41–48 (2004)
9. Fleischmann, M, Krikke, HR, Dekker, R, Flapper, SDP: A characterisation of logistics networks for product recovery. Omega Int. J. Manag. Sci. **28**, 653–666 (2000)
10. Geyer, R, Jackson, T: Supply loops and their constraints: the industrial ecology of recycling and reuse. Calif. Manag. Rev. **46**(2), 55–73 (2004)
11. de Wit, B, Meyer, R: Strategy Process, Content, Context: An International Perspective. Thomson learning, Boston, Massachusetts, United States (2004)
12. Östlin, J, Sundin, E, Björkman, M: Importance of closed-loop supply chain relationships for product remanufacturing. Int. J. Prod. Econ. **115**, 336–348 (2008)
13. HowStuffWorks©. (2013). Pictures downloadable at: http://auto.howstuffworks.com/, accessed 2013-11-15
14. Daugherty, PJ, Richey, RG, Hudgens, BJ, Autry, CW: Reverse logistics in the automobile aftermarket industry. Int. J. Logistics Manag. **14**(1), 49–62 (2003)
15. Spekman, RE, Kamauff Jr, JW, Myhr, N: An empirical investigation into supply chain management: a perspective on partnerships. Supp. Chain. Manag. **3**(2), 53–67 (1998)
16. Cox, A: The art of the possible: relationship management in power regimes and supply chains. Supp. Chain. Manag. **9**(5), 346–356 (2004)
17. Ellegaard, C, Johansen, J, Drejer, A: Managing industrial buyer–supplier relations - the case for attractiveness. Integr. Manuf. Syst. **14**(4), 346–356 (2003)

10

Examination of demand forecasting by time series analysis for auto parts remanufacturing

Mitsutaka Matsumoto[1*] and Akira Ikeda[2]

* Correspondence: matsumoto-mi@aist.go.jp
[1]Center for Service Research, National Institute of Advanced Industrial Science and Technology (AIST), 1-1-1, Umezono, Tsukuba, Ibaraki, 305-8568, Japan
Full list of author information is available at the end of the article

Abstract

Production planning and control in remanufacturing are more complex than those in traditional manufacturing. Developing a reliable forecasting process is the first step for optimization of the overall planning process. In remanufacturing, forecasting the timing of demands is one of the critical issues. The current article presents the result of examining the effectiveness of demand forecasting by time series analysis in auto parts remanufacturing. Most previous studies on demand forecasting in remanufacturing assume that the time distribution of new product sales are known and that the time distributions of product end-of-life and demands for remanufactured products are calculated by adding the product lifespan period to the time distribution of product sales. However, this assumption is not always correct. For example, independent remanufacturers (IRs) do not always have precise information on the time distribution of new product sales, and in this case, a different approach is needed. Based on this background, this study examined the effectiveness of forecasting by time series analysis that does not need those information. To verify the forecasting accuracy, actual data of an auto parts IR was used. The study used the time series data of the shipments of an actual IR of auto parts for a total of 400 types of remanufactured alternators and starters over a period of 12 years. The method was employed on the initial 11 years of data to project the demand over the final year, and the forecasting results provided an average error of 27.2% relative to the actual shipments made over the forecasted year. The factors degrading the accuracy and the means of improving the results are discussed. Also, the implications of the results and future steps regarding the present study are argued.

Keywords: Demand forecasting; Time series analysis; Auto parts; Remanufacturing; Production planning

Background

Remanufacturing is an industrial process that seeks to restore end-of-life products to their original working condition. Remanufacturing closes the materials use cycle and forms an essentially closed-loop manufacturing system; therefore, remanufacturing contributes to material and energy savings. Remanufacturing presently represents a significant economic activity. In the United States, the sales of remanufactured products as of 2011 were estimated to be $43.0 billion annually and supported 180,000 full time jobs [1]. Remanufacturing has spread worldwide to sectors as disparate as auto parts, photocopiers, single-use cameras, construction machines, mining machines, medical equipment, aerospace, military vehicles, heavy-duty engines, computers, vending machines, and so on. Lund [2] has identified 75 separate product types that are routinely remanufactured. Numerous driving forces encourage the use of remanufactured products, although the activity encounters numerous barriers as well [2-6]. Lund [2], for example, has proposed seven criteria useful for establishing the suitability of products for remanufacturing: (1) the product is a durable good, (2) the product fails functionally, (3) the product is standardized and the parts are interchangeable, (4) the remaining value added is high, (5) the cost to obtain the failed product is low compared to the remaining value added, (6) the product technology is stable, and (7) the consumer is aware that remanufactured products are available. Guide [7] focused on aspects of the remanufacturing requirements pertaining to production planning and control of remanufacturing: (1) the uncertain timing and quantity of returns, (2) the need to balance returns with demands, (3) the disassembly of returned products, (4) the uncertainty in materials recovered from returned items, (5) the requirement for a reverse logistics network, (6) the complication of material matching restrictions, and (7) the problems associated with stochastic routings for materials used in remanufacturing operations and highly variable processing times. The present study investigates the first and second requirements, namely, the uncertain timing and quality of returns and the need to balance returns with demands. Forecasting product returns and demands is one of the most crucial issues in inventory management [8]. In addition, forecasts can be applied to used-product acquisition management and capacity planning and are also required for various operational planning activities in remanufacturing [9].

The current paper examines the demand forecasting of remanufactured products using time series analysis. Figure 1 illustrates the time distributions of new product sales and demand for remanufactured products in the case of auto parts. The new product sales shown in the figure are supplied as they are assembled into new automobiles. Over time, all new products reach their end-of-life. A proportion of the products that reach the end-of-life stage generate the demand for spare-parts. If the end-of-life of an auto part occurs earlier than the end-of-life of the automobile, there will be

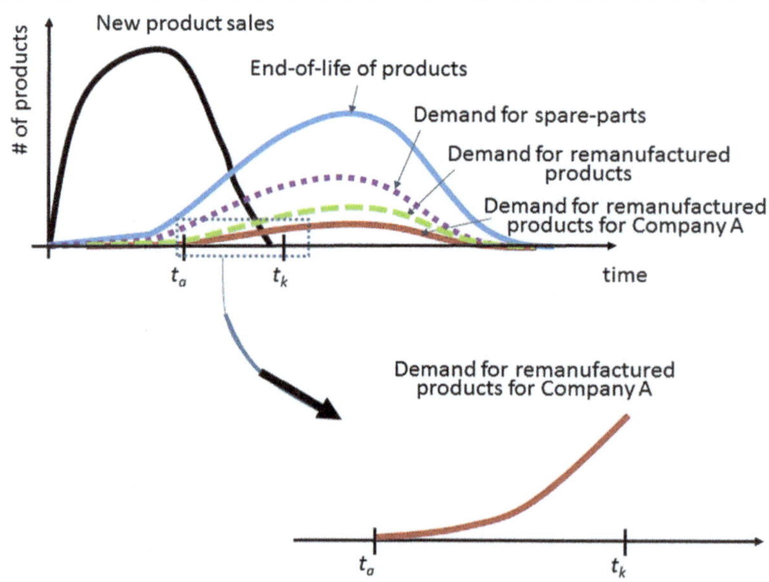

Figure 1 Time distributions of new product sales, end-of-life products, demand for spare parts, and remanufactured product demands.

demand for a spare-part to replace it. A proportion of this demand for spare-parts becomes the demand for remanufactured products. Other parts that reach the end-of-life stage become the demands for new products, reused products, or repairs. In the case of multiple remanufacturing companies, a proportion of the demand for remanufactured products manifests as the demand for remanufactured products from Company A where Company A is the subject company of this study. The section of the plot of demand for remanufactured products from Company A from t_0 to t_k is shown enlarged at the bottom of Figure 1. The demand forecasting that serves as the objective of this paper is to forecast the demand for remanufactured products from Company A.

Most previous studies on demand forecasting in remanufacturing assume that the time distribution of new product sales are known, and the time distributions of product end-of-life and product return are calculated by adding the product use period to the time distribution of new product sales. The assumption that the time distribution is known for new product sales is largely a consequence of the assumption that the original equipment manufacturer (OEM) is the executor of remanufacturing operations. However, an independent remanufacturer (IR) may possess no accurate information of the time distribution of new product sales. In this case, an approach different to the one above is needed. Based on the above background, demand forecasting by time series analysis was employed. To examine the effectiveness of the proposed method, the time series data of the shipments of an actual auto parts IR for a total of 400 types of remanufactured alternators and starters over a period of 12 years was used. Thus far, few studies have investigated the precision of time series analysis demand forecasting using actual data for remanufactured products. In time series analysis, trends are extracted from past historical changes and forecasts are made by extension. The timing of new products sales or product end-of-life is not used, and therefore, the forecast precision obtained by the conventional method where such information is used cannot be expected. However, there are advantages to investigating the effectiveness of forecasting by time series analysis. First, in the case where information on the time distribution of new product sales is unavailable to the remanufacturing company, forecasting by time series analysis may be quite helpful. Second, in the case where partial information on new product sales is available, the combination of forecasts using the product end-of-life model and forecast by time series analysis will increase the accuracy of either method when used separately. However, to conduct such a combined forecast, it is necessary to first clarify the precision and characteristics of forecasting by time series analysis.

The current paper is structured as follows. In the 'Related work' subsection, previous studies on forecasting product return and demand are outlined. In the 'Methods' section, the data, the forecasting method, and the evaluation method used in this study are described. The 'Results and discussion' section presents the results and discussion of this study. The 'Conclusions' section presents the conclusion, where the insights obtained and future issues are summarized.

Related work

Production planning and control in remanufacturing is more complex than those in traditional manufacturing. The complexities derive from the uncertainties in stochastic product returns and imbalances in product returns and demands [7]. Product returns are highly uncertain with respect to timing and quantity, and remanufacturers must balance product returns with demands for remanufactured products to avoid the excessive build-up of inventory or of the inability to meet customer demand where demand exceed returns.

Companies can influence the reliability of product returns to some extent. The common means of reducing the uncertainty of product returns include legislation, incentive systems, contracts, and monitoring and forecasting [10]. Ostlin et al. [11] demonstrated that the ability of the remanufacturer to control product return is characterized according to the relationship between the remanufacturer and the end user. In auto parts remanufacturing, product returns are linked to the demands for remanufactured products. An auto parts remanufacturer supplies remanufactured products, and, in return, it receives used products from the end user. Based on the seven categories proposed by Ostlin et al. [11], the relationship corresponds to either the deposit-based relationship or the credit-based relationship.

There are studies that forecast trends in the amount of product reaching the end-of-life stage over the entire country by building a quantitative model for the product use period. Studies have been conducted for automobiles [12], cathode ray tubes [13], and other diverse products [14]. For remanufacturing, a forecast is generally required for detailed product categories, but a common point is shared for modeling the time distribution of the use period and the product return forecast.

Pioneering works in forecasting product return for remanufacturing include those of Kelle and Silver [15] and Goh and Varaprasad [16]. These researchers developed forecasting models for returns of reusable containers that are typically used in practice to sell or store liquids. Although in strict terms, the return of reusable containers is different from product returns in remanufacturing [7], their works are precursory attempts at product return forecasting. Researchers have sought to expand the model. For example, de Brito and van der Laan [17] investigated the performance of the forecasting methods proposed by Kelle and Silver [15] by considering the impact of imperfect information on inventory related costs.

A case study for building a forecast model designed for specific products in remanufacturing was presented by Toktay et al. [18] who used actual data for returns of Kodak single-use cameras and developed a discrete-time distributed-lag model with dynamic information updates to estimate product returns. Marx-Gomez et al. [19] investigated forecasting models applicable to the remanufacturing of photocopiers. The authors developed a fuzzy reasoning and neuro-fuzzy model to predict the quantity and timing of photocopier returns to the OEM. A Weibullian distribution was employed to describe new product sales and product failure curves, and the return quota was assumed to be uniformly distributed. Umeda et al. [20] presented a model that describes the balance between product returns and demands for single-use cameras, photocopiers, and automatic teller machines based on empirical data. As mentioned previously, conventional forecast studies typically assume that the time distribution of new product sales are known, and the time distributions of product end-of-life and product returns are thereby calculated, which is reasonable when the OEM is the remanufacturer. The condition in which the OEM conducts remanufacturing is applicable to photocopiers, single-use cameras, and other various capital goods. However, IRs may possess no accurate information regarding the time distribution of new product sales. Therefore, the present study investigates demand forecasting by time series analysis.

Studies on demand forecasting by time series analysis have been conducted in other fields. They include demand forecasting for maintenance repair parts [8,21], tourism [22], electricity [23,24], food product sales [25], and some other products and services [26]. Ghobbar and Friend [8], for example, conducted demand forecasting for aircraft maintenance repair parts by time series analysis. Although the demand forecasts for maintenance repair parts and remanufactured products are similar in the objectives, they have different characteristics, where the demand for maintenance repair parts is characterized by its intermittent nature. The intermittent demand tends to be random in terms of time and quantity and has a large proportion of time when the demand is zero. The forecasting accuracies by time series analysis depend, at least partially, on the characteristics of time series of demand. High forecasting accuracies will be expected if the transition curves are stable and periodical, whereas high accuracies cannot be expected if the curves show highly irregular patterns. In this paper, the demand forecasting for auto parts remanufacturing is conducted based on real data, and the precision and characteristics are investigated.

Methods

Data

This study examined the effectiveness of demand forecasting in auto parts remanufacturing. The auto parts remanufacturing industry is reportedly the world's largest remanufacturing sector, accounting for an estimated two-thirds of global remanufacturing activities [1]. Remanufactured auto parts are supplied for automotive after-sales markets. While the targets of auto parts remanufacturing cover a wide range, alternators and starters are the representative products. Among the Automotive Parts Remanufacturers Association (APRA) in the United States, the number of remanufacturers of alternators and starters is highest, representing about 300 companies [27]. The present study employed data obtained from an IR of auto parts that represent shipments of remanufactured alternators and starters.

The authors used data obtained from Shin-Etsu Denso Co., Ltd., which is the largest IR in Japan for alternators and starters [28]. The company has engaged in remanufacturing since the 1960s, but, until the latter 1990s when sales for the Japanese domestic market began to expand, the remanufactured products were almost entirely for export to the United States and European countries. The history of auto parts remanufacturing in Japan is short compared to that of the United States and European countries. Sales of remanufactured alternators and starters for passenger cars in Japan were rare until the late 1990s. The exact figures for the size of the after-sales market for alternators and starters in Japan are not known, and the market scale of the remanufactured products is even more uncertain. The subject company shipped about 160,000 alternators and 170,000 starters in 2013, and about 63,000 alternators and 82,000 starters were for Japanese domestic use, whereas the others were for export. The company remanufactures a combined total of over 7,000 types of alternators and starters. Data collected over a 12-year period reflecting the company's shipments of a total of 400 types of remanufactured products, including 200 types of alternators and starters each, was used in the present study.

Demand for remanufactured alternators and starters has the following features. First, demand is seasonal. Alternators are likely to fail when the temperature is high, and therefore, demand peaks in summer. On the other hand, starters are more likely to fail when the temperature is low, so that demand for starters peaks in winter. Second, the periods of demand differ depending on the product. For some products, the periods are about 10 years, whereas, in other cases, demand may last for decades. This variety is the result of differences of the periods over which an OEM manufactures products. OEMs manufacture some types of alternators/starters for only a few years, whereas other types of products may be manufactured over periods greater than 20 years. Third, for IRs, at least in Japan, it is difficult to accurately determine the time distribution of new product sales for each type of alternator and starter or the number and period over which new products are supplied as they are assembled into new automobiles (i.e., the time distribution of new product sales is shown in Figure 1).[a] Therefore, a method of demand forecasting that does not rely on accurate information regarding the time distribution of new product sales is needed.

If reliable demand forecasting can be realized, the forecasts can be utilized in production planning and control and would result in reduced production costs, product inventory costs, and core inventory costs. Shin-Etsu Denso warehouses a core stock of about 300,000. For each product type, cores representing several months of orders are stocked. In the case of domestic sales, newly returned cores are obtained from the customer at each shipment and returned to the stock. If the number of orders increases, the company increases core stock accordingly to adjust to the situation. The company monitors changes in the number of orders over the past years to determine the stock levels. Reliable demand forecasting could effectively reduce the IRs' core inventory costs.

Demand forecasts, both for the long term, for example over the next year, and for the short term, for example over the next month, are helpful. This study focused on the former and examined the forecasts obtained by time series analysis of demands over a 1-year period.

In this study, the company's shipments per month for a total of 400 types of remanufactured products over a period of 12 years were used. The company handles over 7,000 types of alternators and starters, but only those types of alternators and starters having the highest shipment numbers for the domestic market in the most recent 1-year period were used. The 12-year period was from December 2001 to November 2013.

Examples of the shipment data are shown in Figures 2 and 3. Figure 2 is the record of the number of shipments of a certain alternator (which will be called Alternator A). The number of shipments for this alternator demonstrates an increasing trend. Figure 3 shows the record of the number of shipments of a certain starter (which will be called starter B). This starter reached a peak in its shipment around 2007 and exhibited a decreasing trend thereafter.

To examine the effectiveness of yearly forecasts, the seasonal fluctuations shown in Figures 2 and 3 were eliminated by applying a 12-month moving average to the original time series data. If the original data set is given as x_k ($k = 1, 2, \cdots, 144$), then the moving average time series y_k ($k = 13, 14, \cdots, 145$) becomes $y_k = (x_{k-12} + x_{k-11} + \cdots + x_{k-1})/12$. Therefore, Figures 4 and 5 show the 12-month moving averages of Figures 2 and 3, respectively, where, as an example, the value for December 2013 in Figure 4 is the average of the monthly values from December 2012 to November 2013 shown in Figure 2. In the absence of seasonal fluctuations, Figure 4 shows a relatively simple increasing trend whereas Figure 5 shows an increasing trend followed by a decreasing trend. Forecasts were examined using the trends shown in Figures 4 and 5.

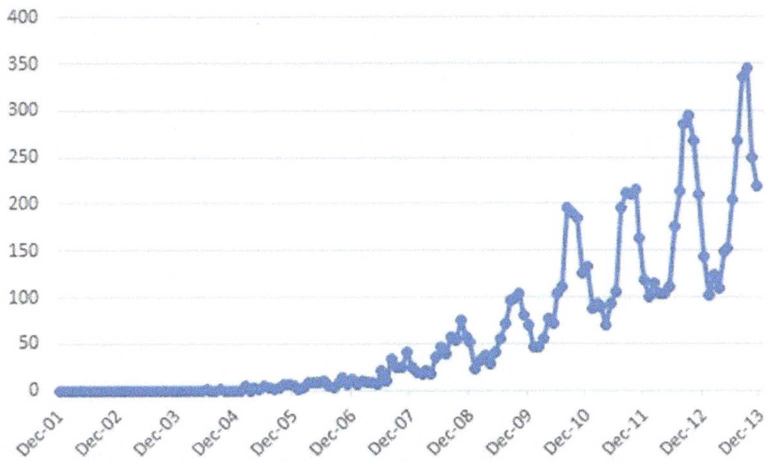

Figure 2 Trend of the monthly number of shipments of one type of alternator (Alternator A).

Forecasting method

This study used the double exponential smoothing method for demand forecasting. Exponential smoothing [26,29] and autoregressive integrated moving average (ARIMA) models [30] are the representative methods in time series analysis. In this study, exponential smoothing was used because exponential smoothing is known to be optimal for a broader class of state-space models than ARIMA models [26].[b] Exponential smoothing is a procedure for continually revising a forecast in the light of more recent experience. In the double exponential smoothing method, two components, namely level and slope, are updated each period. By estimating the level a_T and slope b_T using the observed data (x_1, x_2, $\cdots$, x_T) up to period T, the forecast value s_{T+k} at point $T + k$ ($k = 1, 2, \ldots$) is calculated by the following equation:

$$s_{T+k} = a_T + b_T \times k \tag{1}$$

Holt's method [29] is a representative method of double exponential smoothing. In this method, the estimated values for level a_T and slope b_T are estimated by the following equations:

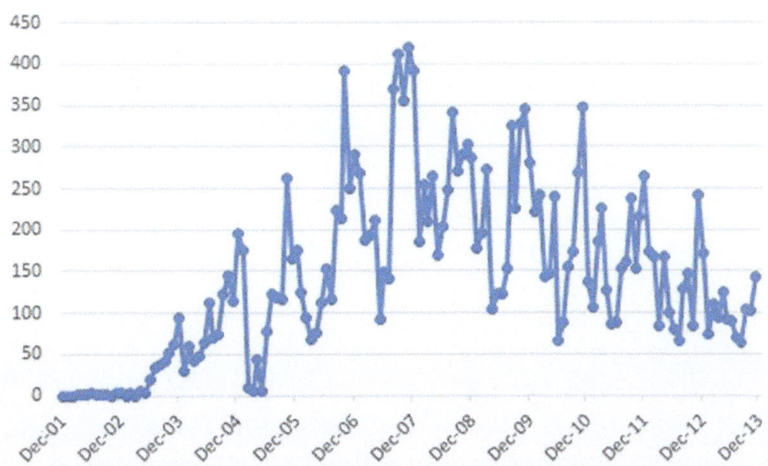

Figure 3 Trend of the monthly number of shipments of one type of starter (Starter B).

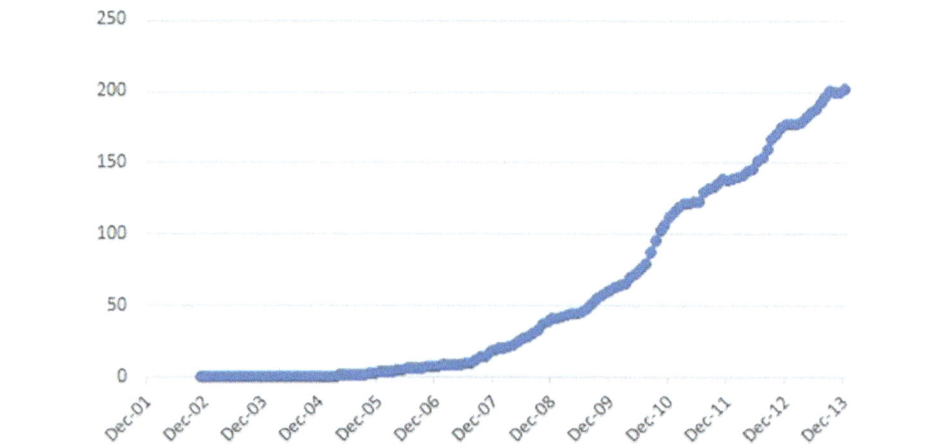

Figure 4 **Trend given by the 12-month moving average of shipments of Alternator A shown in Figure 2.**

$$a_T = \alpha \times x_T + (1-\alpha) \times (a_{T\text{-}1} + b_{T\text{-}1})\ 0 < \alpha < 1 \tag{2}$$

$$b_T = \beta \times (a_T - a_{T\text{-}1}) + (1-\beta) \times b_{T\text{-}1}\ 0 < \beta < 1 \tag{3}$$

The value a_T is estimated by averaging the actual value x_T and the forecast value $(a_{T-1} + b_{T-1})$ at ratio $\alpha : 1 - \alpha$. The value b_T is estimated by averaging the forecast value $(a_T - a_{T-1})$ after obtaining the information during period T and the forecast value b_{T-1} according to information up to point $T - 1$ at ratio β: $1 - \beta$. By supplying the actual value $(x_1, x_2, \cdots, x_T)$, initial values for a_T and b_T and values for α and β, a_T and b_T can be calculated and s_{T+k} obtained. The following are generally used for the initial values for a_T and b_T.

$$a_2 = x_2,\ \ b_2 = x_2 - x_1 \tag{4}$$

The statistic tool *R* was used to calculate the forecast value by Holt's exponential smoothing. Using the *Holt-Winters* () function of *R*, the values for a_T and b_T are

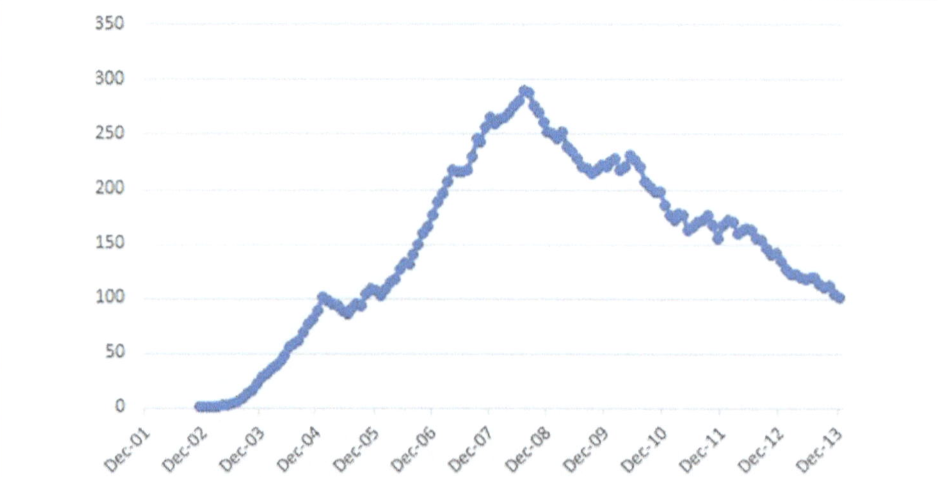

Figure 5 **Trend given by the 12-month moving average of shipments of Starter B shown in Figure 3.**

calculated from the actual value x_T. The values for α and β are successively calculated by taking the least squares sum of the estimated error.

Evaluation method

Demand forecasting was examined using the 12-month moving average data (Figures 4 and 5). Of the 11-year data (from January 2002 to December 2013 for Figures 4 and 5), data for the first 10 years was used (from January 2002 to December 2012), and the value for the last term (December 2013) was forecasted by applying the double exponential smoothing method. The precision of forecasting was evaluated by comparing the forecasted value and actual value for the final term (December 2013). Because of the 12-month moving average, this forecasting corresponds to forecasting the total number of shipments 1 year in the future (in this case, the total shipment occurring over the period December 2012 and November 2013).

Figures 6 and 7 are the results of conducting the forecast calculation for the data shown in Figures 4 and 5. The orange lines overlaying the actual data given by the blue curves in Figures 6 and 7 are the results of the forecast. Comparing the values for December 2013 in Figure 6, the forecasted value is 216.9 and the actual value is 201.3, while the forecast value is 93.2 and actual value is 100.4 in Figure 7.

The error rate of the forecast value is defined by the following equation:

$$\text{Error} = \frac{|\text{Actual value} - \text{Forecasted value}|}{\text{Actual value}} \times 100\ (\%) \tag{5}$$

In Figure 6, the error rate is 7.7% (=|201.3 – 216.9|/201.3 × 100%), whereas, in Figure 7, it is 7.1% (=|100.4 – 93.2|/100.4 × 100%). In addition, as reference value, the difference between the actual value for December 2013 and the actual value for December 2012 is defined as the difference from the previous year as follows:

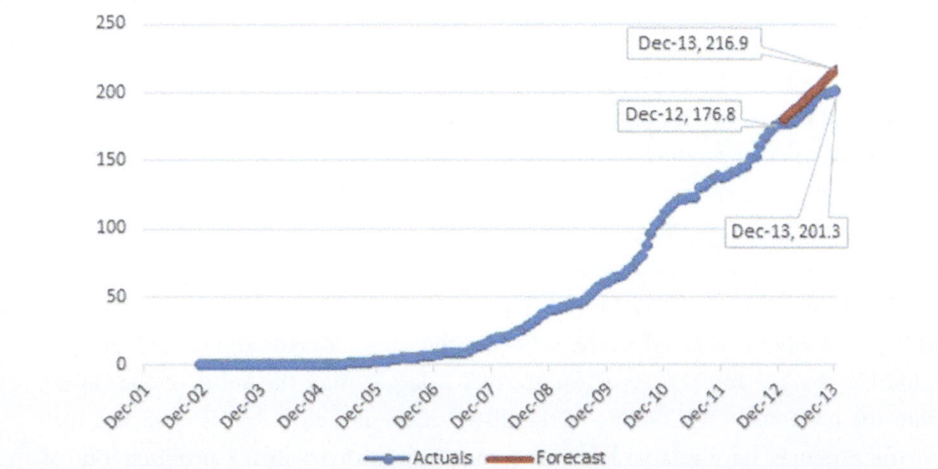

Figure 6 Forecast of number of units of Alternator A to be shipped over the final year based on the previous 11 years.

Figure 7 Forecast of number of units of Starter B to be shipped over the final year based on the previous 11 years.

$$\text{Difference from the previous year} = \frac{|\text{Actual value (in Dec 2013)} - \text{Actual value in previous year (in Dec 2012)}|}{\text{Actual value (in Dec 2013)}} \times 100\ (\%) \tag{6}$$

Referencing the value from previous year is frequently done in the world of business and economics, and this corresponds to the forecast error when forecasting the value of a particular year based on the value of the previous year. In Figure 6, the difference from the previous year is 12.2% (=|201.3 – 176.8|/201.3 × 100%), and, in Figure 7, it is 34.0% (=|134.5 – 100.4|/100.4 × 100%).

Results and discussion

Results

The forecast calculations were conducted for 400 types of products, and the forecast error and difference from the previous year were calculated for each item. Table 1 shows the average forecast error (total row). The average forecast errors were 24.6% for the 200 types of alternator, 29.9% for the 200 types of starter, and 27.2% for the 400 total types of products. The average values of the difference from the previous year were 25.7% for the 200 types of alternator, 29.4% for the 200 types of starter, and 27.6% for the 400 total types of products.

Figure 8 shows a scatter diagram of the forecast and actual values for each product. The line at the center of the diagram represents the condition where forecast = actual. The lines above and below indicate where the error rate is an average of 27.2%, and the region in between indicates an error rate less than 27.2%. The average forecast error value of 27.2% is large compared to the values of the forecast error shown in Figures 6 and 7, and the average forecast error for starters is larger than the value of the difference from the previous year. Looking at the distribution of Figure 8, it is observed that some of the products have a large forecast error. In fact, there are 77 products out of 400 with a forecast error larger than the average of 27.2%. This is slightly less than 20% of the whole, and these degrade the overall average.

Table 1 Average errors of the forecast for 200 types of alternators and 200 types of starters

	Alternator			Starter			Alternator and starter (total)		
	Ratio (%)	Average error (%)	Average difference from the previous year (%)	Ratio (%)	Average error	Average difference from the previous year (%)	Ratio (%)	Average error (%)	Average difference from the previous year (%)
Total	100	24.6	25.7	100	29.9	29.4	100	27.2	27.6
Category A	42	12.9	29.2	43.5	12.6	23.1	42.8	12.7	26.1
Category B	14	52.7	37.1	21	73.3	53.4	17.5	65.0	46.9
Category C	44	26.7	18.9	35.5	25.5	22.9	39.8	26.2	20.6

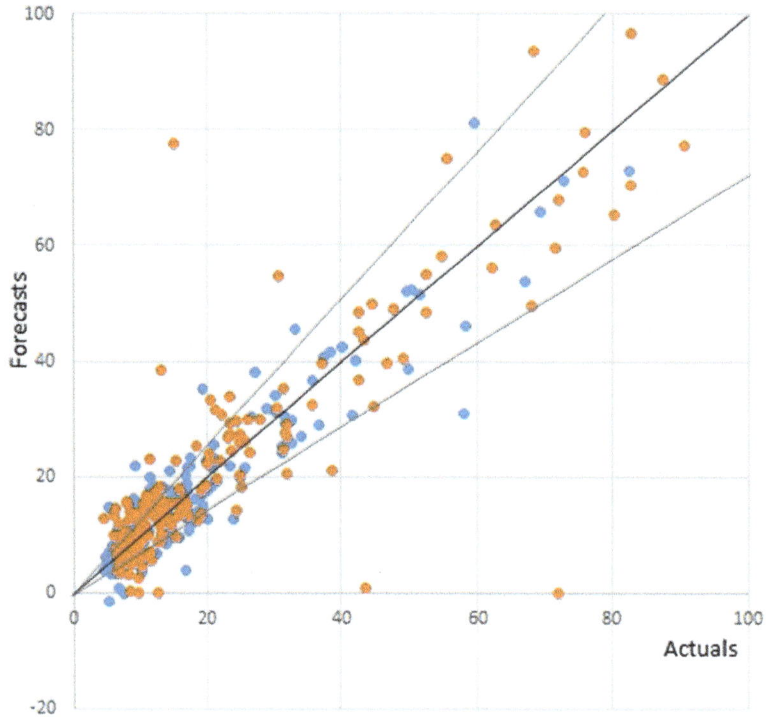

Figure 8 Distribution of forecast and actual values for 200 types of alternators and 200 types of starters. The blue dots represent alternators, and the orange dots indicate starters. The distribution for 385 total products are shown. Fifteen products lie outside of the diagram.

Discussion (1): categorization of transition curves

To address the means to use the proposed method effectively, it is useful to look into the features of the forecasting errors. An examination of the full collection of transition curves allows the curves to be categorized into the following three types.

1. Category A: where the actual data used in the forecast (first 10 years) represents a simple curve with a trend that continues during the forecast period (the last year).
2. Category B: where the actual data used in forecast (first 10 years) represents a simple curve, but the trend changes during the forecast period.
3. Category C: where the transition curve is complex.

A curve is simple if it is either a simple increase or a simple increase followed by a simple decrease. A curve is complex if it exhibits increases and decreases that are repeated.[c]

Figure 9 shows the examples of Category A. The transition curves of Figures 6 and 7 also fall into this category. The forecast errors of the transition curves in this category were small. The examples of Category B are shown in Figure 10. Starter I in the figure is a typical example, where it increases and then decreases during the forecast period, and the forecast reflects the increasing trend up to that time. The average forecast error surpassed 400%. There are also products where the exhibited trend diminished during the forecast period (Alternators F and G) and ones that grew stronger (Starters J and K). The forecast errors were large for these products and degraded the overall precision of

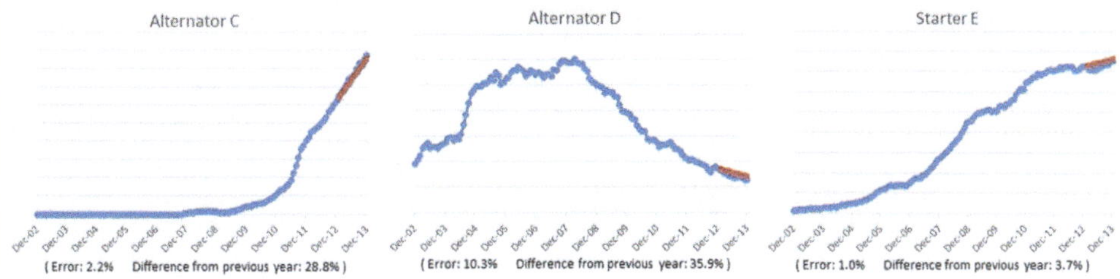

Figure 9 Selected examples of Category A: simple trends.

the forecast. Figure 11 shows three examples representative of Category C. The forecast error exhibits values that may be low, as in the case of Alternator M, or may be high, as for Alternator L.

Table 1 shows the proportions of products for each category. For the alternators, 42% were in Category A (84 products), 14% in Category B, and 44% in Category C, and the forecast errors were 12.9%, 52.7%, and 26.7%, respectively. For the starters, 43.5% were in Category A, 21% in Category B, and 35.5% in Category C, and the forecast errors were 12.6%, 73.3%, and 25.5%, respectively. Figures 12 and 13 show the forecast errors and the values of the difference from the previous year of the 400 transition curves reflecting the 200 alternators and starters, respectively.

The following observations can be made from Table 1 and Figures 12 and 13. First, as seen in the examples of Figure 10, Category B exhibits a simple trend, but the forecast error of the transition curve degrades the overall average forecast error. Particularly, the percentage of transition curves representative of Category B is high for starters, which increases the average error of the starters. Second, for Category C, the overall average forecast error is 26.2%, but the average difference from the previous year is better. Category C curves, as can be seen in Figure 11, repeat increasing and decreasing trends non-periodically. Forecasting by time series analysis is generally not appropriate for such curves to an extent that depends on the frequency of the increase–decrease transitions. In this case, it is safer on average to use the previous year value as the forecast value rather than applying a time series analysis because, in many cases, Category C

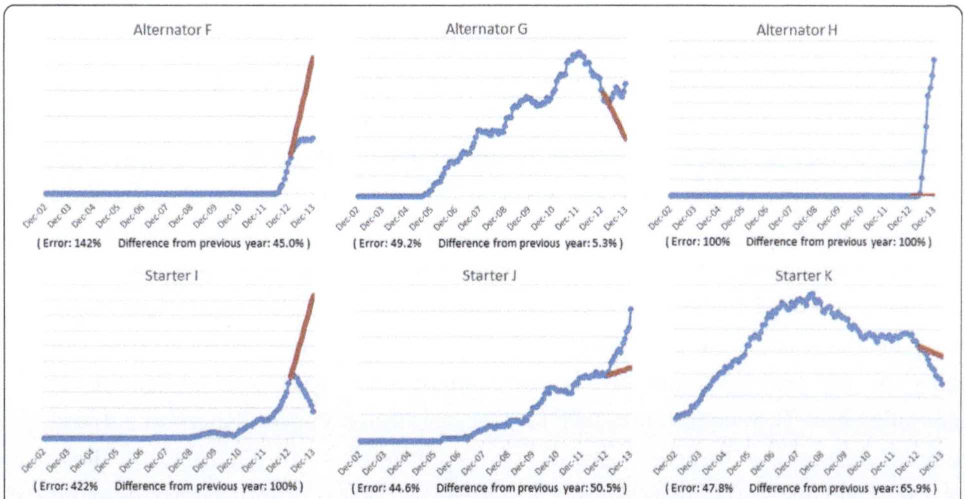

Figure 10 Selected examples of Category B: trend changes over the forecast period (the last year).

Figure 11 Selected examples of Category C: complex trends.

products are long-selling products and the number of shipments throughout the year does not fluctuate to a large extent.

Discussion (2): assessing the possibility of predicting the categories

The results indicated that the forecasts were accurate for the Category A curves, and that the Category B curves increased the overall average error of the forecast. The effectiveness of the proposed method is ensured if Categories A and B could be distinguished without seeing the actual values during the forecast period. Trial investigations were conducted to determine the extent to which Categories A and B can be distinguished through human judgment. In the experiment, three students with no particular knowledge of the subject product or market were shown transition curves categorized as simple (i.e., either in Category A or Category B) during the period excluding the last year and were asked to predict whether the curves were representative of Category A or Category B. The curves for which two or more of the three test subjects estimated to be representative of Category A were categorized as 'Guessed as Category A', and those for which two or more test subjects estimated to be representative of Category B were categorized as 'Guessed as Category B'. The agreements between Category A and 'Guessed as Category A', and Category B and 'Guessed as Category B' were observed.

Table 2 shows the results. Of the 164 Guessed as Category A curves, 132 curves (80%) were correctly evaluated, whereas the other 32 curves (20%) were actually in Category B. On the other hand, of the 77 Guessed as Category B curves, 38 curves (49%) were correctly attributed to Category B, whereas the remaining 39 curves (51%) were actually in Category A. The percentage of correct answers was 71% (= (132 + 38)/(164 + 77) × 100%).

Of these, the curves correctly ascribed to Category B were those where the proximal trend was a rapid increase or decrease, and the forecasts were extensions of such a trend. Alternator G and Starter I shown in Figure 10 fall into these categories. On the other hand, curves incorrectly ascribed to Category B were those that were forecasted to have what appeared to be a too rapid increase or decrease but where such rapid increase or decrease actually followed. For the curves incorrectly ascribed to Category A, the change could not be forecasted, even with the application of human judgment. It was not possible to forecast those curves where the trend moderated during the forecast period (e.g., Alternator F in Figure 10), those where the trend increased (e.g., Starter J and Starter K), or those where demand suddenly emerged (e.g., Alternator H).

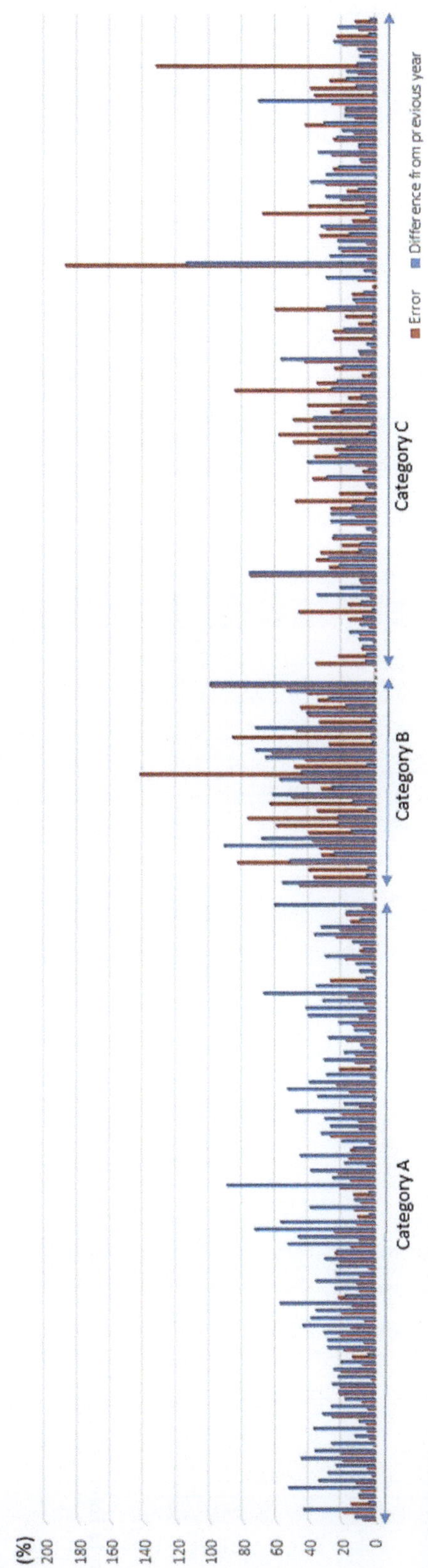

Figure 12 Forecast error rates and the differences from the previous year for 200 alternators.

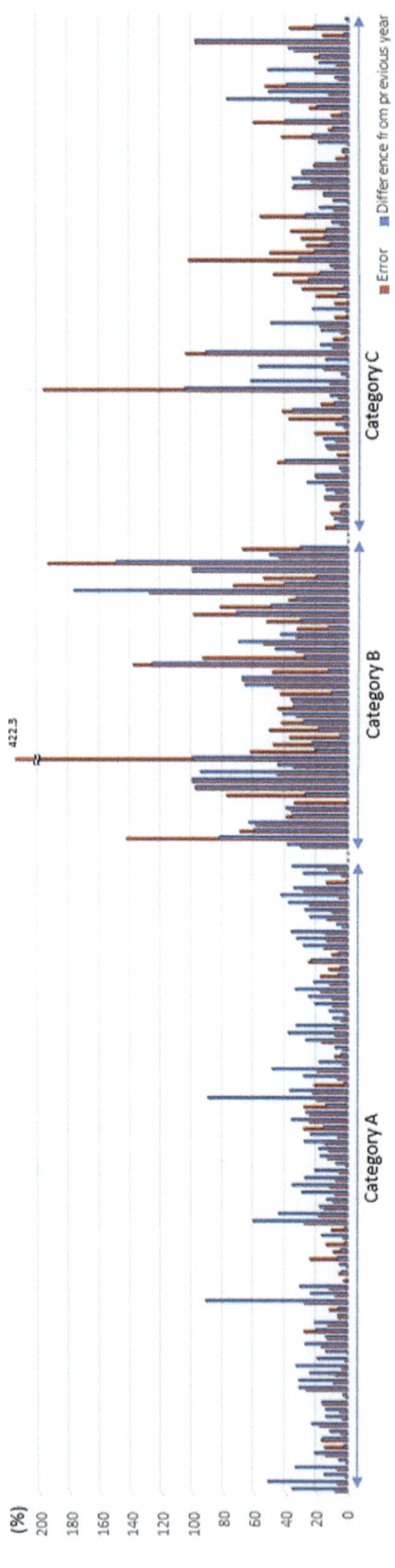

Figure 13 Forecast error rates and the differences from the previous year for 200 starters.

Table 2 Results of the trial to distinguish between Category A and Category B

	Total (alternator and starter)			
	Guessed as Category A		Guessed as Category B	
	Average error (%)	Average difference from the previous year (%)	Average error (%)	Average difference from the previous year (%)
Category A	12.6	21.8	13.0	40.5
	132 types		39 types	
Category B	62.8	49.6	78.8	66.9
	32 types		38 types	
Total	22.4	27.2	45.5	53.5
	164 types		77 types	

Based on this result, the following forecast method was applied. The forecast was conducted by the exponential smoothing method for categories placed in Guessed as Category A, whereas the forecast was based on the previous year value for those placed in Guessed as Category B just as is used for complex curves (Category C). The forecast precision of this method is listed in Table 3. For the alternators, the cover ratio (those curves assigned to Guessed as Category A and Category C) was 79% and the forecast error was 18.8%. For the starters, the cover ratio was 82.5% and the forecast error was 24.3%. In total, the cover ratio was 80.0% and the forecast error was 21.5%. Compared with the average forecast error of 27.2% obtained when the exponential smoothing method was applied to all curves, the average forecast error of 21.5% shows an improvement.

Those curves providing a forecast result with a large error (Category B) were predicted through human judgment. This indicates the possibility for effective use of time series analysis forecasting as a tool supporting human-based decisions for inventory management or for creating production plans.

Conclusions

The current study investigated the effectiveness of forecasting demands in an alternator/starter remanufacturing using time series analysis of actual shipment data from an IR of auto parts. The average error of the forecasts obtained for 400 products using Holt's exponential smoothing method was 27.2%. The result is expected to provide a benchmark for research targeting demand forecasting in auto parts remanufacturing.

Table 3 Forecasts reflecting the distinction among Categories A, B, and C

		Alternator	Starter	Total
Guessed as Category A	Number ratio (%)	35.0	47.0	41.0
	Average error (%)	18.7	25.3	22.4
Guessed as Category B	Number ratio (%)	21.0	17.5	19.3
Category C	Number ratio (%)	44.0	35.5	39.8
	Difference from previous year (%)	18.9	22.9	20.6
Total (Guessed as Category A + Category C)	Cover ratio (%)	79.0	82.5	80.8
	Average error (%)	18.8	24.3	21.5

The results indicated that the forecasts were accurate for products whose demand trends were stable (Category A). Demand trends ascribed to Category A represented 42% of the total, and the demands were forecasted with an average forecast error of 12.7%. On the other hand, the forecasting errors for products whose shipment trends changed during the forecasting period (Category B) were high. Demand trends ascribed to Category B represented only about 18% of the total, but the average error was high at 65.0%, and this degraded the overall average forecasting error. The remaining 40% of the supply curves were those exhibiting complex trends (Category C). While the average forecast error of Category C was 26.2%, better results were obtained by referring to the previous year value. An investigation was conducted to determine the possibility of identifying Category B based on shipment data prior to the forecasting period by human judgment. The experiment indicated that a distinction could be made in part, and the percentage of correct answers for distinguishing between Categories A and B was 71%. As a result, the forecast value using this distinction was assigned to 80.8% of the products with an average forecast error of 21.5%, which showed improvement.

Future steps regarding the present study is proposed as follows. First, the effects of other available information on improving the accuracy of forecasts are worth investigating. As described in the 'Background' and 'Methods' sections, IRs do not have accurate information on the time distribution of new product sales. However, some information is available to IRs in certain cases. For example, information on the number of years that new products were sold is readily available for IRs for some type products although, even in these cases, the number of products sold each year during the period is not available, and thus the time distribution of new product sales is not available. Using this new product sales period information together with data concerning the product lifespan time distribution, the long-term time distribution of demands may become forecastable. In business situations, various information including product-specific market information, word-to-mouth information, etc. are taken into account, in most cases intuitively, to project the trend of demands and to detect the change of the trend of demands. One of the next steps of the present work is to clarify the type of information that may improve the accuracy of forecasts, to develop methods to utilize such information, and to assess the effectiveness of the methods. Second, while this study focused on forecasting demands 1 year in the future, forecasting demands over shorter periods should also be investigated. Both yearly and monthly forecasts are useful for inventory management and production planning in remanufacturing. To conduct shorter-term demand forecasts, seasonal patterns of demands must be extracted. Investigation of monthly demand forecasting is another proposed direction for subsequent work. Third, investigating methods to reflect the forecast results effectively in inventory management and production planning is another significant next step. Depending on the available forecasting accuracy, the effective means of reflecting the results may differ. Also various factors that are specific to the processes such as a feature that underestimation of demands is more costly than overestimation must be taken into account. It is important to develop methods that can accurately reflect demand forecasting in process optimization in remanufacturing based on an understanding of the characteristics of forecasting.

Endnotes

[a]IRs can obtain information on the number of sales of automobile types and can determine the alternator/starter installed in each automobile type. However, because the correspondences between the alternator/starter type and the automobile type is many-to-many, where one type of alternator/starter may be used in several automobile types, and multiple alternator/starter types may be used in a single automobile type according to the engine size and model, it is not feasible for IRs to estimate the time distribution of new alternator/starter product sales from the number of sales of a given automobile type.

[b]In addition, the following are the reasons that exponential smoothing was chosen. First, exponential smoothing is simpler in its formulation, and thus it was expected to be easier to identify the causes when it created an unexpected result. Second, when test calculations were conducted for 30 type products prior to the calculations for 400 type products, Holt's exponential smoothing method outperformed ARIMA model in the average precision.

[c]The categorization to the three categories was supported by human judgment because the intended differentiation was difficult to perform by any other means. For example, if a simple increase or decrease are defined as a monotonic increase or decrease, the definition is too strict and the intended differentiation between simple curves (Categories A and B) and complex curves (Category C) could not be obtained. Three students were asked to choose the appropriate categories for the curves. The categories of the curves were set according to the agreement by two or more students.

Abbreviations

ARIMA model: autoregressive integrated moving average model; IR: independent remanufacturer; OEM: original equipment manufacturer.

Competing interests

The authors declare that they have no competing interests.

Authors' contributions

MM and AI conducted the problem setting for this research. AI compiled and analyzed the data. MM performed the forecasting calculation and drafted the manuscript. Both authors read and approved the final manuscript.

Acknowledgements

This research is partially financially supported by the Grant-in-Aid for Scientific Research (No. 24510220), Japan Society for the Promotion of Science (JSPS).

Author details

[1]Center for Service Research, National Institute of Advanced Industrial Science and Technology (AIST), 1-1-1, Umezono, Tsukuba, Ibaraki, 305-8568, Japan. [2]Shin-Etsu Denso Co., Ltd, 2656-210, Taira, Omachi, Nagano 398-0001, Japan.

References

1. US International Trade Commission: Remanufactured Goods: An Overview of the US and Global Industries, Markets, and Trade. Investigation No.332-525, USITC Publication 4356. (2012). [http://www.usitc.gov/publications/332/pub4356.pdf] (accessed on September 1, 2014)
2. Lund, R: Remanufacturing: An American Resource. In: Fifth International Congress Environmentally Conscious Design and Manufacturing. Rochester Institute of Technology, Rochester (1998)
3. Geyer, R, Jackson, T: Supply loops and their constraints: the industrial ecology of recycling and reuse. Calif. Manage. Rev. **46**, 55–73 (2004)
4. Lundmark P, Sundin E, Björkman M: Industrial challenges within the remanufacturing system. In: Proceedings of Swedish Production Symposium, Stockholm, 2–3 December 2009, 132–139 (2009)
5. Matsumoto, M: Development of a simulation model for reuse businesses and case studies in Japan. J. Cleaner Production. **18**, 1284–1299 (2010)

6. El Korchi, A, Millet, D: Conditions of emergence of OEM's reverse supply chains. J. Remanufacturing **4**, 3 (2014). doi:10.1186/2210-4690-4-3
7. Guide Jr, VDR: Production planning and control for remanufacturing: industry practice and research needs. J. Oper. Manage. **18**, 467–483 (2000)
8. Ghobbar, AA, Friend, CH: Evaluation of forecasting methods for intermittent parts demand in the field of aviation: a predictive model. Comput. Oper. Res. **30**, 2097–2114 (2003)
9. Toktay, LB: Forecasting product returns. In: Guide Jr, VDR, van Wassenhove, LN (eds.) Business Aspects of Closed-Loop Supply Chains. Carnegie Mellon University Press, Pittsburgeh (2003)
10. Pokharel, S, Mutha, A: Perspectives in reverse logistics: a review. Resour. Conserv. Recycling **53**, 175–182 (2009)
11. Östlin, J, Sundin, E, Björkman, M: Importance of closed-loop supply chain relationships for product remanufacturing. Int. J. Production Econ. **115**, 336–348 (2008)
12. Adachi, Y, Daigo, I, Yamada, H, Matsuno, Y: Development of a dynamic model for CO2 emissions from gasoline vehicles in their use stage. Dev. Eng. **11**, 19–29 (2005)
13. Linton, JD, Yeomans, JS, Yoogalingam, R: Recovery and reclamation of durable goods: a study of television CRTs. Resour. Conserv. Recycling **43**, 337–352 (2005)
14. Murakami, S, Oguchi, M, Tasaki, T, Daigo, I, Hashimoto, S: Lifespan of commodities, part I. J. Ind. Ecol. **14**, 598–612 (2010)
15. Kelle, P, Silver, EA: Forecasting the returns of reusable containers. J. Oper. Manag. **8**, 17–35 (1989)
16. Goh, T, Varaprasad, N: A statistical methodology for the analysis of the life-cycle of reusable containers. IIE transactions. **18**, 42–47 (1986)
17. de Brito, MP, van der Laan, EA: Inventory control with product returns: the impact of imperfect information. Eur J Oper. Res. **194**, 85–101 (2009)
18. Toktay, LB, Wein, LM, Zenios, SA: Inventory management of remanufacturable products. Management Science. **46**, 1412–1426 (2000)
19. Marx-Gómez, J, Rautenstrauch, C, Nürnberger, A, Kruse, R: Neuro-fuzzy approach to forecast returns of scrapped products to recycling and remanufacturing. Knowl. Based Syst. **15**, 119–128 (2002)
20. Umeda Y, Kondoh S, Sugino T: Proposal of "marginal reuse rate" for evaluating reusability of products. In: Samuel, A, Lewis, W (eds) Proceedings of International Conference on Engineering Design, Melbourne (2005).
21. Willemain, TR, Smart, CN, Schwarz, HF: A new approach to forecasting intermittent demand for service parts inventories. Int. J. Forecast **20**, 375–387 (2004)
22. Song, H, Li, G: Tourism demand modelling and forecasting - a review of recent research. Tour. Manage. **29**, 203–220 (2008)
23. Taylor, JW, de Menezes, LM, McSharry, PE: A comparison of univariate methods for forecasting electricity demand up to a day ahead. Int. J. Forecast **1**, 1–16 (2006)
24. Weron, R: Electricity price forecasting: a review of the state-of-the-art with a look into the future. Int. J. Forecast **30**, 1030–1081 (2014)
25. Miller, JL, McCahon, CS, Bloss, BK: Foodservice forecasting with simple time series models. J. Hospitality Tourism Res. **14**, 9–21 (1991)
26. Gardner Jr, ES: Exponential smoothing: the state of the art - part II. Int. J. Forecast **22**, 637–666 (2006)
27. Kim HY, Raichur V, Skerlos SJ: Economic and environmental assessment of automotive remanufacturing: alternator case study. In: Proceedings of the 2008 International Manufacturing Science and Engineering Conference (MSEC 2008), Evanston, IL, 7–10 October 2008, 1–8.
28. Matsumoto, M, Umeda, Y: An analysis of remanufacturing practices in Japan. J. Remanufacturing **1**, 2 (2011). doi:10.1186/2210-4690-1-2
29. Holt, CC: Forecasting Trends and Seasonal by Exponentially Weighted Averages. Office of Naval Research Memorandum 52. (1957). Reprinted in: Holt CC: Forecasting seasonals and trends by exponentially weighted moving averages. *International Journal of Forecasting*. 2004, 20:5–10
30. Box, GEP, Jenkins, GM: Time Series Analysis: Forecasting and Control. Holden Day, San Francisco (1970)

11

The manufacturer's value chain as a service - the case of remanufacturing

David Opresnik* and Marco Taisch

* Correspondence:
david.opresnik@polimi.it
Department of Management,
Economics and Industrial
Engineering, Politecnico di Milano,
Piazza Leonardo Da Vinci 32, 20133
Milano, Italy

Abstract

Manufacturing enterprises globally have already largely adopted the product-service strategy into their operations. However, due to gradual commoditization of services, manufacturing enterprises will have to further extend this strategy. One possibility is for manufacturers to servitize, not only their final products, but also a part of their value chain, with the aim of increasing their long-term competitive advantage. In this article, the application of servitization to remanufacturing, as a set of operational and business competences and processes, is conceptualized. By offering remanufacturing as a service, manufacturers will create an additional revenue stream. The synergies created from integrating remanufacturing into an enterprise with a product-service system are scrutinized. The impact of offering remanufacturing as a service (servitizing) is then assessed from the perspective of the competitive advantage of both, the provider and the consumer of the service. Three main sets of implications are identified. The first is that the integration of remanufacturing into a product-service system could increase customer satisfaction through a larger service scope and higher service quality, while decreasing operational costs. Furthermore, it is shown that the higher the level of servitization, the stronger is the positive impact of remanufacturing. The second set of results shows that servitizing remanufacturing can substantially increase the competitive advantage of both the provider and the consumer. While the first two sets of results have industrially oriented implications, the third set constitutes a theoretical contribution through the proposal and conceptual validation of extension of the application of servitization theory. Finally, while the reasoning is of a conceptual nature, it is based on established theories and includes remanufacturing-related industrial cases as a basis for assessment.

Keywords: Remanufacturing; Product-service; Servitization; Value chain; Competitive advantage

Background

This section first depicts two main concepts of the article product-services and remanufacturing. Afterwards, the aim and hypotheses are introduced.

Product-services (P-S) comprise an integrated combination of products and services that deliver value to the customer. The term is found under multiple names; however, Baines [1] proposed a synthesis of those terms, servitization being the broadest one. It is defined as 'market packages or bundles of customer-focused combinations of goods, services, support, self-service and knowledge' [2]. Servitization has become a very successful strategy in manufacturing enterprises, 20% of such enterprises having already integrated some kinds of services [3]. The result of this is that, due to the pervasiveness of servitization as a strong basis for differentiation against competitors, it is going to slowly fade away in the future, implying that product-service bundling will slowly become commoditized or, put differently, customers' expectations will increase with time so that manufacturers will have to find new ways to satisfy them. Consequently, servitization is shifting from representing a distinctive basis for competitive advantage towards being a necessary, though not sufficient, condition for sustaining its competitive advantage. This implies that new business approaches capable of creating synergy with the existing servitization strategy need to be sought. One such path in industry and research that is becoming more and more established is the collaboration between product-service system (PSS) and remanufacturing. The latter is 'the process of transforming durable products that are worn, defective, or discarded to a "like new or better" condition through a production-batch process of disassembly, cleaning, refurbishment and replacement of parts, reassembly, and testing' [4].

Synergies between the two strategies have already been identified. Remanufacturing for instance can be enhanced if selling the functionality of a product and not directly the product itself [5]. In this case, the larger part of the risk associated with a remanufactured product lies with the supplier rather than the user [6]. The synergy also enables the manufacturing enterprises to have better control over the products, as they can more readily plan and manage the manufacturing process [5]. In such cases, profit margins can be sustained or improved compared to the one from new sales [7]. Furthermore, remanufacturing can constitute an essential part of, or even a condition for, economically and environmentally advantageous service selling [8]. However, discussion of only functional sales can be very restrictive, since it represents only one level among the four levels of servitization [9] through which enterprises undertake the shift from products to services. Namely, those levels are also very pervasive among manufacturing enterprises and hence are essential. Consequently, although that there exist some synergies between P-S and remanufacturing strategies, their detailed relation is not yet clear [10]. This is preventing managers of manufacturing enterprises to fully exploit those synergies in order to increase their long-term level of competitiveness.

The aim of this article is to help managers of manufacturing enterprises evolve their service-oriented strategies after integrating a PSS. The evolution of the service-oriented strategies is achieved in two steps, each expressed by a single objective. The first step is to scrutinize the impact of integrating a remanufacturing strategy into a manufacturing enterprise that is already offering P-S, i.e. already servitizing and is aiming to improve its competitive advantage through the evolvement of service-oriented strategies. The focus is thus on manufacturing enterprises that have already servitized their business, as their share has already become substantial, and in which the synergies between P-S and remanufacturing have already partially been identified; this relationship has not been studied in detail, except in some cases like those of Sundin, Bjorkman and Jacobsson [8] and Sakao and Mizuyama [10]. If the relations between P-S and remanufacturing are identified as being positive, then it would imply that the number of manufacturing enterprises that will integrate remanufacturing will increase. Regarding the future potential demand of remanufacturing and its high knowledge intensity, the second objective deals with enterprises that are already providing P-S and have already set up its own remanufacturing activities. The objective has scrutinized the impact of offering remanufacturing-based developed capabilities (competences and processes) to design and set up such remanufacturing activities from an operational and business perspective, as a service to other manufacturing enterprises. Remanufacturing can be offered in a variety of business modes. We have chosen the widely accepted servitization framework to support the analysis. This means that the objective will be to analyse, how remanufacturing activities can be offered as a service in four different modes, where each refers to one of the servitization levels. This will be followed by an analysis of how those remanufacturing activities, offered as a service, impact on the competitive advantage of both the provider and the consumer of such services. Accordingly, the following two hypotheses have been formulated:

H1 - The higher the level of servitization of a manufacturing enterprise, the greater the relevance of its remanufacturing strategy for the enterprise in question.
This hypothesis contributes to defining the relationship between servitization and remanufacturing, which has first to be clarified. As established earlier in the previous paragraphs, multiple synergies exist between functional sales, as the higher level of servitization, and remanufacturing; however, for other levels, the relation is unknown. Due to the characteristics of servitization, however, it is expected that the relevance of remanufacturing increases with the level of servitization, i.e. the higher the servitization level, the more motivated is the manufacturer to minimize costs through the product life cycle. This hypothesis will consequently depict the relevance of remanufacturing as a strategy integrated into the existing business model of an enterprise that has already servitized or that is going to servitize its products.
The second hypothesis is divided into two sub-hypotheses.
H2/A - A manufacturing enterprise that provides P-S and has integrated its own remanufacturing activities can increase its competitive advantage by providing remanufacturing - the competences and processes needed to set up a remanufacturing operation from an operational and business perspective - as a service to other manufacturing enterprises.
H2/B - A manufacturing enterprise that already provides P-S can increase its competitive advantage by acquiring the service of remanufacturing (design and set-up) from another manufacturing enterprise.

The sub hypotheses H2/A and H2/B contribute to the identification of the exploitation possibilities of introducing an integrated remanufacturing strategy into an already servitized business from the perspectives of the provider and consumer of such a service. The first hypothesis, H1, is introduced to explain the relations between remanufacturing and P-S strategies. The second set of hypotheses H2 (A and B) are formulated to investigate the possibility that a manufacturing enterprise can exploit its already developed remanufacturing activities (competences and processes) by offering them as a service (through the four servitization levels) to other manufacturing enterprises. Offering in this way, remanufacturing, which is a specific part of the manufacturer's value chain, is in line with the concept of servitizing the manufacturer's value chain [11], in this way applying the concept of servitization to a specific part of the value chain. The concept of servitizing the manufacturer's value chain thus lies at the core of the proposed concept, although made specific for the case of remanufacturing. This method proposes that manufacturing enterprises offer high added-value, knowledge-intensive parts of their value chain to other manufacturing enterprises. Thus, in this article, the offering of remanufacturing is made through four servitization levels. Servitization [9] is based on the presumption that the core of the offering is a product and that, in each further servitization level, different types of services are added, while the ownership of the product changes. However, in this case, remanufacturing is a service *per se* and not a product. Thus, a service-dominant logic [12] is adopted, implying that the services (the process of providing benefits) are in the centre of the value exchange, using products as units of output that are sometimes used to support the delivery of the service itself [13]. This perspective was adopted because the parts of the manufacturer's value chain (in this case, remanufacturing) are not products but services. In addition, different types of service levels fit some of the main rationales for a service-dominant logic introduced by Vargo and Lusch [14], especially the one stating that skill and knowledge are the fundamental units of exchange. The latter is the case when remanufacturing is offered as a service. Remanufacturing is seen as an interesting part of the value chain to be servitized, since it can be a complex process from the operational standpoint and from the business perspective of making a valid business case. Finally, due to multiple factors (among others also legislative in some industries), remanufacturing activities are becoming more and more pervasive; hence, the demand for such competences and skills will increase.

With the objective of reinforcing the findings of the hypothesis, a research question (RQ) is set. It can be referred to as a productive research question [15], with the aim of reinforcing conceptual findings from the previous hypotheses, thus as follows:

RQ: What are the risks associated with the introduction of offering remanufacturing as a service, as investigated in the previous set of sub-hypothesis H2A and H2/B?

Relation to existing work

Remanufacturing

Remanufacturing preserves most of the added value of a product by giving it a second life and, typically, reduces the energy used by eliminating production steps [16], hence being a sustainable approach in manufacturing. The now already famous case of remanufacturing from Caterpillar that has 17 dedicated manufacturing facilities in eight countries provides remanufactured products of the same quality as new products [17], but with the preservation of approximately 85% of original energy added [18]. The benefits of remanufacturing are not only environmental, but in many cases, it enables the original equipment manufacturer (OEM) to benefit from a secondary market, opening new business opportunities over which the manufacturer previously had no control, since independent operators entered the market and competed for returned 'cores' and thus a share of the market with remanufactured products [19]. These authors established the frequent occurrence of such a situation; namely, based on the assessment of Guide Jr. [20] that 95% of remanufacturing programs are not managed by their OEMs. The potential benefit of remanufacturing is not limited to the remanufacturer but extends also to the consumers, since the remanufactured product can be sold at a price 20% to 30% lower than that of the new product, while the production costs constitute only 35% to 60% of the original manufacturing cost [21]. Regardless of the positive impact of remanufacturing, its integration into existing business models is not always self-evident. Some of the main barriers to integrating remanufacturing into an existing business are consumer acceptance and lack of understanding of the required business model and of a broad enough skill set within a single company to cope with differences of the product entering the business [6]. Such barriers can, to some extent, be alleviated through integration into a P-S strategy. Thus, implementation of remanufacturing practices may be risky without a comprehensive understanding of the factors governing the complex set of remanufacturing activities [22]. For instance, in order to maximize profit, a remanufacturer must be able to balance the return of the product's cores with the demand from customers for remanufactured products [23], which in relation to reverse logistic is a major challenge [24]. However, the barriers are not only those of internal organization and operation, but are also market related. For example, the demand for machine remanufacture is falling because of the introduction of cheaper alternative products into mature markets like the one in UK [25]. Companies find it difficult to implement and consolidate remanufacturing activities [26], being for many still a new topic. Thus, even if remanufacturing has multiple positive impacts, its integration, from a business and operation perspective, is not obvious and can represent a serious challenge to managers. Hence, the research path of integrating remanufacturing, and thus its more pervasive usage, deserves special attention. This article adopts the perspective of Barquet et al. [26] on remanufacturing as an integrated system with its elements and complex set of relationships. Thus, this article studies how to expand remanufacturing activities, seen as a system, through PSS. This also contributes to increasing sustainability in manufacturing.

Service-oriented strategies in manufacturing

For some time now, profit margins from products are not the primary criterion for managers deciding which market to enter with what kind of customer offerings. The more significant profit margins have shifted away from the products, towards the services, one step closer to the customer. For instance, the cost of acquisitioning a train represents only 10% of the product life-cycle costs, while the other 90% represents train-related services [27]. However, the trend of services related to products is not new, although the term servitization was only coined in 1988. Two cases indicated by Schmenner [28] points to that. The first is the company Cyrus McCormick that manufactured reapers that, in the 1860s, were some of the most complex machines. They built an organization through which they reached farmers with services like sales, repair and also credit. The second case - that of Singer who produced a sewing machine that had no technical competitive advantage - still used the English manufacturing system but were nevertheless still the market leaders. This was due to a network of their own retail stores where sales and repairs were made, where inventories were kept, and where they demonstrated the sewing machine for customers. Thus, services in manufacturing were already impacting positively on their competitive advantage. The other mainstream in service-oriented strategies, in addition to servitization, is the PSS. It is a Scandinavian concept offering a promising avenue to achieve, or at least to increase, sustainable consumption and production and to improve the competitiveness of industrial manufacturers [29-31]. In referring to service-oriented strategies in manufacturing, there are various terms like servicizing [32], functional sales [33] and full-service contracts [34].

Servitization is the widest term employed to refer to service-oriented strategies. Thoben's [9] representation of four servitization levels will therefore be used. It is depicted as part of Figure 1 in the form of four blue and white circles, each one representing a single level of servitization. The first level deals with the pure product and does not include any services whatsoever (e.g. no warranty, no maintenance, no consultancy). The second includes services that support the product and can in some cases also generate revenues (e.g. warranty, maintenance), although the product still remains in the ownership of the customer. The third level adds services that enable a manufacturer to differentiate its product from others (e.g. drilling machine leasing). At this level, the product can remain the property of the manufacturer. In the last level, the product is still part of the offering, although it remains in the property of the manufacturer and its characteristics are not essential anymore to the consumer, as here only the functionality is sold and not the product (e.g. the manufacturer sells a constant temperature in a specific place and not the heating system or Hilti International's 'pay per hole model' [35]). This is the highest level of servitization that can be found in the literature and is referred to as a functional sale.

All servitization levels have in common is that they apply to the usage phase of the product, thus the benefits of the service-oriented approaches are intended solely for consumers that use the final products. However, the final product is not the only valuable and tradable resource in manufacturing enterprises and/or along the manufacturer's value

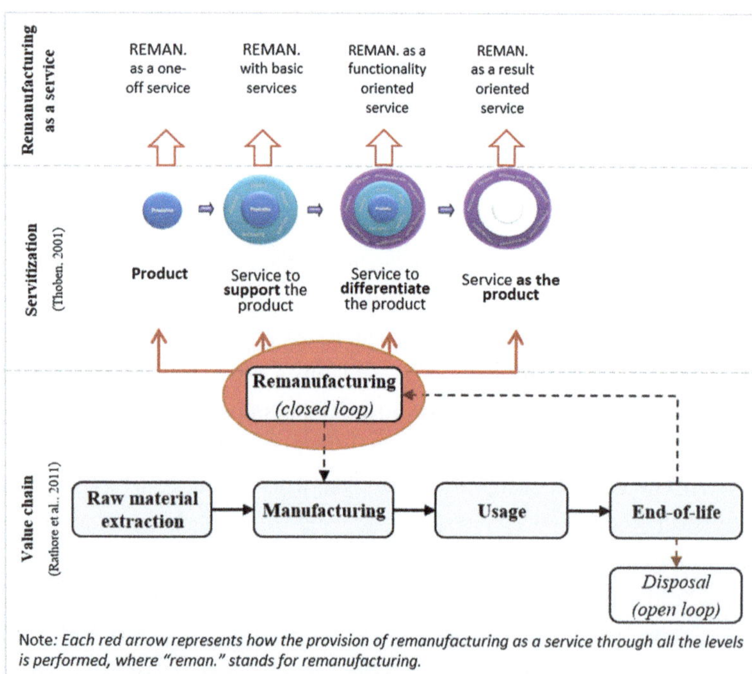

Figure 1 The application of the concept of servitization on remanufacturing. The figure depicts the application of servitization onto another part of the value chain, i.e. remanufacturing. Three main levels: the value chain (with remanufacturing), the servitization framework (with its four levels) and the output of conceptual simulation of servitizing remanufacturing (offered as a service) through four different servitization levels.

chain. Those resources comprise tangible (e.g. a machine) and intangible (e.g. sales and distribution networks or specific competences) ones. Those hold certainly a certain added value. Thus, by applying servitization (as defined in Figure 1) onto high added value parts of the value chain, manufacturers are able to offer those parts as services (through different levels of servitization) to other business entities. Furthermore, the core product, around which servitization revolves, is not necessarily always a 'pure' product but can also be a 'pure' service. For instance, such a service that is offered through servitization can be constituted from formalized relations with different manufacturers with a set of predefined business terms (price, flexible delivery, etc.), which could be termed according to its function (e.g. sales, distribution or development channel) or its form (e.g. network, chain or system). This concept is called servitizing the manufacturer's value chain and was introduced by Opresnik et al. [11], however not yet applied on industrial cases. Nonetheless, this concept was identified in quickly growing innovative manufacturing enterprises; one is textile and the other in the automotive industry. The authors showed that a manufacturing enterprise, by servitizing a part of its value chain, can increase its competitive advantage. Furthermore, it can also increase the competitive advantage of the buyer (consumer) of the part of the value chain in question. Thus, the provider and consumer of the value chain both have benefits. Their business in both cases becomes harder to imitate, their offers become more personalized and both become more flexible on the market (i.e. the seller designed a new product by servitizing its supply chain, while the buyer can more quickly enter new knowledge-intensive markets with smaller risks).

In this article, remanufacturing as a specific part of the value chain is chosen, to which servitization is applied (Figure 1). The figure depicts three main levels: the value chain (with remanufacturing), the servitization framework (with its four levels) and the output of conceptual simulation of servitizing remanufacturing (offered as a service) through four different servitization levels. The four circles represent the four servitization levels that serve as the framework for the process of servitization, based on Thoben [9]. Till now, servitization has been applied during product usage (in Figure 1 depicted as in the value chain in the rectangle 'Usage') while, in the present case, servitization is applied to another part of the value chain, i.e. remanufacturing. In Figure 1, below the servitization framework, the value chain of a manufacturer is depicted together with the process of remanufacturing, based partially on Rathore et al. [36].

The next section provides the main methodological steps upon which the article is based and positions the article's theoretical contribution in the theory building process. Thus, the conceptual application is presented in two main steps. In the first, the impact of integrating remanufacturing into a manufacturing enterprise that is already servitizing; in the second, the impact of 'remanufacturing as a service' is analysed in relation to its competitive advantage. Finally, a discussion is offered, including new questions that have arisen as a consequence of this research.

Methods

The following section is divided into two main parts. In the first, the main research steps are defined. In the second part, the theory presented is positioned within the existing theory building process in the field of servitization and remanufacturing, in order to enable easier and more valid advancement of the overall theory.

Main research steps

Since applying servitization to a specific item on the manufacturer's value chain, in this case remanufacturing, is novel, much information is missing, and there are consequently many uncertainties. The lack of information and uncertainty presents the methodology of conceptual simulation as suitable to be applied in such case [37]. Conceptual simulation is based on three main steps: a) visualizing the situation, b) operating on the situation and c) observing what results, thus enabling causal reasoning [37]. Consequently, we have employed the methodology of conceptual simulation in three main steps. In the first, the impact of integrating remanufacturing into a manufacturing enterprise with a PSS in place is identified. This is based, among other factors, on a meta-analysis of published industrial cases and defines the state of the concept. In the second step, servitization is applied conceptually to a specific part of the value chain, i.e. remanufacturing. This step represents the execution of the application. The third step scrutinizes the impact of the operation on the competitive advantage of the provider and the consumer of such service, while indicating the advantages and risks arising. This is where causal reasoning is performed, based also on established theories from strategic management. Regardless of the fact that, in the first step, sound theories are employed and that, in the second step, the simulation is based on a meta-analysis of cases of manufacturing enterprises providing remanufacturing, the validity of the reasoning being reinforced using Resource-Based View (RBV) [38,39], the article is still presented on the conceptual level, although specific for remanufacturing.

In the first step, multiple industrial cases of remanufacturing are analysed through the lens of servitization. Where possible, all levels of servitization are used in order to identify the benefits that could arise from the collaboration of both strategies. The results are summarized in Table 1. In the second step, servitization of the manufacturer's value chain [11] will be applied to a specific part of the value chain - remanufacturing. The concept of servitizing the manufacturer's value chain has to be put into context in order to able to accept or refute the possibility of applying servitization on the value chain. The application has to depict the main circumstances, hence the contingencies. Thus, the characteristics of the part of the value chain that would be suitable for servitization should be the following:

- The demand for those kinds of services is on the increase;
- It should be highly knowledge intensive;
- It should be complex to integrate into an existing business model; and
- It should create high added value.

Remanufacturing is becoming more and more pervasive (depending on the industries) and relevant but highly complex to manage from operational and managerial perspectives. It thus constitutes a part of the value chain with extremely high added value. Thus, remanufacturing fits those criteria. In order to further increase the validity of the conceptual application, remanufacturing cases have been taken from the literature as the basis for applying the concept of servitization. The conceptual application is thus a step closer to a real industrial environment. In order to identify the impact of the servitization on a specific part of the value chain, some of the most prominent theories and perspectives from strategic management are used. Scrutinizing the impact of servitization

Table 1 Impact of integrating remanufacturing into a manufacturing enterprise with a PSS

Servitization level	Increased customer benefits	Increased P-S provider benefits
Service supporting the product (second servitization level)	Higher customer service quality (e.g. immediate replacement with a less expensive product)	Higher level of customer retention
	Reduction of unexpected product downtime (due to more regular maintenance)	Supplementation of product quality by basic support services (i.e. parts of remanufactured products can be changed more often, due to their lower price)
	Reduced maintenance costs and price of the final product (due to lower prices of spare parts)	More sustainable manufacturing (i.e. remanufactured parts have a much smaller CO2 footprint)
	Wider scope of services supporting the product (e.g. possibility of parts and product substitution by remanufactured ones)	
	More sustainable consumption (i.e. remanufactured parts have a much smaller CO2 footprint)	
Service differentiating the product (third servitization level)	Reduction of unexpected product downtime (with cheaper remanufactured parts, more regular predictive maintenance can be undertaken)	Reduced maintenance costs (e.g. through remanufacturing product costs are decreased throughout the its life cycle)
	Wider scope of services differentiating the product (e.g. customer can choose between new or remanufactured parts or be offered a substituted remanufactured product that otherwise would not be available)	Improved competitive advantage (through reduced price - e.g. MPS performed with lower costs)
	Reduced price of the final product or the price of its usage (due to less costly remanufactured parts)	Larger customer basis (based on a larger service scope and reduced price)
	More sustainable consumption (i.e. remanufactured parts have a much smaller CO2 footprint)	More sustainable manufacturing (i.e. remanufactured parts have a much smaller CO2 footprint)
Product as a service (fourth servitization level)	Higher service reliability (i.e. remanufactured units are always available, especially in the case of unexpected product downtime)	Much lower product life-cycle costs (due to a partially closed loop, the costs are decreased and the provider's profit margin increased)
	Higher service flexibility (i.e. since the product is designed for remanufacturing, its functionalities can be changed faster)	Lower costs of ensuring constant functionality (i.e. during product downtime, it can be substituted with a remanufactured one that is less costly but provides the same functionalities)
	Possibility of reduced service price (due to remanufactured parts)	More sustainable manufacturing
	More sustainable consumption (i.e. remanufactured parts have a much smaller CO2 footprint)	Improved innovation of business model (servitization at this level is in most cases a relevant business innovation that can be optimized through remanufacturing)

through other research disciplines increases the possibility of identifying tensions, thus potentially leading to a more valid theory [40], which is a crucial objective in theory-building [41]. The impact within the field of strategic management will be observed through various lenses within the same field, using multiple theoretical foundations for competitive advantage, thus enabling the results to be meta-triangulated [42].

The main research steps that are undertaken in this article can be useful for managers of manufacturing enterprises needing to be able to exploit better the synergies between remanufacturing and servitization. The process of theory building in the field of servitization is presented in the following section, while positioning the theoretical contribution of this article.

Positioning within the process of theory development

The impact of the present study on the development of the theory of servitization and on remanufacturing is now presented. The impact of servitization is scrutinized through the theory-building process originating from Christensen [40]. He has depicted the process in a pyramid with four main stages of theory building. We populate this structure with the process of theory building for servitization. Described from the top of the pyramid towards the bottom, servitization is not yet a paradigm, since that would require that there are no phenomena within the known field of investigation that do not fall into this category [40], as also multiple perspectives are developed. In the second stage, servitization (as the term encompassing product-service-oriented strategies) could be categorized as a theory as, to some extent, it answers the following questions mandatory for a good theory - what causes what, why and under which circumstances [40]? Hence, it could be said that servitization, thus the integration of services into the customer offering, creates higher added value, as it can more precisely satisfy customers' needs. The third stage involves multiple categorization of servitization, as for the last stage, it refers to different types of studies available for servitization. However, a theory *per se* should not necessarily be accepted as being static. One of the primary purposes of the deductive process of the theory building cycle is to seek anomalies in theories, being the way how a theory is improved [15]. Thus, when observing the manufacturing industry, their strategic positioning and their competitive advantages, an anomaly was identified that the theory of servitization - is not able to explain directly. The theory never explicitly encompassed the possibility of servitizing anything other than the final product (in Figure 1, it is depicted as part of the 'Usage' phase) - although it did not refute it. Hence, by proposing the concept of servitizing the manufacturer's value chain [11], the process of theory building went from the top of the pyramid towards the bottom by noting the anomaly (details see in Figure 2). The research in this article proposes an improvement or extension of the application of the existing theory of servitization, thus going towards the top of the pyramid. However, in order to propose a valid improvement of an existing theory, its proposed application must be clear and sufficiently contingent upon statement of cause and effect [40]; otherwise, it is not falsifiable and cannot be improved or rejected.

A review of the field of remanufacturing from the same structural perspective has to start somewhat differently, from the explicit needs of the manufacturing enterprises asking themselves how to make remanufacturing more profitable and/or how to introduce the process into their existing business model. This would place the topic of the research at the bottom of the pyramid, cycling up towards an attempt to propose a causality, thus having a theory at the end with predictive power. However, regardless of the fact that the theoretical development is positioned into a specific context, which is remanufacturing and substantiated by industrial cases from existing scientific publications, this research is

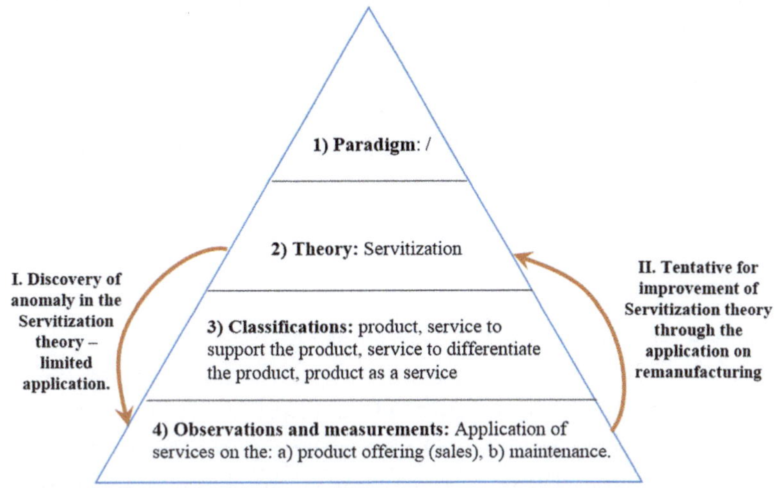

Figure 2 Positioning of the contribution of this article within the theory of servitization. The figure depicts the process of theory building in servitization by proposing the application of servitization onto other parts of the value chain than the product. The process of theory building went from the top of the pyramid towards the bottom by noting the anomaly.

still conceptual in its nature, though based on strong theoretical foundations and clear industrial needs and implications.

Conceptual application

Two main applications are performed. The first subsection studies the impact of integrating remanufacturing into a P-S-oriented manufacturing enterprise, relating to the first hypothesis. In the second subsection, the second application is undertaken, studying the impact of offering 'remanufacturing as a service'. The second application can be seen as the logical continuation of the exploitation of a service-oriented strategy, which is seen in Figure 3. The following model positions all hypotheses within the evolvement of the service-oriented strategies adopted in this article. The initial status of a manufacturing enterprise is as a provider of P-S. In the first step, remanufacturing is integrated,

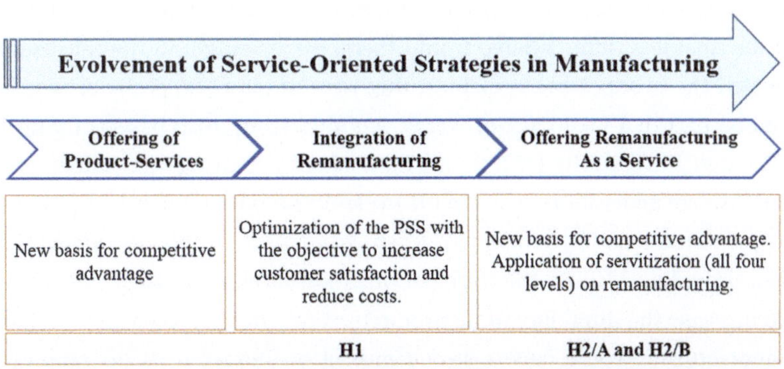

Figure 3 Evolvement of the service-oriented strategies through remanufacturing. The following model positions all hypotheses within the evolvement of the service-oriented strategies adopted in this article. The initial status of a manufacturing enterprise is as a provider of P-S; in a second step remanufacturing is integrated, as a way to create additional added-value; while in the third step, the manufacturers offer remanufacturing as a service to other manufacturing enterprises.

as a way to create additional added value, based on the synergies arising with the PSS. In the second step, the manufacturing enterprise after establishing its own remanufacturing activities and acquiring the related capabilities, it offers those and any potential related infrastructure as a service to other manufacturing enterprises. This step increases the exploitation level of the newly obtained competences from remanufacturing, hence creating the possibility to additionally increase profitability with the existing set of resources (tangible and intangible).

Impact of integrating remanufacturing into enterprises provisioning P-S

This relevance is observed through the impact of remanufacturing on the PSS that is used to provision the P-S, resulting in improved performance indicators like cost efficiency, time to delivery or in P-S quality, customer satisfaction etc. The first level of servitization (pure product) is not part of this analysis, as it represents a pure product offering, without any services. Thus, this level is excluded. Published industrial use cases are used for the analysis.

The first case is that of photocopy machines described by Matsumoto and Umeda [43]. Fuji Xerox started remanufacturing in 1990s, and Ricoh and Canon began selling remanufactured machines ten years later. The price of such remanufactured photocopiers was smaller by 30% or even 50% than that of new products, and profits were larger than those from newly produced machines [43]. As photocopiers need frequent maintenance, such services are classified into the second level of servitization, i.e. offering supporting services like repair and classic maintenance. With a remanufacturing operation in place at this level of servitization, the costs of spare parts could decrease and hence increase customer satisfaction. Abdulrahman et al. [22], who rests on, among others, Hauser and Lund's report [44], describe how the reduced cost of raw materials, energy, manufacturing plants, equipment and extended product life cycles reduce the product price, in turn resulting in a larger customer base. Further on, Fuji Xerox products can be sold, guaranteeing a fixed price per copy from products/processes designed for remanufacturing [1]; in such case, the printers can be completely managed by the provider of the printers including cartridges, maintenance, optimization of use, security, governance, compliance and assuring cost savings as also having a positive environmental impact [45]. Such services, referred to as managed print services (MPS), enable a product to be produced in such way that it is differentiated from others, positioning it in the third servitization level, at which the product can remain in the ownership of the provider (e.g. leased, not sold) or can be part of a take-back program. Another study, also based on Fuji Xerox photocopiers, though in Australia [46], found that remanufacturing can reduce resource consumption and waste generation over the life cycle of the photocopier by up to a factor of 3, with greatest reductions if a product is designed for disassembly and remanufacturing. Thus, when providing MPS in the third servitization level, since the printer provider is motivated to increase the durability of its products, reducing maintenance costs and integrating remanufacturing would have a strong impact on reducing its operational costs throughout its life cycle and/or reducing the final product price (depending on the exact service provided). However, in the third servitization level, there is another possibility in relation to the MPS and remanufacturing, namely, the printer manufacturer could integrate its remanufacturing operations before offering MPS. In this situation, integrating

MPS would mean to provide services at least at the third servitization level. This would make remanufacturing much more beneficial, as the manufacturer has a much greater motivation, than at lower servitization levels (e.g. basic maintenance), to reduce life-cycle costs. In the fourth and highest level of servitization, the manufacturer still owns the machine and provides only the functionality of the product (according, of course, to certain agreed specifications). The latter could be like those offered by MPS. At this level there is also the well-known formula referred to as 'pay-per-print', where the consumer pays for the actual use. At this level, the manufacturer bears all the product life cycle costs and is thus even more motivated to reduce them. Thus, on this level, remanufacturing gains an even greater relevance. Namely, the incentive to provide a certain functionality with minimized costs is even greater. Through remanufacturing, the manufacturer can increase its profit and/or increasing customer satisfaction by delivering a more reliable service with a higher level of quality. It was seen that the impact of remanufacturing clearly increases with the increase of the level of servitization. Thus, if remanufacturing activities are already in place, it would be sound to servitize the products through MPS or *vice versa.* If MPS is already in place, integrating remanufacturing would considerably reduce costs.

The next case is presented by Walsh [7] and deals with game consoles PlayStations. Sony Computer Entertainment Europe (SCEE) is responsible for sales, marketing, distribution and software development for the PlayStation® 2, PSP and PS3 video game and multimedia consoles. Warranty returns are also the responsibility of SCEE. As licensing fees to third party game developers are charged, it is in the interest of SCEE to have a maximum number of end users in order to attract game developers, making customer loyalty vital to the success of the product [7]. Thus, SCEE must offer a high level of quality of the supporting services in addition to their product (console). Basic warranty and maintenance services positions SCEE within the second servitization level. However, by introducing remanufacturing into their operations, SCEE was able to design the following service: when a defective unit is reported to SCEE and a pickup time is arranged for the unit, a remanufactured unit is offered to the customer in exchange for the cost of a service [7]. Consequently, the introduction of remanufacturing not only additionally reinforced customer satisfaction (i.e. minimized the time their customer remains without a working product) but also differentiated additionally their product by introducing this service, that represents also an additional revenue stream. Thus, through the introduction of remanufacturing, SCEE was able to increase customer satisfaction in a market where this indicator is vital, gaining an additional competitive advantage over its competitors.

The third case also comes from Walsh [7] and concerns Edwards Vacuum, one of the leading organisations in the world specialising in the design and manufacture of vacuum products for both general and semiconductor applications. This enterprise established its remanufacturing operations near its larger customers. Even a small downtime, caused in this case by loss of pumping capacity, can far outweigh the cost of repairing or replacing a pump, so the fabricator may opt for an on-site service team. Remanufacturing can play different roles in relation to the product; for instance, it can provide a temporary replacement in order to minimize downtime or as a new (remanufactured) product it decreases the product price for the buyer of the P-S. Hence, when remanufacturing is seen as a service, it is seen to increase customer satisfaction, by decreasing machine downtime and by decreasing costs of the final product (pump), while providing the same product characteristics and quality. The impact of remanufacturing is even more relevant here, where it is integrated into a business model in two ways. Remanufacturing thus enables the assortment of services supporting the product to be broadened, through remanufacturing.

The positive impact of remanufacturing increases throughout the increase of the servitization levels. Thus, at the latest level of servitization (i.e. product as a service, functional sale), the risks of providing the functionality lie completely on the P-S provider, as do the costs of providing it. This requires a fundamental shift not only in the revenue stream but also in the costs structure [47], where remanufacturing can play an important role. For example, the manufacturer has to bear all the costs of maintenance, so it is motivated to minimize the costs and at the same time to provide the widest possible range of services of the highest quality. Both objectives are accessible through remanufacturing which becomes a basis for an innovative paradigm shift in the manufacturing industry from selling physical products to supplying services [5]. The results from the analysis of the industrial cases that have already integrated remanufacturing into their operations are synthesized in Table 1. The first level of servitization is excluded from the analysis, as it is a pure product, while the other three servitization levels are included. The main benefits are categorized in two types of benefits, one is from the customers' perspective, while the other is from the P-S provider's perspective. The benefits from remanufacturing can be direct or indirect.

It can be seen from Table 1 that the higher the servitization level, the more manufacturers bear responsibility for the life-cycle costs. Remanufacturing is contributing considerably to reducing those costs. This analysis strongly indicates that, the higher the servitization level, the greater is the impact of remanufacturing. This benefit is on the side of the provider, who has the possibility of transferring part of this benefit to the consumer of the P-S for instance in the form of lower product, maintenance price, product reliability and/or wider service scope. As for the first level of servitization, i.e. the 'pure' product, manufacturing has a constant level of relevance. This relevance is higher at each succeeding level of servitization. The relevance of remanufacturing is highest in the last level of servitization, the fourth. The main limitation of this representation is that it assumes that the level of intensiveness within the servitization level increases linearly, although this increase is probably case- or industry-specific. Consequently, it can be deduced that the manufacturing enterprises that will servitize, the more relevant and beneficial it becomes for them to include remanufacturing.

The servitization of remanufacturing

The current section aims to analyse the sub-hypotheses H2/A and H2/B, thus studies the impact of offering remanufacturing (i.e. its related processes and competences) as a service to other manufacturers that already servitize and would like to increase their competitive advantage by evolving their current service-oriented strategy.

The application of the concept of servitization onto a part of the manufacturer's value chain is presented. The impact of remanufacturing on the competitive advantages of the provider and the consumer of such service is analysed. Risks that arise are emphasized.

Before commencing with the conceptual application, this paragraph depicts the starting situation. Remanufacturing comprises a very complex set of activities that have to be integrated into an enterprise. They provide considerable value to enterprises from an operational (e.g. patents for a flexible production, product redesign, special skills), and from a business perspective (i.e. how to integrate remanufacturing into an existing P-S business model). The provider and consumer of 'remanufacturing as a service' are first presented. The provider of remanufacturing as a service is a manufacturing enterprise that first has competences in servitization, secondly in remanufacturing and thirdly in exploiting synergy between the two. In order to offer remanufacturing as a service, the provider of such a service has to design it in a way that it can be offered to a new market. From a strategic point of view, the provider is trying to diversifying its business, creating a new revenue stream and entering a new market with high added value that is extremely knowledge-intensive. Offering such service is in line with the strategy of maximal exploitation of valuable internal resources. As for the consumer, buyer of remanufacturing as a service, it is an enterprise that has already servitized its products and is in search for a new basis for competitive advantage, where remanufacturing is one potential solution. Such an enterprise usually would like to move fast through markets, reducing the learning curve and reducing related risks to a minimum, while evolving its service-oriented strategy.

The conceptual application of remanufacturing as a service through all the four servitization levels, critical to H2/A and H2/B, is described as follows. In the first level of servitization, where only the product is present in the customer offering, remanufacturing constitutes a set of one-time activities, including a full (operational and business) or partial plan for integration (operational or business), without any further support. The entire risk, i.e. functionality, reliability and efficiency relies in the hand of the consumer of the service, and the ownership of the remanufacturing system passes over entirely to the consumer. In such case, it is a one-off transaction. In the second servitization level, services are aimed to support the product. Remanufacturing, although it is a service, represents also in this case the 'product' (as already explained, it is actually in this case the service of remanufacturing), the nexus of the offering, while its maintenance, optimisation and modularisation are supporting the service core. Thus, the 'product' is sold with some additional services, which can be either included in the price of the remanufacturing or offered separately; hence the 'product' represents the design and set-up of the remanufacturing service, while further optimisation and maintenance are supporting services. In the third level of servitization, the aim of new services is to differentiate the core offering, which is remanufacturing service. In this level, the ownership of the core offering remains in the hands of the manufacturer that provides remanufacturing. In this case, remanufacturing could be even set up for free (i.e. similar to the product model, that is not actually sold), but the provider of this service can be paid monthly with a fixed or variable amount that could depend on the actual savings achieved by the manufacturer based on using the remanufacturing system. This way, the consumer of remanufacturing as a service does not bear the risk of inefficiency, and the provider of the P-S is actually highly motivated to provide a viable business and operationally efficient solutions. This concept is similar to the energy retrofitting of buildings through the energy service company (ESCO) model, where the provider of the solution is paid according to the savings. As for the fourth servitization level, only the functionality of the product is provided - how it is provided is completely in the domain of the provider, only the result (functionality) is agreed upon. In this way, the functionality of the remanufacturing activities is provided. The consumer of the remanufacturing activities does not own the processes, infrastructure and business network tied to this set of activity but only pays for its use (e.g. per remanufactured piece). The risk lies in the hand of the provider of such service who, in this case, has a strong incentive to fully optimize the functionality of the service since, by creating savings; he can increase its profit. The entire remanufacturing service does not have to be offered as a functionality - only one part may be, for example the development of the network of recovering product cores. Otherwise, providing such a complete service would be to set up the recovery of the cores and its actual remanufacturing process. From the perspective of the consumer, this would mean a complete outsource of the remanufacturing activity.

The concept of servitizing the manufacturer's value chain has been exemplified through the context of remanufacturing. Its impact on the competitive advantage of both provider and consumer are presented, commencing with the former. According to Porter [48] there are two main determinants for relatively long-term performance – the operational effectiveness and the choice of strategic positioning or simply the choice of strategy. The latter defines which activities to perform and in what combinations [49], while operational effectiveness defines how efficiently those activities are performed. The provider, by providing a part of its value chain through servitization, partially changes its strategic positioning. Even though it cannot be said that such a provider is entering or even creating a completely new market, since consulting enterprises do already exist, it is nonetheless an unusual approach, creating new offerings and innovative markets. According to Mintzberg [50], the development of new environments is the role of the entrepreneur, which is the prime basis for economic development. From the perspective of the RBV, the provider develops and frames new, valuable resources to be offered onto the market. However, in the long-term, RBV predicts a competitive advantage only if the enterprises' resources have the following characteristics [51]: a) valuable, b) rare, c) imperfectly imitable and d) non-substitutable. Remanufacturing as a service meets the first three criteria, and is, to some extent, imitable, by consulting enterprises, that become competitors of the provider on this market. It can be concluded that servitizing this part of the value chain improves the competitive advantage of the provider. However, if remanufacturing represents a core element of the provider's basic strategy and if it is hard to imitate, then offering such a service to competitors could have negative consequences. This would be called coopetition, where two enterprises can compete and cooperate simultaneously [52]. In order to avoid that an enterprise offers its strategic resources, it must know very well what its core strengths and weaknesses are. The second risk could be to put too much emphasis on developing the new product and moving away too much from their core business, endangering their current competitive advantage, although, the change of core businesses and basis of competitive advantage is not always negative, as it was seen in the development case of Nokia.

The consumers of the service of remanufacturing will now be scrutinized. By the acquisition of such a service, the possibility of swiftly obtaining valuable, rare and difficultly imitable resources opens up to the consumers of services of remanufacturing. This will enable to make their own business as a whole even more hard to imitate, more efficient and thus more competitive. Furthermore, through the acquisition of such a service, new revenue streams can be created and operational costs lowered, while improving customer satisfaction. Consequently, their long-term competitive advantage is improved. Furthermore, the consumption of service enables also to jump market much more quickly and with smaller risks; for instance, a manufacturer can enter the market of remanufactured parts that was previously in the domain of third party enterprises. However, the impact of acquiring the service of remanufacturing is not without risks. The first risk deals with acquiring a part of the value chain, remanufacturing, that does not fit the current business model. This is referred to as strategic fit. The second risk is related to the lack of experience of the provider in the targeted industrial sector (e.g. the provider has experience in the automotive sector, while the targeted application is in the sector of consumer electronics).

In conclusion, the long-term competitive advantage of both the provider and consumer can increase through the application of servitization to remanufacturing, but both parties run risks that they have to manage carefully. The risk for the provider is highest at the highest servitization level (functional sale), while the risk for the consumer is highest at the first servitization level.

Discussion

Based on the results from both applications in the previous section, each in its independent sub-section, two things are seen. First, multiple synergies between remanufacturing and the PSS can be exploited. Secondly, offering remanufacturing as a service can considerably increase the long-term competitive advantage of the provider as well as the consumer of the service, although while encountering some risks.

The first application demonstrated the impact of integrating remanufacturing into a manufacturing enterprise that is already servitizing. Its integration was observed based on the changes inflicted on the PSS; through the PSS, the P-S is designed, manufactured, delivered and maintained. It has been concluded that the higher the level of servitization, the higher is the impact of remanufacturing on the PSS. Consequently, the first hypothesis is accepted. This indicates that a manufacturing enterprise that is remanufacturing should seriously consider the integration of a PSS and *vice versa*, the enterprise that is providing P-S should seriously review the possibility to integrate remanufacturing. Although the limitation has to be taken into account, this conclusion was based on a limited number of industrial cases. The synergies between remanufacturing and P-S were already introduced by Sundin [53], specifying that remanufacturing can constitute an essential part of, or even a condition for, economically and environmentally advantageous service selling. However, synergies exist also in the lower servitization levels (i.e. services to support the product and services to differentiate the product), that are extremely common in industry. Interestingly, remanufacturing can improve customer satisfaction (as expressed in Table 1) and can decrease the cost of provisioning the P-S. Remanufacturing has often been applied to spare parts and intermediate products but not to final products, which can partially explain the strong impact of remanufacturing on customer satisfaction through improved scope and quality of maintenance services. Table 1 depicts the impact of introducing remanufacturing in an enterprise with a PSS; however, the impact is bidirectional. When a manufacturing enterprise that already remanufactures starts providing P-S at the third or fourth servitization level, thus the product stays in the ownership of the P-S provider, he gains an open access to the 'product's cores' that represent a major challenge in remanufacturing

[54]. Consequently, by starting servitizing, manufacturing enterprises can resolve this challenge. With regard to the characteristics of refurbishment, the results regarding relations with servitization would be similar. However, remanufacturing and refurbishment can also be applied separately or interchangeably.

Furthermore, it has been identified that, in relation to servitization, other motives for remanufacturing can arise for manufacturers - a wider service scope, higher service quality, a higher level of P-S adaptability and thus, higher customer satisfaction, which is also the nexus of servitization. Thus, although this is not the focus of this paper, it can be seen that P-S also makes a positive impact on encouraging remanufacturing activities. This would imply that, if a manufacturing enterprise only remanufactures but does not also servitize, multiple synergies are lost. Understandably, some products cannot be servitized more than on the second servitization level, but even so, remanufacturing still can bring a lot of benefits (e.g. service quality and scope). The most ideal product to servitize and to remanufacture would be one that is expensive, with high maintenance costs and has a long product life-cycle; such a product could be a specific type of a machine tool.

As a synthesized observation from the first application, remanufacturing makes a P-S strategy more competitive, since it:

- Makes the PSS more sustainable in production (through lower needs for resources and energy);
- Enables new and/or more reliable services to be set up;
- Increases customer satisfaction, while decreasing costs;
- Partially closes the product life-cycle loop, making the offering of product functionalities even more efficient.

In the second application, the impact of servitizing a specific part of the manufacturer's value chain, remanufacturing, was assessed. There are two main findings related to the second application. The first finding is that the application of servitizing remanufacturing activities can substantially increase the long-term competitive advantage of the provider. Such a provider develops specific competences when designing a remanufacturing system that exploits synergies with the existing PSS. Those competences and processes in turn represent a new and valuable resource for the manufacturing enterprise despite the fact that they are mostly of intangible nature. Their lack of tangibility does not infer a lack of relevance, since investments in intangible resources in the UK are estimated to have more than doubled between 1970 and 2004 [55], although, according to Zadrozny [56], management awareness about how to take advantage of intangible resources still remains surprisingly low. This is problematic, since it is the role of the strategy of an enterprise to identify the intangible resources that could form the basis for sustainable advantage [57]. From this perspective, servitization serves as a framework with which to review the best possible exploitation opportunities for such intangible resources. Consequently, there is a need for manufacturing enterprises to develop mechanisms for recognizing those parts along the value chain that are valuable and then applying the appropriate exploitation strategy. Nonetheless, there are various possible risks. The manufacturer could risk making public a core part of its value chain that, until then, could not have been imitated by competitors. The provider of such a service would therefore have to define clear limits to what is defined by the 'service', in such way that it does not reveal the strategic core related to this set of activity.

The second finding states that the long-term competitive advantage of the consumer of such a service can also be increased through the acquisition or consumption of such a servitized part of the value chain. This is due to the ability to rapidly set up a knowledge-intensive (at the operational and managerial levels) activity through which new synergies can be created in the core business. Furthermore, this enables manufacturing enterprises that were not previously able to servitize their products (e.g. spare part manufacturer) to now use the concept of servitization, however not on their core product (e.g. fuel pump), but onto their value chain, in this case remanufacturing. Consequently, the servitization framework and its related strategies can be seen as a framework for transferring a structured set of competences to other enterprises, thus making intangible resources transferrable. In the case of integrating remanufacturing, such enterprise can enter new markets with smaller risks and quicker.

Both presented applications had not only managerial implications but also theoretical. Namely, the possibility to apply servitization to other parts of the value chain, besides the final product, was the theoretical anomaly discovered already in Opresnik et al. [11] and visible in Figure 2. The applications analysing the impact of remanufacturing and then the possibility to offer this as a service were steps undertaken with the aim to try to extend the application of the theory of servitization. The feasibility of servitizing a part of the value chain was seen to be conceptually valid. More exactly, this means that the proposal has been conceptually verified, but not validated through industrial application. Hence, we cannot talk about 'real' validation, because this would demand further investigations. If the findings from this article will be found valid in future researches, it would indicate that the theory of servitization was enriched by applying to other parts of the value chain then the product (for details, see Figure 2).

Finally, another synergy between remanufacturing and PSS has been identified. Remanufacturing is a kind of sustainability initiative (from an environmental perspective at least), and its business benefits are strongly reinforced when products are servitized. This means that servitization contributes considerably in making a sound business case for sustainability initiatives. This contribution becomes more relevant the higher the servitization level. Such a relation is quite natural, since the third and, especially, the fourth servitization levels (where the product actually remains in the ownership of the manufacturer, who has to take care of all the costs throughout the product life-cycle) force the manufacturer to adopt a life-cycle perspective over the product. Hence, the costs during this time are born by the manufacturer, creating a strong incentive to optimize all processes, which is done in this article through remanufacturing. Thus, for manufacturing enterprises that are already servitized, remanufacturing represents an excellent example of such synergies. However, further research is needed to identify in which industries those effects are the strongest.

Although this article proposed answers to the introduced hypotheses, it opened even more questions leaning towards operationalization of the concept. The questions are the following:

- How to identify which part of the value chain to offer through servitization?
- How to delineate the core part of the 'product' that will be offered, so not to reveal the strategic core of its primary business? The response is most certainly

multi-layered, for instance taking into consideration the maturity of the industry, the level of technological advancement, the position of the provider of the service in the market (leader, niche innovator etc.).
- How can a consumer of a servitized part of the value chain trust the provider? Trust can be a barrier, especially if the provider can be seen as a competitor. For that purpose, the concept of business ecosystems are set up, with the aim of increasing trust and knowledge sharing, among others, by setting clear IP rules.

The final question that arose in this article is the following. Are the synergies created between PSS and remanufacturing and the additional benefits created from providing or consuming remanufacturing as a service strong enough to disrupt the classical idea or business model to sell as much products as possible, which is not being adequate in terms of sustainability? Thus, does the proposed approach in this article represent a potential solution or mean to the need of carefully revising such (unsustainable) models [58]? Namely, from the synergies of remanufacturing and PSS arise the benefit of increased customer satisfaction and/or reduced operational costs. While from providing the remanufacturing as a service, additional revenue stream that is sustainable (from the environmental perspective) is created. While consuming such service, it enables the manufacturer to obtain access to the first set of synergies described, as also to enter into new markets (e.g. market of used parts). Are the visible impacts sufficient to develop models that could stipulate more sustainable consumption and production?

To synthesize, the manufacturing enterprises that servitize their business, should strongly consider the integration of remanufacturing. Secondly, as the number of enterprises that servitize is increasing around the world, this means that the market for offering remanufacturing as a service is also developing itself.

Conclusion

An important share of manufacturing enterprises is already servitizing, which is slowly making their additional competitive advantage fading away, thus awakening the need to evolve the servitization strategy further. In this article, such a development path was proposed, based on the concept of servitizing the manufacturer's value chain, being in this case remanufacturing, as a knowledge intensive part of the value chain. Its use is becoming more and more pervasive, but still presents operational and managerial challenges to manufacturing enterprises. To test the applicability and impact of applying servitization (through all four levels) on remanufacturing, the methodology of conceptual simulation was employed. At first, it was scrutinized whether remanufacturing has a positive impact on the PSS, since otherwise a proposal to servitize remanufacturing to manufacturing enterprises would not make sense. The results show that considerable synergy is achieved by integrating remanufacturing into an existing PSS, but through all the servitization levels and not just in the last one, i.e. functional sale. The two most significant benefits were shown to be increased customer satisfaction and lowered operational costs related to the P-S provision. Furthermore, the higher the servitization level, the greater the benefits arising from the synergy between remanufacturing and PSS are. The main incentives for remanufacturing are long-term economic and environmental. When remanufacturing is integrated into an enterprise with PSS, the economic incentives become much stronger. The business case for remanufacturing has, therefore, been strongly reinforced because of the way products are offered to customers, through servitization. Thus, this reinforcement is amplified, through the increase of the servitization level.

Secondly, the impact of servitizing remanufacturing activities through all four servitization levels has been analysed. The impact on both partakers (provider and consumer of remanufacturing activities) was positive, and it was found that both could increase their long-term competitive advantage while, however, encountering certain risks. The higher the servitization level, the more risk was born by the provider of the service of remanufacturing. Although of conceptual nature, the paper is based on established theories of servitization, remanufacturing and competitive advantage. Furthermore, examples from journal-based industrial cases based on meta-analysis were used. The results are relevant for managers of manufacturing enterprises that already recognize the value of servitization as also its limitations. It is proposed that remanufacturing should first be integrated, and then offered as a service. Finally, it is concluded that the application of servitization strategy should be extended onto the value chain, as well as the final product. Real industrial validation is still needed, while also conceptual application on other parts of the value chain.

Abbreviations
MPS: managed print services; P-S: product-service; PSS: product-service system; RBV: Resource-Based View.

Competing interests
The authors declare that they have no competing interests.

Authors' contributions
DO and MT conceptualized the application of servitizing the value chain on the case of remanufacturing and designed in detail the application. D.O. prepared the drafted manuscript. Both authors reviewed critically the manuscript and gave the final approval of the version to be published.

Acknowledgements
This work has been partly funded by the European Commission through the FoF-ICT Project MSEE: Manufacturing SErvice Ecosystem (No. 284860).

References
1. Baines, TS, Lightfoot, HW, Evans, S, Neely, A, Greenough, R, Peppard, J, Roy, R, Shehab, E, Braganza, A, Tiwari, A, Alcock, JR, Angus, JP, Bastl, M, Cousens, A, Irving, P, Johnson, M, Kingston, J, Lockett, H, Martinez, V, Michele, P, Tranfield, D, Walton, IM, Wilson, H: State-of-the-art in product-service systems. Proc. Inst. Mech. Eng. Part B J. Eng. Manuf. **221**, 1543–1552 (2007)
2. Vandermerwe, S, Rada, J: Servitization of business: adding value by adding services. Eur. Manag. J. **6**, 314–324 (1988)
3. Santamaría, L, Jesús Nieto, M, Miles, I: Service innovation in manufacturing firms: evidence from Spain. Technovation **32**, 144–155 (2012)
4. Ijomah, WL: Addressing decision making for remanufacturing operations and design-for-remanufacture. Int. J. Sustain. Eng. **2**, 91–102 (2009)
5. Sundin, E, Bras, B: Making functional sales environmentally and economically beneficial through product remanufacturing. J. Clean. Prod. **13**, 913–925 (2005)
6. Rosamond, E: The application of product services system in remanufacturing activities. In: IMS2020-Proceedings from the IMS2020 Summer School on Sustainanable Manufacturing (2010)
7. Walsh, B: {PSS} for product life extension through remanufacturing. In: CIRP IPS2 Conference. pp. 261–268. Linköping (2010)
8. Sundin, E, Bjorkman, M, Jacobsson, N: Analysis of service selling and design for remanufacturing. In: Proc {IEEE} Int Symp Electron Environ, pp. 272–277. (2000)
9. Thoben, K-D, Eschenbächer, J, Jagdev, H: Extended products: evolving traditional product concepts. In: 7th Int Conf Concurr Enterprising Bremen. (2001)
10. Sakao, T, Mizuyama, H: Understanding of a product/service system design: a holistic approach to support design for remanufacturing. J. Remanufacturing **4**, 1 (2014)
11. Opresnik, D, Zanetti, C, Taisch, M: Servitization of the manufacturer's value chain. In: Prabhu, V, Taisch, M, Kiritsis, D (eds.) Advances in Production Management Systems. Sustainable Production and Service Supply Chains. IFIP Advances in Information and Communication Technology, pp. 234–241. Springer, Berlin Heidelberg (2013)

12. Vargo, SL, Lusch, RF: From goods to service(s): divergences and convergences of logics. Ind. Mark. Manag. **37**, 254–259 (2008)
13. Lusch, RF, Vargo, SL, O'Brien, M: Competing through service: insights from service-dominant logic. J. Retail. **83**, 5–18 (2007)
14. Vargo, SL, Lusch, RF: Evolving to a new dominant logic for marketing. J. Mar. Sci. **68**, 1–17 (2004)
15. Christensen, CM: The ongoing process of building a theory of disruption. J. Prod. Innov. Manag. **23**, 39–55 (2006)
16. Bernard, S: Remanufacturing. J. Environ. Econ. Manage. **62**, 337–351 (2011)
17. Fiksel, J: A systems view of sustainability: the triple value model. In: Sustainable Futures: Multi-Disciplinary Perspectives on Multi-Level Transitions, 2nd edn, pp. 138–141. (2012)
18. Cat® Reman Products - Sustainability [http://parts.cat.com/en/catcorp/cat-reman-products] (accessed 5.20.2015)
19. Bulmus, SC, Zhu, SX, Teunter, R: Competition for cores in remanufacturing. Eur. J. Oper. Res. **233**, 105–113 (2014)
20. Guide Jr, VDR: Production planning and control for remanufacturing: industry practice and research needs. J. Oper. Manag. **18**, 467–483 (2000)
21. Giutini, R, Gaudette, K: Remanufacturing: the next great opportunity for boosting US productivity. Bus. Horiz. **46**, 41–48 (2003)
22. Abdulrahman, MD-A, Subramanian, N, Liu, C, Shu, C: Viability of remanufacturing practice: a strategic decision making framework for Chinese auto-parts companies. Journal of Cleaner Production, 1–13 (2014). doi:10.1016/j.jclepro.2014.02.065 IN PRESS
23. Sundin, E, Dunbäck, O: Reverse logistics challenges in remanufacturing of automotive mechatronic devices. J. Remanufacturing **3**, 1–8 (2013)
24. Östlin, J, Sundin, E, Björkman, M: Importance of closed-loop supply chain relationships for product remanufacturing. Institutional. Perspect. Supply. Chain. Manag. **115**, 336–348 (2008)
25. Butler, P: Machine Tools: A Report on the Remanufacture of Manual and CNC Tooling Machinery in the UK. Centre for Remanufacturing and Reuse, UK (2006)
26. Barquet, AP, Rozenfeld, H, Forcellini, FA: An integrated approach to remanufacturing: model of a remanufacturing system. J Remanufacturing **3**, 1 (2013)
27. Wise, R, Baumgartner, P: Go downstream. Harv. Bus. Rev. **77**, 133–141 (1999)
28. Schmenner, RW: Manufacturing, service, and their integration: some history and theory. Int. J. Oper. Prod. Manag. **29**, 431–443 (2009)
29. Goedkoop, MJ, te Riele, HRM, Rommens, PJM: Product Service Systems, Ecological and Economic Basics. Report for Dutch Ministries of Environment (VROM) and Economic Affairs (EZ). PricewaterhouseCoopers NV/Pi!MC, Storrm CS/PRé consultants (1999)
30. Tukker, A, Tischner, U: Product-services as a research field: past, present and future. Reflections from a decade of research. J. Clean. Prod. **14**, 1552–1556 (2006)
31. Tukker, A: Eight types of product-service system: eight ways to sustainability? Experiences from {SusProNet}. Bus. Strateg. Environ. **13**, 246–260 (2004)
32. Rothenberg, S: Sustainability through servicizing. MIT SLoan. Management. Review. **48**, 83–89 (2007)
33. Markeset, T, Kumar, U: Product support strategy: conventional versus functional products. J. Qual. Maint. Eng. **11**, 53–67 (2005)
34. Stremersch, S, Wuyts, S, Frambach, RT: The purchasing of full-service contracts. Ind. Mark. Manag. **30**, 1–12 (2001)
35. Reinartz, W, Ulaga, W: How to sell services more profitably. Harv. Bus. Rev. **86**, 91–96 (2008)
36. Rathore, P, Kota, S, Chakrabarti, A: Sustainability through remanufacturing in India: a case study on mobile handsets. J. Clean. Prod. **19**, 1709–1722 (2011)
37. Trickett, SB: Trafton, JG: {"What} if…": the use of conceptual simulations in scientific reasoning. Cogn. Sci. **31**, 843–875 (2007)
38. Peteraf, MA: The cornerstones of competitive advantage: a resource-based view. Strateg. Manag. J. **14**, 179–191 (1993)
39. Wernerfelt, B: A resource-based view of the firm. Strateg. Manag. J. **5**, 171–180 (1984)
40. Christensen, CM: The ongoing process of building a theory of disruption. J. Prod. Innovat. Manag. **23**(1), 39–55 (2006)
41. Yin, RK: Case Study Research: Design and Methods. Volume 5. Sage (2009)
42. Lewis, MW, Grimes, AJ: Metatriangulation: building theory from multiple paradigms. Acad. Manag. Rev. **24**, 672 (1999)
43. Matsumoto, M, Umeda, Y: An analysis of remanufacturing practices in Japan. J. Remanufacturing. **1**, 1–11 (2011)
44. Hauser, WM, Lund, RT: The Remanufacturing Industry: Anatomy of a Giant: A View of Remanufacturing in America Based on a Comprehensive Survey Across the Industry, p. 165. (2003)
45. Managed Print Service Solutions - Xerox [http://services.xerox.com/managed-print-services/managed-print-service-solutions/enus.html], accessed: 7.5.2014
46. Kerr, W, Ryan, C: Eco-efficiency gains from remanufacturing. J. Clean. Prod. **9**, 75–81 (2001)
47. Johnson, MW, Christensen, CM, Kagermann, H: Reinventing your business model. Harv. Bus. Rev. **86**(12), 57–68 (2008). Harvard Business Publishing.
48. Porter, ME, Kramer, MR: The competitive advantage of corporate philanthropy. Harv. Bus. Rev. **80**, 56–68 (2002)
49. Porter, ME: What is strategy? Harv. Bus. Rev. **57**(2), 61–78 (1996)
50. Mintzberg, H: The strategy concept 1: five P's for strategy. Calif. Manag. Rev. **30**(1), 11–24 (1987). Berkley.
51. Barney, J: Firm resources and sustained competitive advantage. J. Air Waste Manage. Assoc. **17**, 99–120 (1991)
52. Nalebuff, BJ, Brandenburger, A: Co-Opetition. Harper Collins. Business, New York (1996)
53. Sundin, E, Bjorkman, M, Jacobsson, N: Analysis of service selling and design for remanufacturing. In Electronics and the Environment, 2000. ISEE 2000. Proceedings of the 2000 IEEE International Symposium on (pp. 272–277). IEEE, San Francisco
54. Östlin, J, Sundin, E, Björkman, M: Product life-cycle implications for remanufacturing strategies. Journal of Cleaner Production **17**, 999–1009 (2009)

55. OECD: New sources of growth: intangible assets. OECD. http://www.oecd.org/sti/inno/46349020.pdf. Accessed 15 May 2014.
56. Zadrozny, W: Leveraging the power of intangible assets. MIT Sloan. Manag. Rev. **48**, 85–91 (2006)
57. Hall, R: A framework linking intangible resources and capabilities to sustainable competitive advantage. Strateg. Manag. J. **14**, 607–618 (1993)
58. Garetti, M, Taisch, M: Sustainable manufacturing: trends and research challenges. Prod. Plan. Control. **23**, 83–104 (2012)

12

Sustainable manufacturing for Indonesian small- and medium-sized enterprises (SMEs): the case of remanufactured alternators

Yun Arifatul Fatimah[1*], Wahidul Biswas[1], Ilyas Mazhar[2] and Mohammad Nazrul Islam[2]

* Correspondence: yun.arifatul@postgrad.curtin.edu.au
[1]Sustainable Engineering Group, Curtin University, Perth, Australia
Full list of author information is available at the end of the article

Abstract

Achieving sustainability is a great challenge for most of the Indonesian manufacturing small- and medium-sized enterprises (SMEs). Remanufacturing has been considered to be a key strategy to attain sustainable manufacturing by maximising the use of old components and minimising landfill size and energy usage. However, SMEs, which are undoubtedly the engine of the Indonesian manufacturing industry, do not have adequate experience, skill, resource, technology and financial support in the remanufacturing area. This paper proposes a new concept for sustainable manufacturing assessment framework through remanufacturing strategies in Indonesian SMEs. In this sustainable manufacturing assessment framework, the existing remanufactured products are assessed using sustainable manufacturing criterion (e.g. reliability, life cycle cost, employment opportunity and greenhouse gases). This framework identifies improvement opportunities, including eco-efficiency, cleaner production and green technology to make existing remanufactured products technically, economically, environmentally and socially sustainable. The sustainability of remanufactured alternators produced by Indonesian SMEs has been assessed to validate the aforementioned sustainable manufacturing assessment framework.

Keywords: Remanufacturing; SMEs; Sustainable manufacturing

Introduction

The manufacturing industry is a wealth-producing sector of the Indonesian economy accounting for 24% of gross domestic product (GDP) followed by agriculture (15%) and other economic sectors [1]. Large enterprises and small- and medium-sized enterprises (SMEs) accounted for 43.28% and 56.72% of the total GDP in the manufacturing sector, respectively, while SMEs alone represent 99.96% of total number of manufacturing industries (3.27 million companies) and also absorb employment for 87.47% of the total industry workers [2].

However, the proportion of SMEs for total manufacturing export and GDP is still low compared to the large number of enterprises due to a lack of innovative technology, inefficient production processes, limited skilled workers, insufficient capital investments and unqualified and unclassified standardisation products [3]. Most of SME exports are dependent on large enterprises, and their contributions are usually unrecorded and undermined. As a result, SMEs cannot expand their market independently

[4]. Furthermore, many Indonesian manufacturing SMEs are not environmentally conscious and contribute a large amount of pollution and resource depletion due to inefficient equipment usage [5].

Thus, Indonesian manufacturing SMEs do not comply with the economic, social and environmental objectives of sustainable development. This is because the objective of sustainable manufacturing is to develop and improve human life continually over time through the optimization of production and consumption activities by conducting efficiency on material and energy consumptions, focusing on poverty reduction and maintaining the resources for human beneficial reasons [6].

The main strategies for attaining sustainable manufacturing are remanufacturing, reuse, recondition and recycling [7]. Remanufacturing which is defined as a series of manufacturing steps acting on an end-of-life part or product in order to return it to like-new or better performance, with warranty to match [8], appears to be the most appropriate strategy to attain sustainable manufacturing in Indonesia due to the following reasons.

Firstly, it fulfils three objectives of sustainable manufacturing. Remanufacturing is economically viable by maximising the use of old components or product, and it is environmentally friendly by reducing the size of landfill, minimising the energy usage, and it is socially viable by providing employment opportunities and developing prosperity flows [9-11]. For example, an automotive engine remanufacturing company reduced metal consumption by 7,650 tonnes, conserved energy of 16 million kW h and decreased emission of about 11,300 to 15,300 tonnes CO_2 equivalent (CO_2eq) [12]. In the UK, remanufacturing contributes to the workforce around 50,000 employees and provides £2.4 billion GDP in 2009 [13], while in the USA, 73,000 remanufacturing industries employed about 480,000 people [14]. In addition, the increased employment opportunity, job satisfaction, income and clean environment will improve the quality of human life. These advantages significantly place remanufacturing as the main contributor to the sustainability of prosperity [10].

Secondly, since 1997 - when the economic crisis took place Indonesia - the new products became unaffordable and expensive for majority of the Indonesian people. Consequently, refurbished, reconditioned, cannibalised, reused and remanufactured products such as electronic, household appliances, automotive components and office furniture have become the usual products in the Indonesian market [15].

Unfortunately, the development of the remanufacturing industry which is mainly held by SMEs is still undercover, neglected and environmentally unfriendly. Many SMEs feel doubtful that their remanufacturing business will continue to grow due to high competency in the global market. Only few giant companies (e.g. PT Sanggar Sarana Jaya, PT Komatsu Remanufacturing Asia) have recognised the value of remanufacturing strategies in Indonesia as they can offer economic, social and environmental benefits to the manufacturing sector of Indonesia [16,17].

State of the art of Indonesian-remanufactured auto parts

The automotive industry sector is growing alarmingly in Indonesia in the recent years. The production growth was about 35.21% from 2009 to 2011, while the market growth was 37.54% [18]. The total number of automotive industries was around 445 which are mainly held by SMEs absorbing around 185,000 to 204,596 employees [19]. In the case of motor vehicle industries, sales were 894,164 units in the last 2011, and the prediction shows that the percentage will be increased to 50% in the next 5 years [18,20]. As a result, the use of auto parts such as engine, transmission, steering gear, starter and alternator will increase.

Alternator is part of the automotive components which can potentially be remanufactured. The remanufactured alternator dominates the remanufacturing market around 22% [21]. The remanufactured alternator offers 50% cheaper costs than the new ones and consumes 60% less energy and 70% less material compared to the new product. Also about 80% of the alternator parts can potentially be reused [22,23].

The legislations (e.g. End-of-Life Vehicle Directive and Energy-Using-Product Directive), which are crucial drivers for a successful implementation of the remanufactured products in developed countries (e.g. UK, Japan) [10], have been found absent in Indonesia. The next important driver is market demand. The Indonesian automotive component market is divided into original equipment market and market for low-cost and minimum standard product which are manufactured mainly by SMEs. Due to the affordable price of remanufactured products [24], the remanufactured alternators are expected to dominate the market.

Except for a few (e.g. Galeri alternator), most SMEs are not effectively producing remanufactured alternators. The Indonesian SMEs producing remanufactured alternators experienced a number of challenges. Firstly, the lack of financial mechanism has been perceived to be an obstacle to expand SMEs. Although the Indonesian government offers business credit programme (KUR-Kredit Usaha Kecil) for SMEs, it is mostly applied to few specific locations in Indonesia (e.g. Java, Bali) [25]. Secondly, the lack of knowledge and skilled worker has led to ineffective and inefficient production activities. Thirdly, the orientation of product standardisation based on the domestic market generates low quality, less reliability and short warranty period which reduce the attractiveness of the remanufactured alternators in the market. Fourthly, most of the SMEs are not environmentally efficient and produce more pollution due to lack of advanced technology and waste management.

Thus, this paper provides a remanufactured alternator as a case study to test the sustainable manufacturing framework for both assessing and improving the SMEs' remanufacturing auto parts in Indonesia and seeks to come up with the best solutions and strategies to overcome the aforementioned problems.

The model for sustainable manufacturing

Based on Figure 1, the steps for assessing the sustainability of remanufactured products are stated.

Firstly, the existing remanufactured products are assessed using the sustainable manufacturing criterion, including technical (i.e. reliability), economic (i.e. life cycle cost, sales) environmental (i.e. solid waste, GHG emission) and social aspects (i.e. employment opportunity, warranty). Secondly, the calculated values of these criterions are compared with the threshold values of the sustainability criterion. These threshold values are derived from the standard Indonesian and international literature discussing Indonesian and global remanufacturing issues. Examples of the threshold values of technical, economic, social and environmental aspects are reliability (99%), life cycle cost (50% cost of new product), warranty period (2 to 3 years) and greenhouse gas

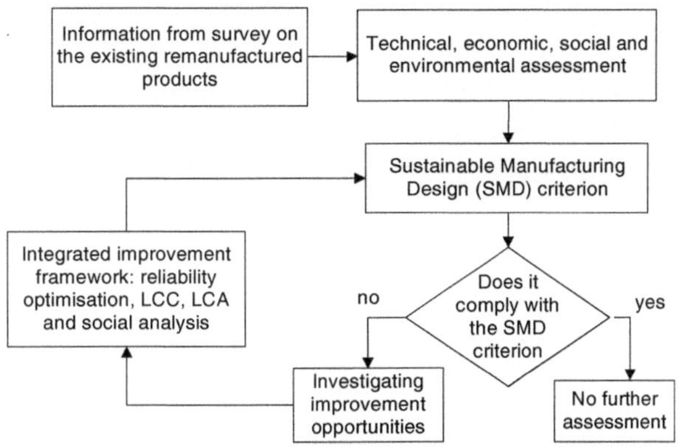

Figure 1 Model for sustainable manufacturing assessment.

emission (2.58 kg CO_2eq) which appeared in the remanufacturing sectors [23,24]. Remanufactured products not complying with any of these sustainability criteria are considered as unsustainable remanufactured product and are then redefined by investigating improvement opportunities through an integrated improvement framework involving reliability optimization, life cycle cost (LCC), life cycle assessment (LCA) and social assessment. The objective of this integrated improvement is to determine technically feasible solutions which are economically, socially and environmentally feasible and meet the sustainable manufacturing criterion. Figure 2 shows the detailed version of the integrated improvement framework.

Firstly, technical criterion involving reliability of the remanufactured products is assessed. Using this criterion, the material, method, man, machine, energy and information are analysed, and appropriate technical feasible solutions involving the best available technologies, processes, technical skills and energy and material consumption are proposed. Along with technical feasibility studies, LCA analysis of the technically feasible solutions is carried out following the ISO 14040–43 guideline [26] to determine environmental criteria including greenhouse gas (GHG) emissions and solid waste. If technically feasible solutions are not environmentally viable, 'hotspot' or the process producing the most pollution is identified to apply mitigation strategies (e.g. cleaner production, eco-efficiency, green technology and industrial symbiosis).

These two analyses are carried out simultaneously until both technically feasible and environmentally friendly solutions are obtained. Once technically and environmentally feasible solutions are obtained, the socio-economic viabilities of these options are

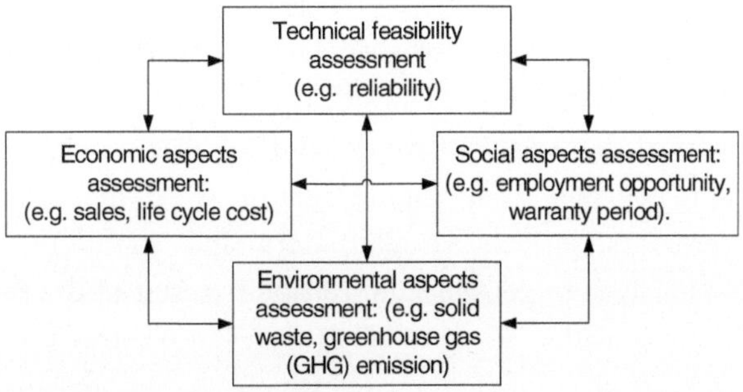

Figure 2 An integrated improvement framework.

assessed involving sales and life cycle cost assessments. The economic viability is attained when sales and life cycle cost are equal or greater than the threshold values. Similarly, social criteria such as employment opportunity and warranty period are required to meet the threshold value. If the revised version of technical option is not socio-economically viable, different socio-economic policies including rebate, subsidy and education are considered to attain socio-economic feasibilities.

Once technically, economically, socially and environmentally feasible solutions are determined, some appropriate institutional framework, including policy instruments, stakeholder responsibilities, key actions, targets and key performance indicator, are developed to implement sustainable manufacturing in SMEs. The institutional framework takes into account the constitution of direct (e.g. remanufacturer, supplier and consumer) and indirect (e.g. government, research institution and bank) stakeholders who participate in the decision-making process for sustainable remanufacturing.

The case study of remanufactured alternators

The existing situation of SME remanufactured alternators

A survey has been conducted in an SME producing remanufactured alternators in Central Java, Indonesia to test the proposed model and the framework. Remanufacturing operations have begun with a recovery process where alternators were collected directly from used alternator collection centres (e.g. service centre and auto junkyard). It should be notable that this research is limited to factory gate, and therefore, downstream activities such as customer satisfaction have been excluded.

In the factory line, disassembly is the first step where housing, stator, rotor, regulator, rectifier, brush and bearings are dismantled using simple tools and manual methods, and then all parts have been cleaned and dried by gas heating. Following this, testing including inspection and sorting of parts has been done for reconditioning on the basis of the reusability of parts. The reconditioned and new parts have been reassembled to make the remanufactured alternator. The alternator was tested before they were packed and ready to be delivered to the customer. The assessments of the existing remanufactured alternators have been done in the following ways.

Technical assessment

In this case study, the number of alternators that failed in 3 years (2008 to 2010) was 3,838. Failure of alternator components happened mainly due to failure of regulator (63.68%), followed by rectifier (13.25%), brush (11.97%) and other components such as stator and rotor (11.11%).

The reliability of the remanufactured alternator was calculated using the Weibull ++8 software. Intensive data including failure identification of the alternator, time of failure of the alternator and sales of the alternator were identified. The purchasing date, failure date and failed parts were collected to determine the time to failure. Data related to the number of failed alternators, suspended alternators and status (e.g. failure/F or suspended/S) were also involved to get the valid reliability result of the remanufactured alternator. Based on the calculation using Weibull distribution, the Weibull reliability plot in Figure 3 presents the shape parameter (β) which was 2.4 and the scale parameter (η) which was 1,288 days.

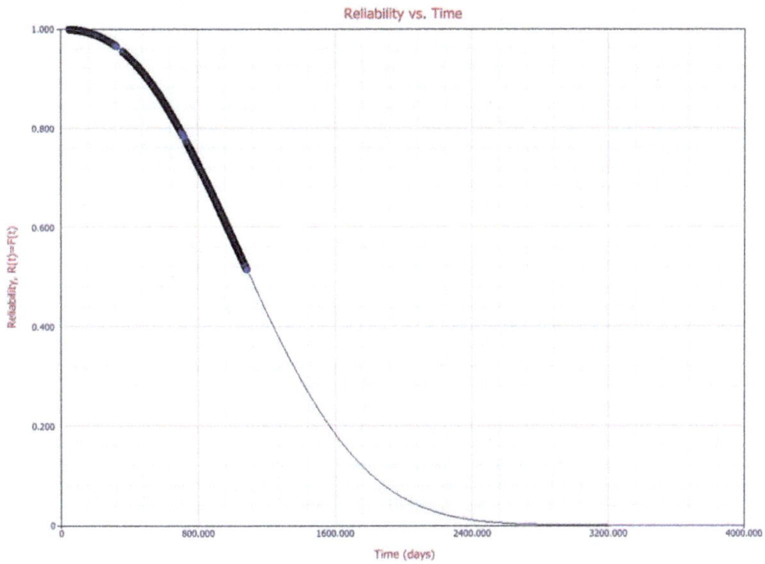

Figure 3 The reliability vs. time of the remanufactured alternators.

However, the reliability of new product is 100% in the early part of its life and gradually declines during its lifetime [8]. The mean time to failure of the alternator was 1,142 days, and the reliability model of the remanufactured alternator is presented in the following equation:

$$R(t) = exp\left[-\left(\frac{t}{1288}\right)^{2.4}\right] \tag{1}$$

The threshold value of the remanufactured alternator reliability is estimated to be about 99% which is close to 194 days (6.5 months), while the reliability of the remanufactured product in 1 year is 95%.

Economic assessment

The lack of investment on SMEs, inefficiency and ineffective production and the lack of quality and reliability made difficult situations for the SMEs to compete in the global market. This can be seen in the sales performance of the SMEs which showed that the number of sales decreased gradually from 2008 to 2010. In 2009, the sales decreased by 7.20% and dropped again in 2010 by 21.7% which led to a financial crisis in the SMEs.

The life cycle cost analysis of the remanufacturing process that includes the cost of product recovery, disassembly, cleaning and washing, inspection and sorting, reconditioning, replacement, reassembling, final testing, packaging and warranty have been performed. The acquisition cost was not considered in this LCC because the remanufacturing process used old infrastructure and machinery.

Based on the calculation done, the unit cost of the remanufactured alternator product was 183,000.00 IDR (USD = 9,700 Indonesian Rupiah (IDR) in 2010). The threshold value of LCC that was used to evaluate the economic criterion of the remanufactured product was 50% of the new product [25]. The breakdown of the remanufacturing operation costs of the alternator is shown in Figure 4. The product recovery cost contributed the largest portion (34.84%) of total unit cost, followed by replenishment cost

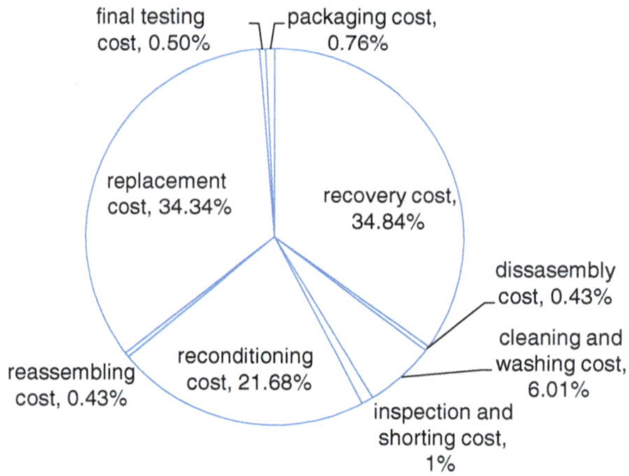

Figure 4 The life cycle cost of the remanufactured alternator.

(34.34%), reconditioning cost (21.68%), cleaning and washing cost (6.01%), inspection and shorting cost (1%), packaging cost (0.76%), final testing cost (0.50%), reassembling cost (0.43%) and disassembling cost (0.43%).

Social assessment

Warranty is one of the key social indicators of remanufactured product which represents an industry for its corporate social responsibilities [27]. The warranty of remanufactured product is proposed to maintain the performance of the product to be as good as new, which offers a social benefit to the customer. However, the remanufactured alternator SMEs could only provide a 1-year warranty against a 2- to 3-year warranty for new alternators. The next social indicator is employment opportunity because a higher employment opportunity could alleviate poverty and enhance social equity by reducing the gap between the rich and the poor [28]. The survey conducted in the SMEs' remanufacturing alternators found that they could hire nine to ten employees per year. However, the financial crisis faced by the SME has decreased the number of employee to about 24% per year in 2008 to 2010.

Environmental assessment

Environmental assessment includes the determination of solid waste and GHG emissions. After the disassembling, washing and testing processes, 59% of components are reused, 22% of components are replaced with used component and 19% of components are replaced with new components. Therefore, the remaining parts have been categorised as solid waste. About 41% (0.83 kg) of the parts have been replaced with used and new parts consisting of steel (24%), cast iron (26%), copper (33%), plastic (11%) and carbon (1%). The threshold value used to evaluate the solid waste contributed from the remanufactured alternator was 21% [24].

The GHG emission was calculated using an LCA software known as Simapro 7.3 [29]. As the remanufacturing processes were conducted in Indonesian SMEs, Indonesia's emission factors for electricity generation were used to determine the CO_2 equivalent emission from the remanufacturing process. The emission from the production of new components which replaced the worn-out components had also been

Table 1 The existing situation and the threshold value

Assessment	KPI	Threshold value	Existing situation	Yes/No
Technical	Reliability	99%	95%	No
Economic	LCC	172,500 IDR	183,000 IDR	No
	Sales	Increase	Decrease 21.70%	No
Social	Employment opportunity	Increase	Decrease 24%	No
	Warranty	2 to 3 years	1 year	No
Environment	Solid waste	0.43 kg	0.83 kg	No
	GHG emission	2.58 kg CO_2eq	5.98 kg CO_2eq	No

considered. GHG emissions associated with transportation for product recovery and shipment of new components from China to Indonesia were estimated using emission factors obtained from the Simapro 7.3 software database. The results show that the remanufactured alternator contributed 5.98 kg CO_2eq of GHG emissions, while the threshold value of the GHG emissions was 2.58 kg CO_2eq for a remanufactured alternator [24]. Table 1 represents the existing scenario of the remanufactured alternator in comparison with the threshold values.

The result indicates that the remanufactured alternator industry does not comply with the SMD criterion as all key performance indicators do not satisfy the threshold value.

The integrated improvement framework

The processes, material, labour and machinery from the existing situation were assessed using the sustainable manufacturing framework to determine the best possible technical solution meeting the sustainability criteria (or threshold values).

Scenario I

In this scenario, the technical feasibility of the replacement of all worn-out components (41%), including voltage regulator, rectifier, insulator, brush, bearings, pulley and rotor, with original equipment components was assessed in order to increase the reliability of the alternator. This means that 22% of the components, which were previously replaced with the used components, have now been replaced with new components.

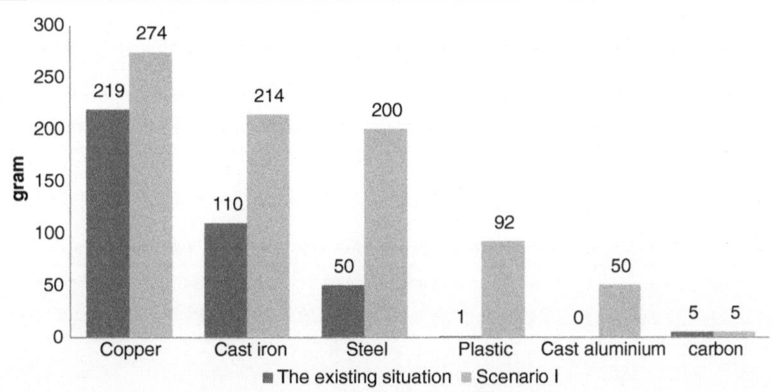

Figure 5 The materials used in the existing situation and scenario I.

Table 2 The environmental impacts of scenario I

	GHG emission (kg CO_2eq)	Solid waste (kg)
Existing situation	5.98	0.83
Threshold values	2.58	0.43
Scenario I: replacement with new components	14.6	0.83

The reliability of the alternator in scenario I was assumed to be the same as a new alternator because this scenario had considered the use of new parts instead of reconditioned parts. However, some key information such as time to failure, failure data and suspension data associated with the use of new components could have been considered in order to calculate the exact value of reliability of the improved version, but this was beyond the scope of this paper. Figure 5 shows the comparison of material intensities between the existing situation and scenario I.

The environmental impact assessment for scenario I in Table 2 shows that the replacements of worn-out components with new ones do not satisfy the threshold value of the environmental indicators, so an alternative scenario (scenario II) was conducted.

Scenario II

This scenario involves the use of recycled components. The following assumptions and considerations were applied to this scenario:

- The components including housing and pulleys have been reconditioned. The only recycled item considered was the coil of stator and rotor.
- The use of recycled components could reduce the GHG emission to about 83% and waste to about 88% [24].
- Since the brush and plastic cannot be recycled in an economically viable way, they were assumed to be sent to a landfill.
- The electricity consumption for reconditioning, recycling and transportation was considered for assessing environmental aspects of this mitigation strategy [24,30].
- It was assumed that there is a collaboration between the remanufacturing SMEs and the recycling industry to exchange worn-out and recycled components.

Table 3 shows that the technical and environmental criteria have been met by applying scenario II. The next step was to assess the socio-economic viabilities of the remanufactured alternator.

Table 4 shows that the LCC (170,600 IDR) of scenario II had met the threshold value (172,500 IDR). Following Cook and Ali [31], the reliability of the remanufactured alternator was assumed to increase the customer's satisfaction by 0.2%. This means that the increase in market response was predicted to increase the sales of the remanufactured

Table 3 The technical aspect and environmental impact assessment of scenario II

	Reliability (%)	GHG emission (kg CO_2eq)	Solid waste (kg)
Existing situation	95	5.98	0.83
Threshold values	99	2.58	0.43
Scenario II: replacement with recycled components	99	2.48	0.10

Table 4 The socio-economic analysis of scenario II

Indicator	Existing situation	Scenario II	Threshold value
LCC[a]	183,000 IDR	170,600 IDR	172,500 IDR
Sales	Decrease by 21.7%	Increase by 0.2%	Increase
Warranty	1 year	2 years	2 to 3 years
Employment opportunity	Decrease 24%	Increase 0.3%	Increase

[a]LCC analysis was calculated based on the additional variable cost for replacement (15,000 IDR), sustaining cost (237,000 IDR) and cost saving (61,000 IDR).

alternator by about 0.2%, which had in turn increased the employment opportunity by around 0.3%. The warranty period of a remanufactured alternator was considered to be the same as that of a new alternator. The increase in reliability was expected to extend the warranty period and to reduce warranty cost. Nominal customer's risk analysis was used to assess the warranty of the remanufactured alternator. Based on this analysis, the warranty of the proposed scenario was about 2 years which is the same as the threshold values.

However, these strategies can only be applied through the implementation of appropriate policies by the Government of Indonesia and stakeholders' involvement. Socio-economic policies, including rebate, subsidy and education, are some possible policy instruments. In addition, the stakeholder responsibilities, key actions, targets and key performance indicators can be developed to implement the mitigation scenario for attaining sustainable manufacturing.

Conclusions

The development of sustainable manufacturing is a great challenge for Indonesian SMEs. Remanufacturing has the potential to achieve sustainable manufacturing in Indonesian SMEs. This research proposes a sustainable manufacturing assessment model and integrated improvement framework to address socio-economic and environmental objectives of remanufacturing SMEs. Following the integrated sustainable framework, it was found that the existing remanufacturing operation in Indonesian SMEs is unsustainable. The use of recycled components in the remanufactured alternator could help attain the threshold values for sustainable manufacturing, including reliability (99%), GHG emission (2.48 kg CO_2eq), solid waste (0.10 kg), LCC (170,600.00 IDR), sales (increase 0.2%), warranty period (2 years) and employment opportunity (increase 0.3%). Last but not least, an institutional framework, including remanufacturers, suppliers, consumers, government, research institutions and financial institutions, needs to be established in order to promote sustainable manufacturing strategies in Indonesian SMEs.

Competing interests

The authors declare that they have no competing interests.

Authors' contributions

YAF collected data, carried out literature review, conducted technical, environmental, economic and social analyses and finally completed this manuscript. WB reviewed the environmental, economic and social analyses. IM and MNI reviewed the technical analysis and participated in the consultation process. All authors read and approved the final manuscript.

Acknowledgements

Yun Arifatul Fatimah sincerely appreciates the Indonesian Directorate General of Higher Education (DIKTI) for the financial support granted through her doctoral scholarship, and Muhammadiyah University of Magelang for the support and encouragement with regard to her doctoral study.

Author details
[1]Sustainable Engineering Group, Curtin University, Perth, Australia. [2]Department of Mechanical Engineering, Curtin University, Perth, Australia.

References

1. Bureau of East Asian and Pacific Affairs: Background note: Indonesia. http://www.state.gov/r/pa/ei/bgn/2748.htm. Accessed 20 March 2012
2. Indonesian Bureau of Statistics: Trends of the Selected Socio-economic Indicators of Indonesia. Indonesian Bureau of Statistic, Jakarta (2009)
3. Tambunan, T: SME Capacity Building Indonesia. Kadin Indonesia-JETRO, Jakarta (2006)
4. Tambunan, T, Xiangfeng, L: SME Development in Indonesia and China. Kadin Indonesia-JETRO, Jakarta (2006)
5. Dhewanthi, L: Addressing Financial Obstacles of Micro, Small, Medium Enterprises (MSMEs) for Environmental Investment in Indonesia, in Greening the Business and Making Environment a Business Opportunity. United Nation, ESCAP, Bangkok (2007)
6. Azapagic, A, Perdan, S: Indicator of sustainable development for industry: a general framework. Process. Saf. Environ. Prot. **78**, 243–261 (2000)
7. Anityasari, M, Kaebernick, H: A concept of reliability evaluation for reuse and remanufacturing. Int. J. Sustain. Man. **1**, 3–17 (2008)
8. Centre for Remanufacturing and Reuse: Remanufacturing and reuse. http://www.remanufacturing.org.uk. Accessed 20 May 2010
9. Gupta, SM, Kamarthi, SV: LRM-Engineering Solutions for Evolving Customer and Environmental Need. Research Publications Related to Environmentally Conscious Manufacturing. Northeastern University, Boston (2002)
10. Gray, C, Charter, M: Remanufacturing and Product Design: Designing for the 7th Generation. University College for Creative Arts, Farnham (2007)
11. Bernard, S: Remanufacturing. J. Environ. Econ. Manag. **62**, 337–351 (2011)
12. Shi-can, L: Benefit analysis and contribution prediction of engine remanufacturing to cycle economy. J. CSUT **12**, 25–29 (2005)
13. Chapman, A, Barlett, C, McGill, I, Parker, D, Walsh, B: Remanufacturing in the UK: a snapshot of the remanufacturing industry in the UK in 2009. Center for Remanufacturing and Reuse, Aylesbury (2009)
14. Smith, VM, Keoleian, GA: The value of remanufactured engines: life-cycle environmental and economic perspective. J. Ind. Ecol. **8**, 193–221 (2004)
15. Ramdansyah: Komputer Layak Pakai Senjata Melawan Kebodohan - Gerakan Pemberantasan Buta Komputer di Masyarakat Bawah. Rumah Demokrasi, Jakarta (2011)
16. Komatsu Reman Indonesia: Reuse and recycling activities. http://www.Komatsu.com. Accessed 20 July 2012
17. Statistics Indonesia: Statistics Indonesia. http://www.datastatistik-Indonesia.com. Accessed 20 November 2010
18. GAIKINDO: Statistic Data. GAIKINDO, Jakarta (2012)
19. Layton, C, Rustandie, J: Gambaran rantai nilai komponent otomotif, Justifikasi pasar dan strategi peningkatan pasar komponen dalam negeri. SENADA, Indonesia Competitiveness Program, Jakarta (2007)
20. Aftermarket Australian Automotive Association: Indonesian Automotive Aftermarket Opportunities. www.aaaa.com.au.files/IndonesiaReport12.pdf. Accessed 2 July 2013.
21. Jung, D, Seo, Y, Chung, W, Song, H, Jang, J: The quality stability improvement for remanufactured alternator. In: The Global Conference on Sustainable Product Development and Life Cycle Engineering: Sustainability and Remanufacturing VI, Busan, 29 September-1 October 2008. (2008)
22. Asif, FMA, Simere, DT, Nicolesu, CM, Hauman, M: Methods analysis of remanufacturing option for repeated lifecycle of starters and alternators. In: 7th International DAAAM Baltic Conference Industrial Engineering, Tallinn, 22–24 April 2010. (2010)
23. Zhang, T, Chu, J, Wang, X, Lie, X, Cui, P: Development pattern in enhancing system of automotive components remanufacturing in China. Resour. Conserv. Recycl. **5**, 613–622 (2012)
24. Kim, H, Severengiz, S, Skertos, SJ, Selinger, G: Economic and environmental assessment of remanufacturing in the automotive industry. 15th CIRP International Conference on Life Cycle Engineering, Applying Life Cycle Knowledge to Engineering Solution, pp. 195–200. CIRP, New South Wales, Sydney (2008)
25. Maurougane, A: Promoting SME Development in Indonesia. OECD Economics Department Working Papers no. 995. Organisation for Economic Co-operation and Development, Paris (2012)
26. ISO 14040: Environmental Management - Life Cycle Assessment - Principles and Framework (ISO 14040). International Organization for Standardization, Geneva (1997)
27. Anityasari, M: Reuse of industrial product - a technical and economic model for decision support. Thesis. The University of New South Wales, Sydney (2008)
28. Biswas, W, Bryce, P, Diesendorf, M: Model for empowering rural poor through renewable energy technologies in Bangladesh. J. Environ. Sci. Policy **4**, 333–344 (2001)
29. PRe Consultants: Simapro 7.2. PRe Consultants, Amersfoort (2010)
30. Jolly, JL, Maryland, D: The U.S. Copper - base Scrap Industry and Its By-Products - An Overview. Copper Development Association Inc, New York (2012)
31. Cook, VGC, Ali, A: Using net present value methods to evaluate quality improvement projects. Int. J. Quality and Reliability **27**, 333–350 (2010)

13

Reverse logistics network design: a holistic life cycle approach

Joanna Daaboul*, Julien Le Duigou, Diana Penciuc and Benoît Eynard

* Correspondence:
joanna.daaboul@utc.fr
Department of Mechanical Systems Engineering, Université de Technologie de Compiègne, UMR CNRS 7337 Roberval, rue du Dr Schweitzer, CS60319, 60203 Compiègne Cedex, France

Abstract

When aiming for a more sustainable world, enterprises such as aircraft and automobile industries are highly interested in lightweight components and solutions. Among these solutions are aluminum wrought alloys that offer high potentials for dramatic weight reduction of structural parts. Nevertheless, the production of virgin aluminum is highly energy consuming. Major energy reduction may be achieved if recycled aluminum is used. The aim of the European project SuPLight is to address new industrial models for sustainable lightweight solutions - with recycling in high-end structural components based on wrought alloys. This article addresses the issue of designing the reverse logistics network assuring the needed volume of recycled aluminum for the production of L-shaped front lower control arms for personal cars. It details the developed method and demonstrator for designing a reverse logistics network based on a life cycle assessment.

Keywords: Reversed logistics design; LCA; Wrought aluminum

Introduction

Until recently, reverse logistics (RL) was not given a great deal of attention in organizations. Actually, implementing reverse logistics programs to reduce, reuse, and recycle wastes from distribution and other processes generates tangible and intangible value and can lead to better corporate image [1]. Its main drivers are legislations and directives, consumer awareness, and social responsibilities towards the environment [2-4]. Another motivating driver of reverse logistics is economic factors. Reverse logistics can generate profits by reselling valuable components or products [5]. Reverse logistics is of high importance for aluminum-based products, since the production of virgin aluminum is highly energy consuming [6]. It is also proven that remanufacturing is beneficial for many industries in general [7] and for the automobile industry specifically [8].

Reverse logistics operations and chains they support are significantly more complex than traditional manufacturing supply chains [9-11]. Therefore, and similar to how companies develop efficient logistics processes for new goods, it is necessary to plan operations for returned goods, taking into consideration that the processes are most probably quite different from those defined for forward distribution [12]. Reverse logistics is not a symmetric picture of forward distribution [13] and requires different management and planning approaches. For example, it is difficult, in the case of RL, to estimate supply-related parameters such as the unit operational costs directly from reported statistical data. The design of a reverse logistics network should be based on an integrated approach [14].

The aim of this article is to design an effective reverse logistics network for the L-shaped front lower control arms (FLCA) (Figure 1) for personal cars [15] within the SuPLight project which is a multidisciplinary research project, combining physics at the atomic scale level, metallurgy, continuum mechanics, structural mechanics, optimization algorithms, tolerance analysis, life cycle analysis, manufacturing, and business modeling. The project addresses new industrial models for sustainable lightweight solutions - with 75% recycling in high-end structural components based on aluminum wrought alloys.

The designed reverse logistics network has the following characteristics:

1. The reverse logistics network should be environmentally friendly as much as possible. Thus, the reverse logistics design method should integrate environmental impact as decision criteria.
2. The different facilities (collection locations, remanufacturing facilities, etc.) are not necessarily owned by one company.
3. The reverse flow has different sources and might have depending on these sources different disposal routes (reselling, remanufacturing, recycling, etc.).

Background and related works

A reverse logistics network is formed by four main activities: gatekeeping, collection, sorting, and recovery (disposal) [16]. These are shown in Figure 2.

Many recent works propose models and methods for the design of a reverse logistics network. These mainly propose the use of mathematical programming, and more precisely mixed-integer linear programming models are used [17]. According to the recent literature review in [17], the only objectives considered in reverse logistics or forward/reverse network design are cost, profit, responsiveness (service level), source balance,

Figure 1 Front lower control arm (FLCA) [14].

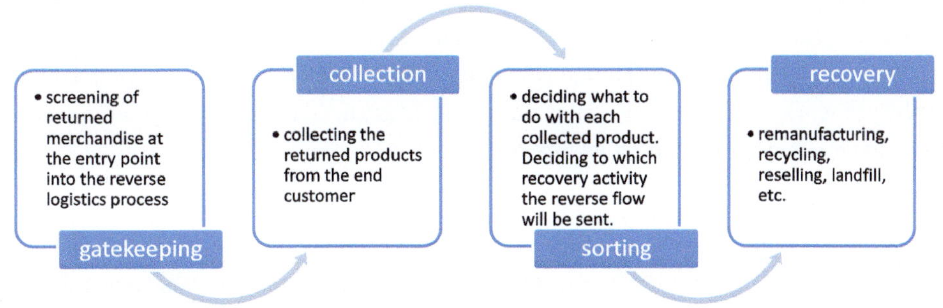

Figure 2 **Reverse logistics network main activities.**

and quality. The most considered decision criterion is the total cost of the network [18-22], followed by the service level [23]. Zhou and Wang [24] propose a generic model for reverse logistics design considering only costs as decision criteria. Lee and Dong [25] propose a two-stage stochastic programming model for multi-period reverse logistics network design with the different investment and operational costs as the main decision criterion. Lambert et al. [26] consider the environmental regulations as a constraint to their decision model but not as decision criteria. Cardoso et al. [27] proposed a mixed-integer linear programming model for supply chain planning including reverse logistics activities. Yet they consider the expected net present value (ENPV) as the decision criterion. In this case study and research work, the environmental impact of the network is an important design factor; therefore, it should be considered as a main decision criterion in the design of reverse logistics network. This means considering not only the environmental impact of transportation activities, but also the remanufacturing and recycling activities.

The main decisions of the proposed models in the literature are location/allocation of facilities and transportation values [18-23]. Most of these models consider the network design problem with collection, sorting, and recovery (disposal) facilities owned by one company which is collecting the reverse flow and reusing it. This is not always the case. Reverse logistics networks include not only facilities owned by the company itself, but also facilities owned by other partners in the network. This means that designing the best reverse logistics network includes three main decisions: 1) making/buying, 2) supplier selection, and 3) location/allocation of facilities.

These three decisions should be made while taking into consideration the environmental performance of the reverse logistics network. This means integrating environmental indicators as decision criteria.

Few works considered the recycling option of the recovery step [20,28]. Most works found in the literature consider the remanufacturing option [18,19,21-23]. In this case study, the interest is in using recycled aluminum. Thus, the main considered recovery activity is recycling. This means that the reverse flow will probably have different sources, which might not all have the same recovery route (related recycling activities).

A framework for the design of reverse logistics

Network design method

Since the main aim of this project is to be more sustainable and more precisely to have a better environmental performance, a life cycle approach seems adequate to analyze the performance of the chosen network. It permits assessing the environmental performance of the reverse logistics network alone and also its impact on the whole life cycle assessment of the product. This permits integrating the environmental performance of the network as a decision criterion for its design. The proposed method is formed by five main steps presented in Figure 3.

Step 1: Context definition

In this step, the company and the product(s) are described. The constraints of the company are listed as well as its preference for outsourcing/insourcing reverse logistics-related activities.

Step 2: Parameter definition

The parameter definition, as shown in Figure 4, consists of identifying the different roles in a reverse logistics network, the responsibilities and activities per role, the requirements for these activities, and the performance indicators per requirement per activity. Four roles were identified:

1. The collector: responsible for the gatekeeping and collection activities
2. The sorter: responsible for sorting activities
3. The recoverer: responsible for recovery activities (recycling, remanufacturing, landfill, etc.)
4. The end collector of reverse flow: the user of the final recovered reverse flow.

The same partner may have more than one role, and different partners might have the same role. For example, the end collector of a reverse flow might also collect the

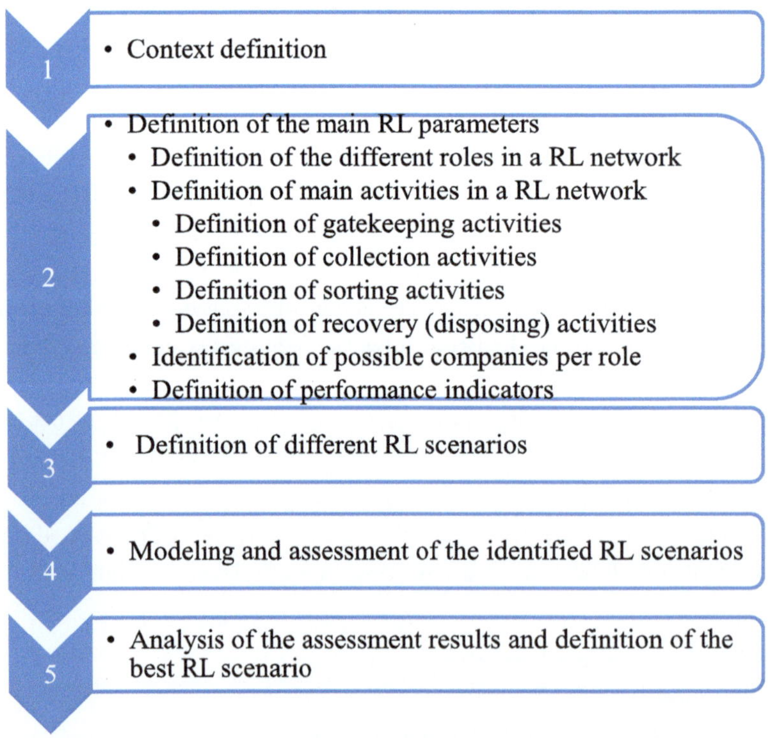

Figure 3 **Proposed method for reverse logistics network design.**

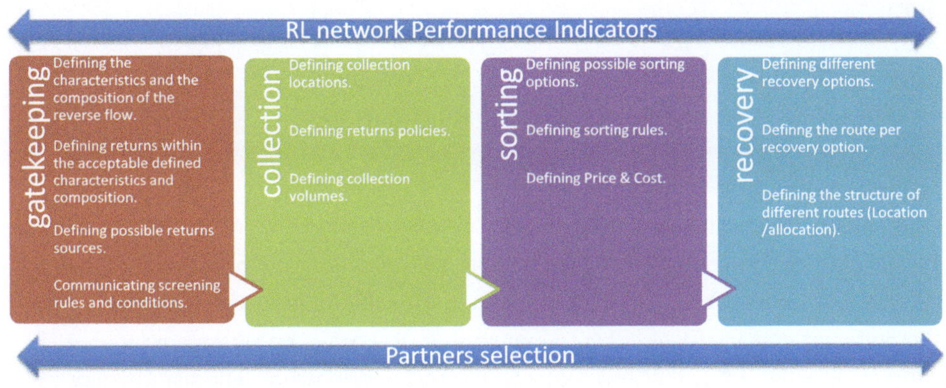

Figure 4 Parameter definition step.

returns, and a sorter might also recycle or remanufacture the reverse flow. One partner may have all roles. In this case, the same company owns the entire reverse network.

The requirements per activity are case dependent. Most of them depend on the type of the product. The recycling processes for example differ based on the product type, whereas the general performance indicators are the same no matter the product type.

Concerning the performance indicators, these are:

- Total cost of the network including cost of transportation, facilities, purchasing, activities, and stock: This is the most used reverse logistics performance indicator in the literature (see the 'Background and related works' section).
- Recycling rate: This indicator reflects the service level of the entire network. It reflects the percentage of recycled material for the product using the recovered reverse flow.
- Life cycle assessment-related indicators: These indicators permit evaluating the environmental impact of the reverse logistics network. Considering the environmental performance of the reverse logistics network is the main goal of this work. The most used life cycle assessment (LCA) indicators are climate change (or carbon footprint of product), energy-related indicators, chemicals used, PFC/GHG emissions, fresh water consumption, and use of renewable/nonrenewable energy.

The above presented indicators are decision criteria for the design of the reverse logistics network. Other indicators are needed for the supplier selection decision. According to [29], the most used criteria for the supplier selection decision are price, deliver on time, quality, equipment and capability, geographic location, technical capability, and finally management and organization. To these criteria, the environment-related indicator is added: the green processes. This indicator reflects whether the processes used are environmentally friendly or not. This indicator is of outmost importance for the sorting and recycling activities where the environmental impact depends on the technology used.

In this step, candidate companies for playing one or more roles in the reverse logistics network are also defined. For each company, a description sheet is filled containing information on the company, performance indicators, and constraints, such as the minimum quantity of reverse flow and replenishment lead time.

Step 3: Scenario definition

The scenario definition is achieved by the expert user in one of two ways:

1. Defining any reverse logistics network scenario, by choosing for each role one or more companies as well as the quantity of reverse flow and then by choosing the routes between these companies.
2. Defining the reverse logistics network step by step, using decision criteria per role. In this case, and for every role, the user may ask to identify the best company based on many decision criteria, identified in step 2. This option is based on the analytical hierarchy process (AHP). AHP is the most used decision support system for supplier selection [30]. This technique which was developed by Thomas L. Saaty [31] relies on the expertise of the user in order to generate the weights of attributes. It is based on the comparison of pairs of options and criteria. It has found widespread application in decision-making problems, involving multiple criteria in systems of many levels [32]. Its main advantages are:
   - Consideration of non-tangible subjective attributes
   - Ability to structure a complex multi-criteria, multi-person, and multi-period hierarchical problem
   - Capability to investigate each level of the hierarchy separately and to combine the results as the analysis progresses
   - Binary comparison of elements (alternatives, criteria, and sub-criteria)
   - Availability of IT support.

The steps that need to be followed in this option are:

1. Fill for every potential partner the measured performance for all decision criteria identified previously.
2. Fill the criteria comparison matrix for all criteria identified in the 'Background and related works' section.
3. Choose a specific role of the reverse logistics network. The design may be made in a forward or backward manner. This means the allocation of partner per role may be realized by starting at the end of the network or at its beginning.
4. Choose a specific type of reverse flow.
5. Identify all candidates (company/facility/supplier) for the chosen role.
6. Select the decision (selection) criteria of interest (from the pool of all available criteria).
7. Calculate the weights, using AHP.
8. Rank all candidates, using AHP.
9. Select the candidate ranked first.
10. Allocate order quantities, choose transportation mode, and validate.

The user may choose many scenarios as desired and compare their performance in steps 4 and 5. In other words, the user may compare on the basis of all environmental and economic criteria the scenarios built with the decision aid option and with any other scenario built by the user alone without any aid of the system.

Steps 4 and 5: Assessment of reverse logistics scenarios and result analysis

After one or more scenarios are chosen, the different performance indicators are calculated. A comparison between the scenarios is provided.

The evaluation of the reverse logistics network is achieved via life cycle assessment software (SimaPro) [33]. This is possible within the SuPLight project which aims at providing a collaborative platform [34] allowing the communication and exchange of information between different softwares such as the LCA software and the reverse logistics demonstrator.

After the life cycle assessment of the different reverse logistics scenarios is received from the LCA software, a full analysis combing this assessment and the one achieved via the reverse logistics prototype (total cost of the network, quantity of recycled material, etc.) is provided to the user. A comparison of all created reverse logistics scenarios is then delivered to the user.

The AHP method is reused at this stage to identify the best-fitting reverse logistics network.

Reverse logistics prototype (demonstrator)

The proposed method was the basis of the development of the software prototype. It has been designed with a flexible architecture in order to allow portability and interoperability with any external application. The software prototype implementation is based on the MVC software architecture pattern, using HTML5 and CSS to create the view and for the application logic the JAX-RS Java API for RESTful web services which are controlled by JavaScript (JQuery).

The reverse logistics prototype is focused on reverse logistics scenario definition allowing the user to choose the partners of the reverse logistics network and their roles and visualize them on a map; define reverse logistics routes - a single route is characterized by flow and transportation data (the flow type and the quantity exchanged between two partners, transportation mean and distance); visualize the summary of the created scenario and the final route on the map; display LCA indicators for the current scenario; compare several scenarios on a radar graph based on the indicator value and the total RL cost; use the decision aid system to choose partners based on input criteria; determine recommended quantities for each route; and compute RL indicators for each scenario (total cost of the network, quantity of recycled material, etc.). Google services are used for automatic distance calculation and map display based on data provided by the user. The prototype main page and the route definition interface are presented in Figures 5 and 6, respectively. On the main page, the prototype shows the reverse logistics scenario on a map with a description of the different routes including the node names (companies' names) with the type of reverse flow, quantity of flow, transportation mode, and distance between the two nodes. The different enterprises or facilities chosen for the reverse logistics network are presented on the map. This map is dynamic and shows the chosen facility locations or the chosen suppliers as well as all related costs. This map serves only as a visualization tool for the chosen scenario.

This prototype is not intended for advanced planning of a supply chain such as the advanced planning and scheduling (APS) software. But its main function is to provide a support for decision making in designing a reverse logistics network. Even though the APS software may provide 'What if' analysis, it does not offer a decision aid method

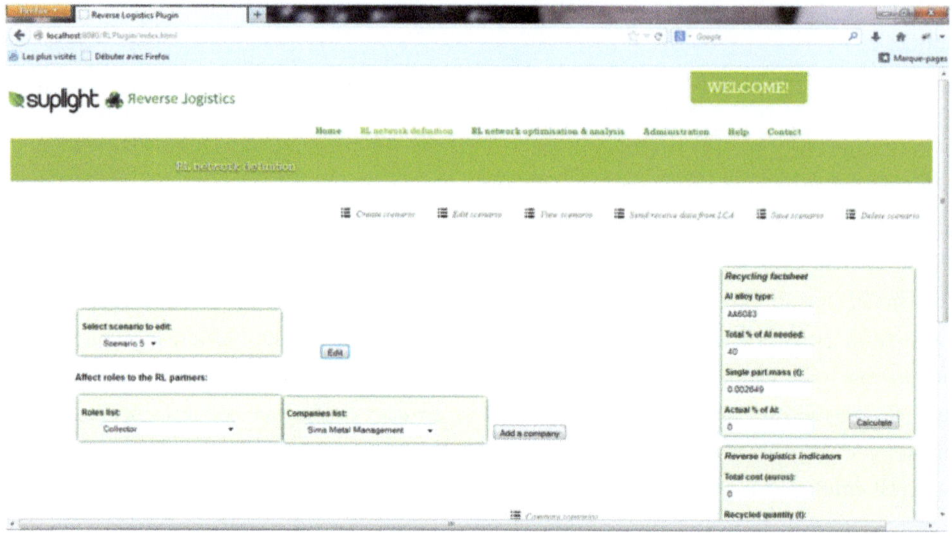

Figure 5 **Reverse logistics demonstrator main page.**

such as AHP nor does it provide the optimal reverse logistics network among all possible scenarios; it permits only to compare between manually created scenarios. The use of the proposed demonstrator does not eliminate the need to use the APS software which is necessary to manage a reverse logistics network. The demonstrator may be considered as a plug-in for the APS software.

Application of the proposed framework to an industrial case study

Step 1: Context definition

The case study concerns a front lower control arm of GM Opel Insignia. This component was chosen by Raufoss Technology (RT) as a case study. RT is the end collector of reverse flow in this case study. RT wishes to only have the role of the end collector of reverse flow. This means all other reverse logistics activities are outsourced.

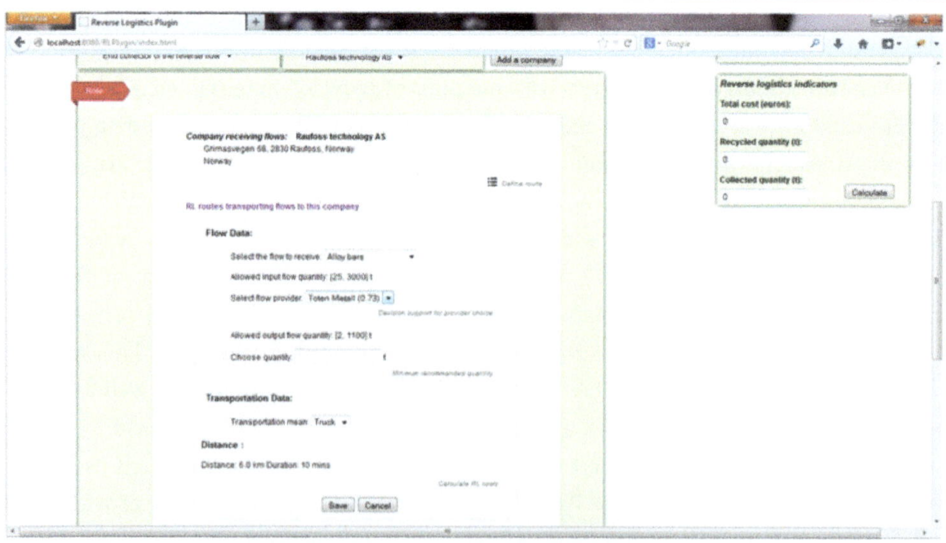

Figure 6 **Route definition interface of the reverse logistics prototype.**

The control arm which is an important part of a vehicle's suspension system consists of a bar used to attach the suspension members to the vehicle's chassis. It is produced from a high-strength wrought aluminum alloy: AA6082. The control arm is produced by hot forging and machining from extruded aluminum billets. Heat treatment is necessary to obtain the necessary strength.

Step 2: Defining reverse logistics parameters

Defining the gatekeeping activity

The acceptable characteristics and chemical compositions are defined as well as possible returns (specific products/alloys) and their sources. These are presented in Table 1.

Defining the collection activity

Possible collection locations were identified (Table 1). At first, the type of these collection locations was defined. Then, specific enterprises of the identified types were determined.

Defining the sorting and the recovery/treatment activities

The aluminum recycling industry includes a wide range of processes, which require a wide range of control techniques. These different processes have different environmental performance and produce different qualities of reverse flow. Moreover, they are not all suited for all types of returns containing aluminum. The only recovery option considered in this case study is recycling for acceptable returns. Thus, the sorting activity consists of selecting acceptable returns, shredding it, and separating aluminum alloys

Table 1 Acceptable returns and their sources

Products/alloys	Product/source	Possible collection locations
Bumper	Vehicle (dismantling)	Car scrap yard, automotive dismantlers, and recyclers
Wheels	Vehicle (dismantling)	Car scrap yard, automotive dismantlers, and recyclers
Suspension, chassis, etc.	Vehicle (dismantling)	Car scrap yard, automotive dismantlers, and recyclers
Closures, skin sheet, and inner stiffness	Vehicle (dismantling)	Car scrap yard, automotive dismantlers, and recyclers
Body structure	Vehicle (dismantling)	Car scrap yard, automotive dismantlers, and recyclers
Seats	Vehicle (dismantling)	Car scrap yard, automotive dismantlers, and recyclers
AA3003	Vehicle (shredding, alloy separation)	Shredding plant
AA5754	Vehicle (shredding, alloy separation)	Shredding plant
AA6016	Vehicle (shredding, alloy separation)	Shredding plant
AA6060	Vehicle (shredding, alloy separation)	Shredding plant
Wrought 1	Vehicle (shredding, alloy separation)	Shredding plant
Wrought 2	Vehicle (shredding, alloy separation)	Shredding plant
Siding	Building (demolition)	Demolition contractors
Window frames	Building (demolition)	Demolition contractors
3004 corrugating roofing	Building (demolition)	Demolition contractors
Foils	Packaging (sorting)	European recycling centers and companies
Rigid containers	Packaging (sorting)	European recycling centers and companies
Closures	Packaging (sorting)	European recycling centers and companies

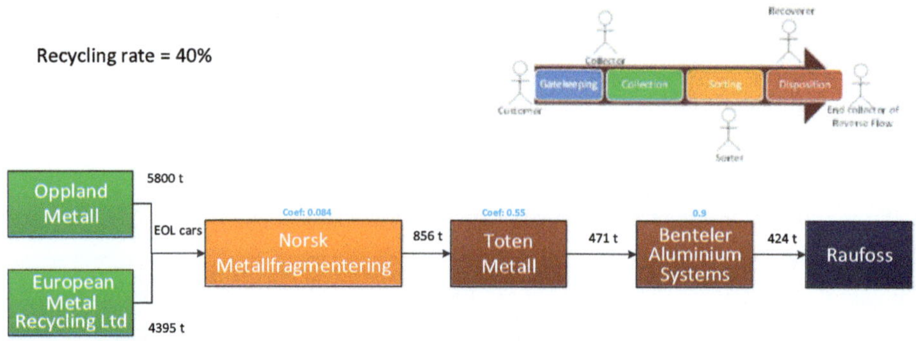

Figure 7 Scenario 1.

from other materials including other metals. This step will output the necessary scrap that will be used in the recycling activity. In summary, the sorting and recovery activities are the following:

1. Shredding
2. Sorting: electromagnetic separation, eddy-current separation, sink-float, and color sorting/etching
3. Refining: distillation, electrolytic separation, hot crush, unidirectional solidification, and flux separation
4. Remelting
5. Casting aluminum
6. Extrusion.

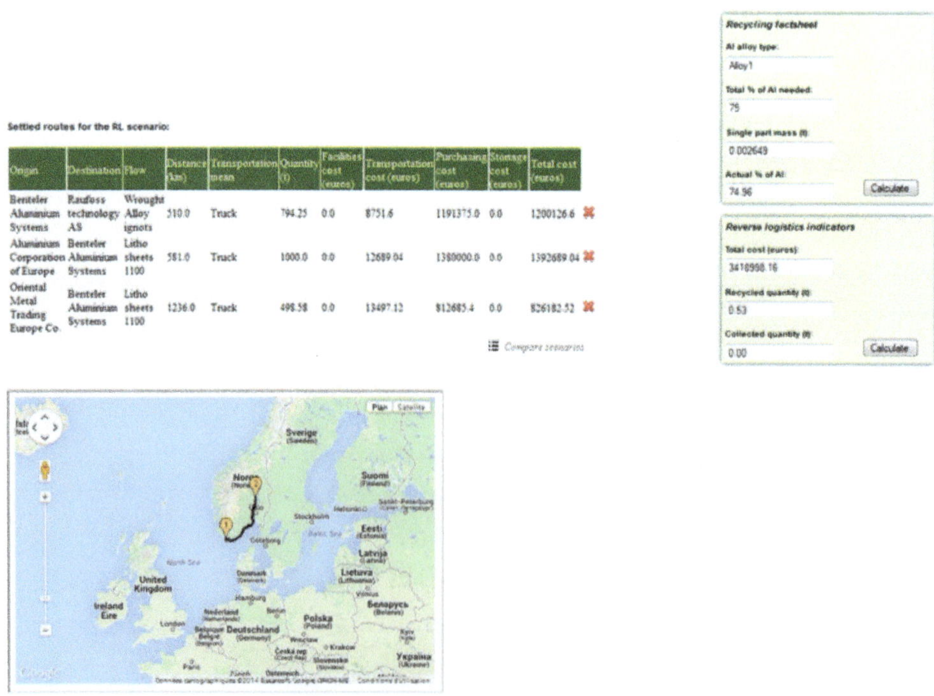

Figure 8 Scenario 1, plug-in interface.

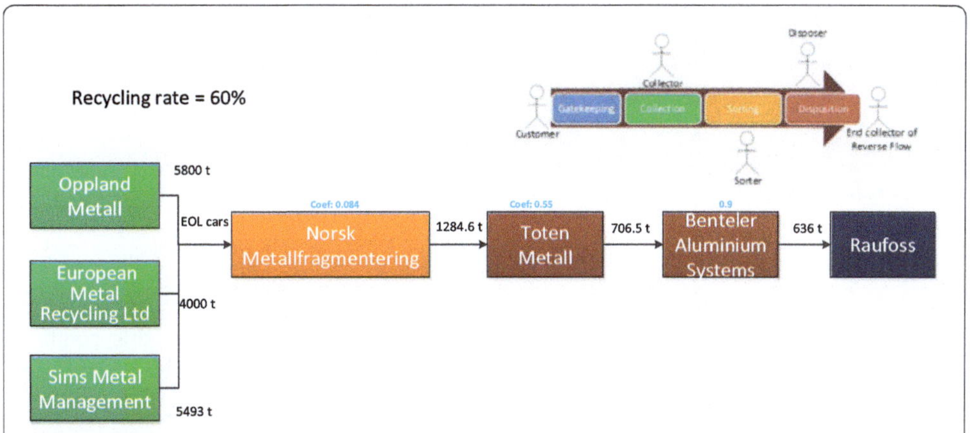

Figure 9 Scenario 2.

Identifying potential partners

Potential partners were identified for the three roles: collector, sorter, and recoverer. These and the related collected data are not shown in this article due to confidentiality issues. In total, 20 potential partners were identified.

Step 3: Defining different reverse logistics scenarios

Using the reverse logistics plug-in, two main RL scenarios were identified. These are presented in Figures 7, 8, 9, and 10. In the figures, a coefficient is indicated for every selected partner. This is equal to the coefficient of output/input. It represents the percentage of the covered input at the end of the activity achieved at this partner. The collector coefficient corresponds to the average percentage of Al in a car. The coefficient of sorting and recycling activities is recovered from AL recycling literature. The

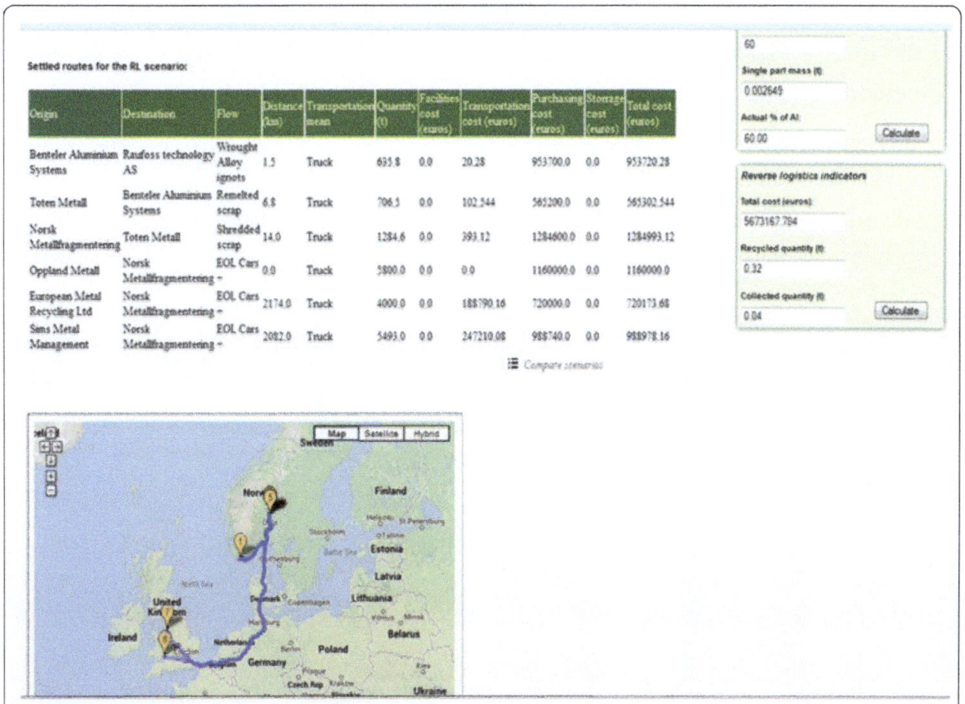
Settled routes for the RL scenario:

Origin	Destination	Flow	Distance (km)	Transportation mean	Quantity (t)	Facilities cost (euros)	Transportation cost (euros)	Purchasing cost (euros)	Storage cost (euros)	Total cost (euros)
Benteler Aluminium Systems	Raufoss technology AS	Wrought Alloy ignots	1.5	Truck	635.8	0.0	20.28	953700.0	0.0	953720.28
Toten Metall	Benteler Aluminium Systems	Remelted scrap	6.8	Truck	706.5	0.0	102.544	565200.0	0.0	565302.544
Norsk Metallfragmentering	Toten Metall	Shredded scrap	14.0	Truck	1284.6	0.0	393.12	1284600.0	0.0	1284993.12
Oppland Metall	Norsk Metallfragmentering -	EOL Cars -	0.0	Truck	5800.0	0.0	0.0	1160000.0	0.0	1160000.0
European Metal Recycling Ltd	Norsk Metallfragmentering -	EOL Cars -	2174.0	Truck	4000.0	0.0	188790.16	720000.0	0.0	720173.68
Sims Metal Management	Norsk Metallfragmentering -	EOL Cars -	2082.0	Truck	5493.0	0.0	247210.08	988740.0	0.0	988978.16

Figure 10 Scenario 2, plug-in interface.

Table 2 Results for the reverse logistics scenarios considering environmental performance for EOL

Indicator	Unit	Plug-in	Scenario 1	Scenario 2
Nonrenewable energy (resources)	MJ	LCA	57.2	72.46
Climate change	kg $CO_2 \cdot$ eq.	LCA	7.74	8.66
Fresh water consumption (water withdrawal)	m^3	LCA	0.058	0.065
Recycling rate	%	RL	40	60
Total reverse logistics cost	Euros	RL	3,848,108.92	5,673,167.78

recommended quantity calculated via the plug-in takes into consideration this coefficient. The choice of a partner is made based on different criteria such as the geographical location, the transportation mean, the transportation cost, the quality of the reverse flow, the price of the reverse flow, the order delay, the partner reputation, the partner flexibility, and green processes used. The final criterion is very important since the aim of the designed reverse logistics is to reduce environmental impacts.

Step 4: Assessment of the different reverse logistics scenarios

Using the reverse logistics plug-in and via its communication with the SuPLight platform, the performance of the designed RL networks is calculated. Considering environmental performance, the user may choose to view the environmental impact of the designed reverse logistics network, either for the end-of-life (EOL) phase or for the whole life cycle. The results presented in Table 2 show the performance of reverse logistics scenarios considering only the EOL phase. Figure 11 compares the environmental impact of reverse logistics scenarios to the baseline scenario which is the current state with 0% recycled aluminum.

Step 5: Analysis of the assessment results and definition of the best reverse logistics scenario

Using the reverse logistics plug-in, scenarios may be compared on a radar graph; the AHP method is used in this step in order to define the best scenario. It permits

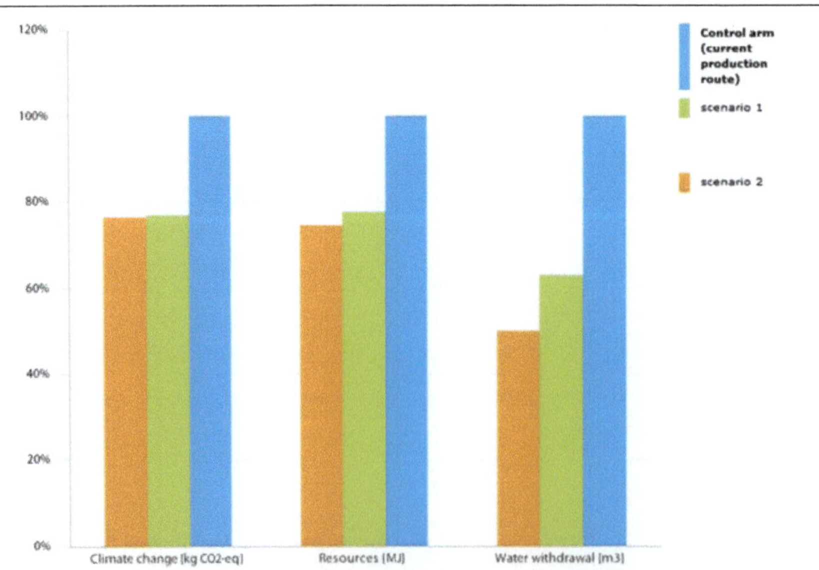

Figure 11 Environmental impacts of the scenarios compared to the baseline (merged life cycle stages).

Table 3 AHP scores for the two reverse logistics scenarios, considering EOL environmental performance

AHP score	Nonrenewable energy		Climate change		Fresh water consumption		Recycling rate		Total RL cost		Final score
	Weight	Score	Weight	Score	Weight	Score	Weight	Score	Weight	Score	
Scenario 1	0.316	0.833	0.286	0.750	0.185	0.750	0.094	0.167	0.119	0.875	0.736
Scenario 2	0.316	0.167	0.286	0.250	0.185	0.250	0.094	0.833	0.119	0.125	0.264

calculating a final score for every scenario in order to identify the best-fitting scenario for Raufoss. These scores are presented in Tables 3 and 4 for EOL and the whole life cycle, respectively.

Results and discussions

The obtained results in Figure 11 show a significant difference (of more than 20%) between the baseline and scenarios 1 and 2 for climate change, resource consumption, and water withdrawal. This means that introducing recycled scrap is beneficial in terms of environmental impact reduction. Aluminum old scrap recycling is considered as less polluting than extracting virgin aluminum. This general point of view has been proved in this case study.

It is important to consider environmental impact of a reverse logistics network for its design. If only reverse logistics cost was considered as the decision criterion, scenario 1 would have been the best scenario. Nevertheless, considering environmental impact shows that scenario 2 is more beneficial. Scenario 2 has better environmental impact for the whole life cycle of the FLCA.

According to the results obtained and shown in Table 4, scenario 2 has a higher score and thus is the best-fitting scenario for the FLCA. This is in contradiction with the results obtained comparing only the environmental performance of the two scenarios for the EOL phase. Considering only the EOL environmental impact, scenario 1 received a final score of 0.736, while scenario 2 received a score of 0.264. Considering these results, scenario 1 is by far the best scenario and should thus be chosen for the reverse logistics network. Nevertheless, when considering the environmental impact for the entire life cycle, scenario 1 received a final score of 0.301, while scenario 2 received a final score of 0.699. Scenario 2 is in this case the best scenario and should be chosen. The total cost of the reverse logistics network and the recycling rate do not change. But the environmental performance changes when considering either one phase or the entire life cycle. This proves that the impact of the reverse logistics network on the whole life cycle of the product should be considered and not only for the EOL phase. In other words, the design of the reverse logistics network should take into consideration the forward supply chain in order to ensure a better performance for the entire chain (forward and reverse supply chain).

Table 4 AHP scores for the two reverse logistics scenarios, considering environmental performance for whole life cycle

AHP score	Nonrenewable energy		Climate change		Fresh water consumption		Recycling rate		Total RL cost		Final score
	Weight	Score	Weight	Score	Weight	Score	Weight	Score	Weight	Score	
Scenario 1	0.316	0.250	0.286	0.250	0.185	0.167	0.094	0.167	0.119	0.875	0.301
Scenario 2	0.316	0.750	0.286	0.750	0.185	0.833	0.094	0.833	0.119	0.125	0.699

Conclusion and perspectives

In this paper, a method for designing a reverse logistics network was presented. It was the basis for the design of a reverse logistics and development of a plug-in. Both were used to design the reverse logistics network for the left front control arm for Raufoss Technology within the European project, SuPLight. The case study validated the method as well as the plug-in developed. They are both designed based on industrial needs. The AHP method was used to compare different reverse logistics scenarios. The aim of the proposed method is to design a reverse logistics network while taking into consideration all possible recovery activities, the possibility of having other partners in the network, and its environmental impact. The results obtained validate the necessity to consider other criteria than cost when designing a reverse logistics network. The results also show the necessity of considering the environmental impact of the reverse logistics network for the whole product life cycle.

The developed method and plug-in are based on human intervention and thus cannot be fully automated. Future works will focus on integrating a closed-loop supply chain approach which might offer several advantages in terms of improving the performance of the whole network and not only the RL network [35].

Competing interests

The authors declare that they have no competing interests.

Authors' contributions

JD carried out the literature review and case study. JD and JLD developed the reverse logistics network design methodology and drafted the manuscript. DP developed the plug-in. JLD and BE supervised the work. All authors read and approved the final manuscript.

Acknowledgements

The authors would like to acknowledge the European Commission and the Picardie French Region for their financial support through the SuPLight FP7 Project (grant agreement no. 263302). They also wish to express their gratitude and appreciation to all the project partners for their contribution during the development of various ideas and concepts presented in this paper.

References

1. Carter, CR, Ellram, LM: Reverse logistics: a review of the literature and framework for future investigation. J. Bus. Logist. **19**(1), 85–102 (1998)
2. Ferrer, G, Ayres, R: The impact of remanufacturing in the economy. Ecol. Econ. **32**(3), 413–29 (2000)
3. Castell, A, Clift, R, France, C: Extended producer responsibility policy in the European Union—a horse or a camel? J. Ind. Ecol. **8**(1–2), 4–7 (2004)
4. Ravi, V, Shankar, R: Analysis of interactions among the barriers of reverse logistics. Technol. Forecast. Soc. Chang. **72**(8), 1011–29 (2005)
5. Toffel, MW: The growing strategic importance of end-of-life product management. Calif. Manag. Rev. **45**(3), 102–129 (2003)
6. DAS, SK, Yin, W: The worldwide aluminum economy: the current state of the industry. JOM. **59**(11), 57–63 (2007)
7. Östlin, J, Sundin, E, Björkman, M: Importance of closed-loop supply chain relationships for product remanufacturing. Int. J. Prod. Econ. **115**(2), 336–348 (2008)
8. Sundin, E, Dunbäck, O: Reverse logistics challenges in remanufacturing of automotive mechatronic devices. J. Remanufacturing. **3**, 2 (2013)
9. Amini, MM, Retzlaff-Roberts, D, Bienstock, CC: Designing a reverse logistics operation for short cycle time repair services. Int. J. Prod. Econ. **96**(3), 367–380 (2005)
10. Rogers, DS, Tibben-Lembke, RS: Going Backwards: Reverse Logistics Trends and Practices. RLEC Press, Pittsburgh (1999)
11. Mitra, S: Revenue management for remanufactured products. Omega. **35**(5), 553–562 (2007)
12. Stock, J, Speh, T, Shear, H: Many happy (product) returns. Harv. Bus. Rev. **80**(7), 16–8 (2002)
13. Fleischmann, M, van Wassenhove, LN, van Nunen, JAEE, van der Laan, E, Dekker, R, Bloemhof-Ruwaard, JM: Quantitative models for reverse logistics: a review. Eur. J. Oper. Res. **103**(1), 1–17 (1997)
14. El Korchi, A, Millet, D: Conditions of emergence of OEM's reverse supply chains. J. Remanufacturing. **4**, 3 (2014)
15. SuPLight: Document Deliverable 1.5: SuPLight Application Scenarios. SuPLight, EU FP7 Project, N°263302, (2012). http://cordis.europa.eu/project/rcn/99059_en.html
16. Stock, J: Product Returns/Reverse Logistics on Warehousing. WERC, IL, Oak Brook (2004)

17. Ramezani, R, Bashiri, M, Tavakkoli-Moghaddam, R: A new multi-objective stochastic model for a forward/reverse logistic network design with responsiveness and quality level. Appl. Math. Model. **37**(1–2), 328–344 (2013)
18. Min, H, Ko, CS, Ko, HJ: The spatial and temporal consolidation of returned products in a closed-loop supply chain network. Comput. Ind. Eng. **51**(2), 309–320 (2006)
19. Üster, H, Easwaran, G, Akçali, E, Çetinkaya, S: Benders decomposition with alternative multiple cuts for a multi-product closed-loop supply chain network design model. Nav. Res. Logist. **54**(8), 890–907 (2007)
20. Listes, O, Dekker, R: A stochastic approach to a case study for product recovery network design. Eur. J. Oper. Res. **160**(1), 268–287 (2005)
21. Demirel, ON, Gökçen, H: A mixed-integer programming model for remanufacturing in reverse logistics environment. Int. J. Adv. Manuf. Technol. **39**(11–12), 1197–1206 (2008)
22. Pishvaee, MR, Kianfar, K, Karimi, B: Reverse logistics network design using simulated annealing. Int. J. Adv. Manuf. Technol. **47**, 269–281 (2010)
23. Du, F, Evans, GW: A bi-objective reverse logistics network analysis for post-sale service. Comput. Oper. Res. **35**(8), 2617–2634 (2008)
24. Zhou, Y, Wang, S: Generic model of reverse logistics network design. J. Transport. Syst. Eng. Inform. Tech. **8**(3), 71–78 (2008)
25. Lee, D-H, Dong, M: Dynamic network design for reverse logistics operations under uncertainty. Transport. Res. E Logist. Transport. Rev. **45**(1), 61–71 (2009)
26. Lambert, S, Riopel, D, Abdul-Kader, W: A reverse logistics decisions conceptual framework. Comput. Ind. Eng. **61**(3), 561–581 (2011)
27. Cardoso, SR, Barbosa-Póvoa, APFD, Relvas, S: Design and planning of supply chains with integration of reverse logistics activities under demand uncertainty. Eur. J. Oper. Res. **226**(3), 436–451 (2013)
28. Wang, HF, Hsu, HW: A closed-loop logistic model with a spanning-tree based genetic algorithm. Comput. Oper. Res. **37**(2), 376–389 (2010)
29. Chen, YJ: Structured methodology for supplier selection and evaluation in a supply chain. Inf. Sci. **181**(9–1), 1651–1670 (2011)
30. Van der Rhee, B, Verma, R, Plaschka, G: Understanding trade-offs in the supplier selection process: the role of flexibility, delivery, and value-added services/support. Int. J. Prod. Econ. **120**(1), 30–41 (2009)
31. Saaty, TL: The Analytic Hierarchy Process. McGraw- Hill, New York (1980)
32. Liu F-H, F, Hai, HL: The voting analytic hierarchy process method for selecting supplier. Int. J. Prod. Econ. **97**(3), 308–317 (2005)
33. SuPlight, Deliverable 4.3: LCA/LCC Tools for Lightweight Solutions. SuPLight, EU FP7 Project, N°263302, (2013)
34. SuPLight, Deliverable 3.2: Generic Framework for Simulation-Based Optimization. SuPLight, EU FP7 Project, N°263302, (2012)
35. MA Asif, F, Bianchi, C, Rashid, A, Mihai Nicolescu, C: Performance analysis of the closed loop supply chain. J. Remanufacturing. **2**, 4 (2012)

14

Forecasting product returns for remanufacturing systems

Xavier Liang, Xiaoning Jin* and Jun Ni

* Correspondence: xnjin@umich.edu
Department of Mechanical Engineering, University of Michigan, 1031 H.H. Dow, 2300 Hayward Street, Ann Arbor, MI 48109, USA

Abstract

One of the major challenges that a remanufacturer faces at strategic planning level today is to match its supply (returned items) with demand due to the inherited uncertainties and variations on both sides. Forecasting product returns is one of the most important tasks of this matching process. Unlike forecasting for traditional manufacturing systems, both quantity and quality forecasts are critical since return timing, quantity, and the quality of returned products can all vary dramatically. This research develops a forecasting method which incorporates knowledge from related sales, product usage, customer return behavior, and product life expectancy information to provide a more accurate prediction of product returns. The models are validated using Monte Carlo simulations. Numerical cases are also presented to illustrate its usage and some important insights.

Keywords: Remanufacturing; Forecast; End-of-life; Simulation

Background

As the global manufacturing environment becomes increasingly competitive, more and more manufacturers view remanufacturing as an important opportunity for profit generation. Remanufacturing provides a means for a society to treat product life cycle from a more holistic perspective, offers an alternative to traditional recycle and reuse, and increases resources utilization. Remanufacturing is the process that recovers residual value from used or degraded products by disassembly and recovery at module level or at component level. This is accomplished by restoring used products to like-new condition by replacing broken or degraded components or by reprocessing used components [1,2]. Focusing on value-added recovery is the main difference between remanufacturing and other types of end-of-life (EOL) treatments, such as recycling [3,4]. This is also the reason that remanufacturing systems are viewed and treated more similar to traditional manufacturing systems than simple recycling.

Remanufacturing is a fast growing industry. In the United States alone, remanufacturing operations (excluding military) has a $53 billion per year market share [5], and it is growing from between 10% to above 50%, depending on industry and product types [6]. Automotive Parts Remanufacturers Association estimates that the remanufactured units were roughly 10 million in 1995, 15 million in 2000, 20 million in 2005, and 30 million in 2015 [7]. Remanufacturing businesses are driven by different incentives in different regions, which, in turn, give different management styles and focus on different metrics and sets of key performance indicators (KPI). Unlike remanufacturers in Europe, who face more rigorous government regulations, North American counterparts are more profit driven. Because the targeted remanufacturers of this research are U.S. based, this paper is more focused on providing economical competitive advantage, such as matching supply with demand to maximize profit, and less on governmental regulation factors or environmental issues.

The fluctuations of both supply and demand sides are the most critical issues facing industry today. Many other remanufacturing topics, such as inventory management, are designed to either isolate or mitigate these fluctuations. In a survey by Hammond et al. [8], automotive part remanufacturers considered that parts availability was the number one difficulty for their operation (43% of them), and 41.2% of them used availability of parts as the main criterion to decide whether or not to remanufacture a given product. Marx-Gmez et al. [9] and Guide et al. [10] stated that the uncertainty in time and amount was the single most important factor that influences remanufacturing system planning. Guide et al. [3] listed seven major problems that remanufacturing systems faced today, and three of them were related to this problem. Uncertainty from supply side (returned products) is the most crucial characteristic of remanufacturing problems, and it distinguishes remanufacturing from traditional manufacturing systems. Unlike traditional manufacturers, remanufacturers usually have less or no direct control of the returned parts. Figure 1 illustrates an example of variations in both supply and demand sides. From a remanufacturer s perspective, supply is the volume of returned used products, and demand is how many remanufactured products are desired. The overlapping area denotes remanufacturable quantity when delays in inventory, remanufacturing time, and other factors are neglected. As the figure shows, remanufacturing operations are essentially matching processes which intend to maximize the overlapping region under demand and supply curves, and in order to match, forecasts of both supply and demand curves are necessary. Moreover, in addition to the variations in quantity and arrival times, quality variation is equally important since returned products can have a wide range of conditions, but finished products usually need to meet the same quality specification. This poses a unique problem which traditional manufacturing systems do not encounter, so new techniques are needed to predict and actively change the shape of both supply and demand curves. Some researches take incentives, such as discount and advertisements, into account to optimize the overall remanufacturable volume [11]. Our study is more focused on the prediction of the supply curve (see Figure 1).

There exists a spectrum of remanufacturing scenarios which have very different types of return characteristics. For example, the return process of leased printer is very

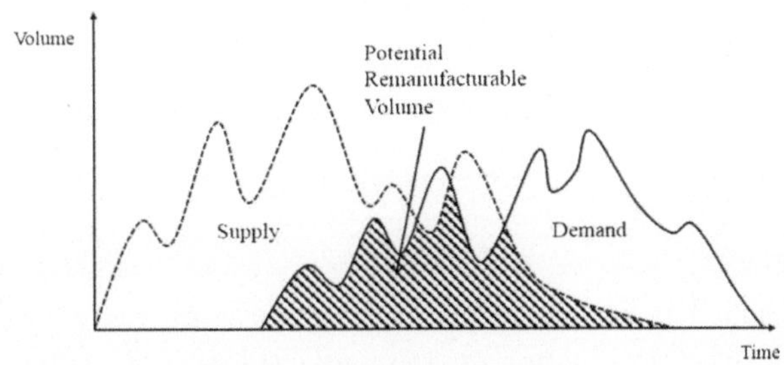

Figure 1 Remanufacturable area under supply and demand.

predictable since the return date and/or quality are predefined by the leasing contracts. However, the return characteristics of many fashionable goods and consumer electronics are much more random since the sales of the items, usage pattern, and customer return behavior are all fairly unpredictable. Because of this variation in predictability, different forecasting approaches are required for different business settings. This paper targets the recently developed electrical vehicle (EV) battery return applications. The urgency and importance of EV battery remanufacturing is because of limited and costly resources, the modular nature of lithium batteries, and potential resale market. Current lithium ion (Li-ion) battery-powered vehicles, such as Chevy Volt and Nissan Leaf, whose batteries are warranted for 8 10 years or 100,000 150,000 miles, are likely to fail before the expected life of the vehicles. The returned battery may still have significant residual value. The cost of Li-ion battery is still very high, usually at $1,000 $1,200 per kWh. Therefore, remanufacturing has great potential to dramatically reduce the total life-cycle costs of Li-ion batteries [12,13]. Industries also propose different types of purchase/warranty schemes in order to allow customers to have operational batteries for the entire vehicle lifespan. That is, customers are now able to purchase not only a single battery pack with the EV but also the warranty of battery use-time, or the combination of them. In this type of business scenario, broken or degraded batteries can be returned or traded in to dealers or battery collectors and be replaced with a new battery before the predefined total use-time expires. The new batteries can then be made with components from returned batteries. This creates strong incentives for customers to return and also for manufacturers to remanufacture. Since this business scenario is completely new, the goal of this research is to develop a long-term forecasting method to remanufacturers and related suppliers to assist their operational decisions.

The challenges for forecasting of product returns are mainly from two sources: lack of quality/credible data and unproven assumptions. There is no or very limited historical data since this type of remanufacturing business scenario is never before seen in the industry and data collecting is also limited or of low accuracy/credibility in similar fields. These challenges seriously limit the type of forecasting techniques we can implement. With these constraints, we propose to build a physical model-based forecasting method instead of traditional data-driven methods which heavily rely on data quality and sophisticated statistical models.

The objective of this research is to provide a methodology to forecast both quantity and quality of returned products, such as EV batteries, based on previous/expert knowledge on indirect information rather than direct historical data. The rest of the paper is organized as follows. Literature review reviews previous literature regarding remanufacturing return forecasting. Forecasting of product return quantity presents the methodology and main factors for quantity forecasting of product returns, and Quality forecasting of returned products is the formulation for quality forecasting and numerical cases. The final section summarizes this research work and points out possible future research directions.

Literature review

Although sale forecasting has been studied for many decades, there are few scientific papers regarding return item forecasting for remanufacturers, and majority of them are focused on short-term tactical and operation level mainly for inventory management and production planning [14]. For some scenarios, simple classical forecasting techniques, such as moving average and exponential smoothing, are sufficient [15]. However, often, there are more valuable information that people can take advantage of, and a variety of forecast outputs are required for different situations [16]. Therefore, specific techniques are developed to tailor to those specific needs. In some cases, periodical information, such as monthly volume, is available and the need is to predict in the future volume [17]. In other cases, historical return dates are available and other characteristics, such as return lead times, are predicted. Kelle and Silver [18] developed four different forecasting methods for expected value and variance of return lead times of containers. Goh and Varaprasad [19] used a transfer function model which included factors, such as previous returns, sales, and time lag, to predict the timing and quantity of returns of Coca-Cola bottles. Toktay et al. [20] developed a Bayesian estimation-based distributed lag model which used newly collected data to update estimated parameters. As listed above, the majority of existing studies use statistics-based method for prediction with historical data.

Other types of forecasting methods including previous knowledge, simulation, or known sub-models are often used. Marx-Gomez et al. [9] combined simulation and fuzzy logic models to forecast the quantities and timing of returns of photocopier. Simulation was used to obtain sales, failures, usage intensity, return quotas, and other so-called impact factors. Then, fuzzy controller was used to combine these impact factors and give one-period prognosis and neuro-fuzzy network was used to provide multi-period prognosis. Similarly, Hanafi et al. [21] used fuzzy-colored petri nets to combine different sub-models, such as technology development, consumer demands, and product reliability to forecast returns at different locations over a specific time period. For others, non-parametric models are more suitable. For example, Monte Carlo simulation is employed to estimate CRT television sales. The gap between existing literature and our current need is that the above methods are used for prognosis and for relatively short-term prediction but not for lifespan planning of the business or facility. Furthermore, the existing literature has not addressed much on the quality variation of returned products which is critical for battery remanufacturing. To fill the gap in the existing literature, this paper develops a new forecasting tool for end-of-life product returns in terms of *timing*, *quantity*, and *quality* to support the remanufacturing strategic planning and decision making.

Forecasting of product return quantity

Quantity forecasts provide information of how many product return a remanufacturer expects to obtain at a future time. However, unlike most forecasts which only deal with one-time customer decision, such as buying or not buying, returning is determined by a series of cascaded events, i.e., product purchase decision, product usage, and product return decision. In this section, a new method is presented for predicting product returns in terms of quantity only. This is critical to determine the availability of remanufacturable products. The key to the new method is to utilize an effective characterization of three main influence factors sales, life expectancy, and return behavior to facilitate an accurate forecasting of return timing, quantity, and quality. More specifically, the reasons of product return considered in this paper are limited to the failure-induced return and end-of-life return. The product technology upgrading is not considered as a cause of return.

Sub-models for influence factors

Sales distribution, S(t)

Sales forecasting has been studied for many decades, and each manufacturer or third party consultant usually has its own forecasting models. Moreover, it is common in industry practice that non-equation form of modeling is used, and people tend to focus more on market research perspective of the forecast. Techniques, such as concept test, focus groups, perceptual mapping, conjoint analysis, and consumer clinics, are more often seen in order to have a better understanding of the customers preference and to adjust future plans. For example, the information acceleration (IA) methods are used for GM s new electric vehicle sales prediction [22]. On the other hand, academic researchers are still interested in more mathematics-orientated approaches because they can provide a more direct relationship among different influential factors. For this reason, this paper provides both analytical and numerical methods in order to accommodate current industry practice.

For analytical-based forecasts, there are generally two categories: *aggregated* and *disaggregated* models. For the aggregated or top-down approaches, only the cumulated behavior of a group of people is studied. On the other hand, disaggregated or bottom-up models study individual decision makers that underlie market demand or supply and integrate them. Although the emphasis in forecasting and econometrics has generally shifted from aggregated to disaggregated models in the past decades, most forecasting models people use in industry today are still of aggregated form simply due to the difficulty and expense of collecting data on individual consumers. Out of all aggregated models, *Bass diffusion model* is the most often used [23].

Diffusion of products or innovation is the theory that seeks to explain how, why, and at what rate a new idea spreads through societies. The parameter that characterizes this process is called the *rate of adoption*, and it is defined as the relative speed in which members of a social system adopt an innovation. This rate is usually measured by the length of time required for a certain percentage of the members to adopt. Customers can be divided into many categories, such as innovators, early adopters, early majority, late majority, and laggards. Groups are different in how they perceive different innovation factors, such as relative advantage, compatibility, and complexity of the product. Bass diffusion model is chosen because it has all the essentials of diffusion models yet is the most simplified and intuitive version. In this model, only two customer groups, early adopters and followers, are considered. The sales only include the originally manufactured products and the remanufactured products are only used for warranty repair but not for original sale.

Mathematically, we use the following differential equation to represent the Bass diffusion

$$\frac{f\ t}{1-F\ t}\quad p\quad qF\ t \qquad 1$$

where $F(t)$ is the base function, and $f(t)$ is the rate of change or derivative of $F(t)$. p is the coefficient of early adopters, advertising effect, or innovators in Bass s original model. It describes how quickly early adaptors are willing to purchase or to enter a new market. q is the coefficient of followers, internal influence, word-of-mouth effect, or imitator factors in the original model. Sales volume at time t, $S(t)$, is the rate of change of installed base $f(t)$ multiplied by the market potential, m, and has the form of

$$S\ t \quad mf\ t \tag{2}$$

The solution to Equation 1 is

$$S\ t \quad mf\ t \quad m\frac{p\ \ q^{2}}{p}\frac{e^{-p\ \ q\ t}}{\left(1\ \ \frac{q}{p}e^{-p\ \ q\ t}\right)^{2}}, m,p,q>0 \tag{3}$$

with its maximum occurring at

$$t^{*} \quad \frac{lnq-lnp}{p\ \ q} \tag{4}$$

In real practice, p and q are set equal since early adopters usually act much quicker than followers. The choice of p and q depends on many other social factors and industry [24].

Bass diffusion model is a generalized form of other more commonly used s-curve functions. When q =0, it reduces to standard exponential distribution with $\lambda = p$. When p = 0, it reduces to logistic distribution. Bass diffusion model is also a special case of the Gamma/shifted *Gompertz* distribution.

Product breakdown distribution, B(t)

Besides sale information, another essential influencing factor for battery return forecasting is battery life expectancy or the degree of quality degradation during the usage stage. Currently, battery condition monitoring typically refers to the evaluation of battery state of charge (SOC), or the state of health (SOH). SOC is defined as the amount of remaining charge in a battery before a recharge is required, and SOH is the potential chargeable capacity of a battery compared to the original unused one. For electric vehicle batteries, consumer usually does not return when the battery is completely dead but rather until certain degradation criteria are reached, such as SOC drops below 75% ~85%. Therefore, the return time prediction is usually to predict the time length that a battery can last until certain manufacturer predefined threshold is reached.

The conventional approach for life expectancy prediction is to model product condition or degradation by an appropriately chosen random process, and the occurrence of failure or reaching of certain threshold is modeled by a Poisson process. Weibull distribution is chosen for this study because its failure rate function is only current state dependent. The system condition, or degradation state, is modeled by a Brownian motion with positive drift. Under this assumption, the time to failure corresponds to the first passage time of the Brownian motion and follows an inverse Gaussian distribution. Weibull distribution is well studied for reliability engineering, and reference can be found in many textbooks (i.e., [25]).

Weibull distribution is one of the solutions that assume the degradation process is deterministic. The probability density function $g(t)$, the hazard function $h(t)$, and cumulative distribution function $G(t)$ of Weibull distribution are:

$$g\ t|a,b,c \quad B\ t \quad \frac{c}{b}\left(\frac{t-a}{b}\right)^{c-1} e^{-\frac{t-a}{b}\ c}; t \geq a; a \in \mathbb{R}; b,c \in \mathbb{R} \tag{5}$$

$$h\ t|a,b,c \qquad \frac{c}{b}\left(\frac{t-a}{b}\right)^{c-1};\ t\geq a;\ a\in\mathbb{R};\ b,c\in\mathbb{R} \qquad 6$$

$$G\ t|a,b,c \qquad 1-e^{-\ \frac{t-a}{b}\ ^{c}};\ t\geq a;\ a\in\mathbb{R};\ b,c\in\mathbb{R} \qquad 7$$

Here, the breakdown function $B(t)$ is exactly the probability density function, i.e., $B(t) = g(t)$ [24]. Weibull distribution is a very versatile reliability distribution and widely used in many reliability engineering applications. By changing the shape parameter, the failure rate $h(t)$ can be changed directly. It can be increasing, constant, or decreasing over time. Piecewise curve fitting is commonly used to model the classic bell curve for failure rate [26,27].

Customer return function, C(t)

Customer return behavior is the third influencing factor that needs to be considered in the forecasting model. The warning indicator of a battery failure is usually signaled on the dashboard, very similar to maintenance reminder signals. Also similar to the maintenance behavior, people usually do not bring back the vehicle for battery treatments immediately when they see the warning signal. This is probably due to the fact that the vehicle is still perfectly drivable and usually no noticeable changes are detected. Another reason is that SOC or SOH aspects of the battery usually degrade very gradually over a period of years without noticeable abrupt changes.

To the authors knowledge, people do not return immediately, and sometimes the time delay between product failure and the return action can be as long as half a year or more. The return function is usually heavily skewed to the left as illustrated in Figure 2. This long-tailed distribution indicates that the majority of people will return damaged or unwanted product in a short period of time. However, there are also a considerable amount of people who will return after a relatively long period of time. This kind of characteristics can be modeled by *inverse Gaussian functions* due to its skewness, positive support, and relatively easy expression (see Figure 2). Choosing inverse Gaussian is also because its flexibility in modeling the following three characteristics: (1) the majority of people return within certain time period, (2) only a small portion of them return whenever they found convenient, and (3) the rest will never return.

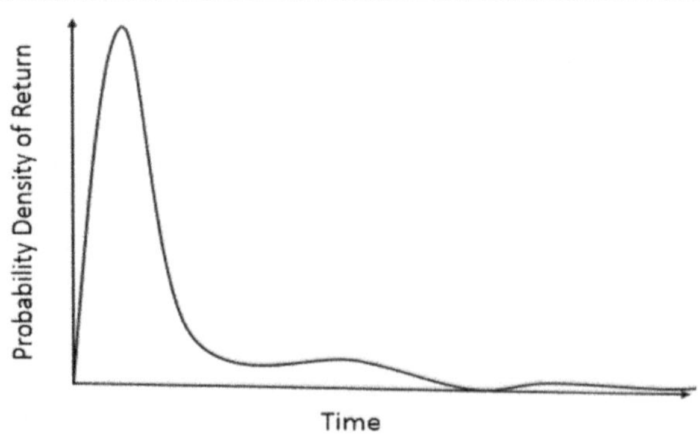

Figure 2 Typical customer return behavior distribution.

Inverse Gaussian functions have the general probability distribution function of

$$C\ t\ \ \left[\frac{i}{2\pi t^3}\right]^{1/2} e^{\frac{-i\ t-j\ 2}{2j^2 t}};\ i,j > 0 \qquad (8)$$

where j >0 is the mean and i >0 is the shape parameter. $C(t)$ represents the probability that the customer returns the battery t time units after its failure.

Although there are many other factors that may influence the quantity and quality of the returned products such as government regulations, they are either not appropriate for the EV battery return in North America or not proven to have significant influence on return quantity or quality. Therefore, the three main factors listed above are the focus in this study.

Return quantity forecasting in continuous case

As illustrated in Figure 3, the return quantity at any point in time, such as at point R, is the summation of different sale quantity and product life expectancy combination. That is, returned product at R may be purchased at time S_1 and used for B_1 years, or purchased at time S_2 and used for B_2 years, and so on.

One way to characterize the relationship of these three influential factors is to use convolution. Convolution for two continuous functions $f(t)$ and $g(t)$ is defined as

$$f\ t*\ \ g\ t \equiv \int_{-\infty}^{\infty} f\ \tau\ g\ t-\tau\ d\tau \qquad \int_{-\infty}^{\infty} f\ t-\tau\ g\ \tau\ d\tau \text{forf},g:\ 0,\infty \in\mathbb{R} \qquad (9)$$

For this problem, the volume of broken products at time t can be represented as the convolution of the sales volume $S(t)$ and the breakdown probability $B(t)$. Then, the total volume of product returns at time t; $R(t)$ is the convolution of the consumer return behavior $C(t)$; the previous convolution result $S(t)*B(t)$ as follows:

$$R\ t \qquad S\ t* \quad B\ t* \quad C\ t \qquad (10)$$

where $R(t)$ denotes the return quantity at time t, $S(t)$ is the probability function of sales quantity, $B(t)$ is the probability function of breakdown time, and $C(t)$ is the probability function of customer return. Note that $B(t)$ and $C(t)$ may not be normalized to 1. The integration of $B(t)$ is less than 1 because not all batteries are degraded to certain

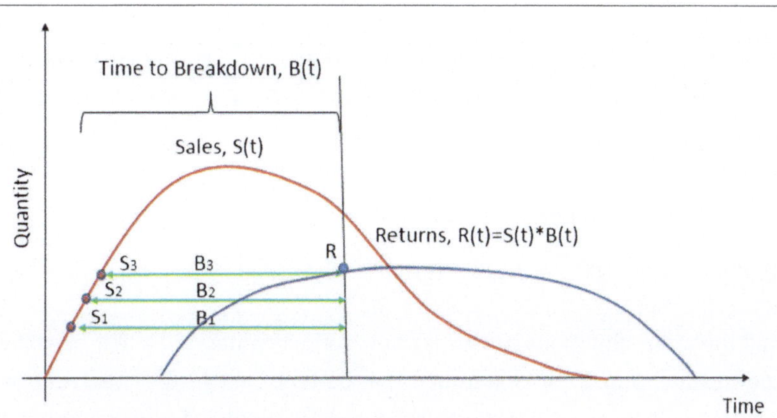

Figure 3 Relationship between sales, breakdown, and return.

threshold during the lifespan of an electric vehicle. The integration of $C(t)$ is less than 1 because not all failed batteries are returned. Some customers may choose not to return or return to third party collectors.

However, due to the complexity of the functions $S(t)$, $B(t)$, and $C(t)$, there is no closed-form solution for the convolution. Because only a finite range is needed, these functions can be approximated over this desired range by some polynomial functions. The coefficients of the polynomial are chosen such that the weighted error between the approximation and original function is minimized in a least squares sense. For example,

$$S\ t \quad \theta_0 \quad \theta_1 t \quad \theta_2 t^2 \quad \theta_3 t^3$$
$$where\,\theta_i \quad argmin \int_a^b \left[R\ t - \tilde{R}\ t, \theta_i\ \right]^2 dt \tag{11}$$

Return quantity forecasting in discrete form

As mentioned, the majority of forecasting models used in industry are not analytical based but are most likely produced and derived from different forms of marketing researches. Periodical, such as monthly or yearly, data instead of equation form of $S(t)$, $B(t)$, and $C(t)$ are provided. The advantage of using a discrete form is that the raw data form from reliability tests can be used directly without fitting into a model first. Similar as continuous convolution, a discrete form of convolution can be defined as

$$f\ n * g\ n \equiv \frac{1}{l}\sum\nolimits_{m\ -\infty}^{\infty} f\ m\ g\ n-m \quad \frac{1}{l}\sum\nolimits_{m\ -\infty}^{\infty} f\ n-m\ g\ m \tag{12}$$

where both $f[n]$ and $g[n]$ have finite positive support, and l is a scale factor which is related to number of samples for both $f[n]$ and $g[n]$ and is used for normalization. Due to the simplicity of discrete convolution computation, many continuous functions are first discretized then are taken convolution in discrete domain. In this way, no symbolic integral of continuous functions is needed.

Numerical examples

This section uses a numerical example to demonstrate the modeling procedure. The sales volume of battery packs are the same as the sales volume of the electric vehicles. A typical EV model usually has a life of 3 ~ 4 years. Hence, it is reasonable to assume 95% of the Bass diffusion function to be in the range of [0, 4]. Substituting p =0.08, q =2, and m =1 into Equation 3, we obtain $S(t)$ and its polynomial approximation $\tilde{S}\ t$ as follows:

$$S\ t \quad 54.08\frac{e^{-2.08t}}{1 \quad 25e^{-2.08t\ 2}}$$
$$\tilde{S}\ t \quad 0.011t^5 \quad -0.101t^4 \quad 0.511t^3 - 1.373t^2 \quad 1.631t - 0.3952 \tag{13}$$

Note that all the approximations in this section are over interval [0, 10] or 10 years.

For breakdown function $B(t)$, it is assumed that batteries are in the wear-out zone in around 3 4 years. Failures most likely occur around the sixth years of its usage. Therefore, we have the following $B(t)$ in the form of Weibull distribution:

$$B\ t \quad 2\left(\frac{t-3.5}{2}\right)^{3} e^{-\ t/2\ ^{4}}$$
$$\tilde{B}\ t \quad 0.005t^{6}-0.046t^{5} \quad 0.249t^{4}-0.713t^{3} \quad 1.007t^{2}-0.588t \quad 0.089 \qquad 14$$

For return behavior, assuming μ =0.5 and λ =0.2, we have the following return function

$$C\ t \quad \left(\frac{0.2}{2\pi t^{3}}\right)^{1/2} e^{\frac{-0.2\ t-0.5\ ^{2}}{2*0.5^{2}t}}$$
$$\tilde{C}\ t \quad -0.0170t^{6} \quad 0.1515t^{5}-0.7493t^{4} \quad 1.9745t^{3}-2.2338t^{2} \quad 0.1179t \quad 1.409 \qquad 15$$

From Equation 17, the resulting polynomial approximation of the return function $R(t)$ is obtained:

$$\tilde{R}\ t \quad 0.0006t^{5}-0.0074t^{4} \quad 0.0445t^{3}-0.1108t^{2} \quad 0.0867t-0.0084 \qquad 16$$

The result is shown in Figure 4.

Monte Carlo simulation verification

Another way to predict the quantity of returned product is using Monte Carlo simulation. Monte Carlo simulation is a family of computational algorithms that rely on repeated random sampling in order to obtain the final results. Assume there total customer to be 100,000, so 100,000 samples will be used. For each sample, its sale date, expected product life, and customer returns are generated according to the functions $S(t)$, $B(t)$, and $C(t)$, respectively. For sales date, it follows Bass diffusion model as in previous sections. Technically speaking, it is not strictly a probability distribution since the integration of the function may not add up to 1. Therefore, normalization is needed before sampling. For each sample, the deterministic calculation is then calculated:

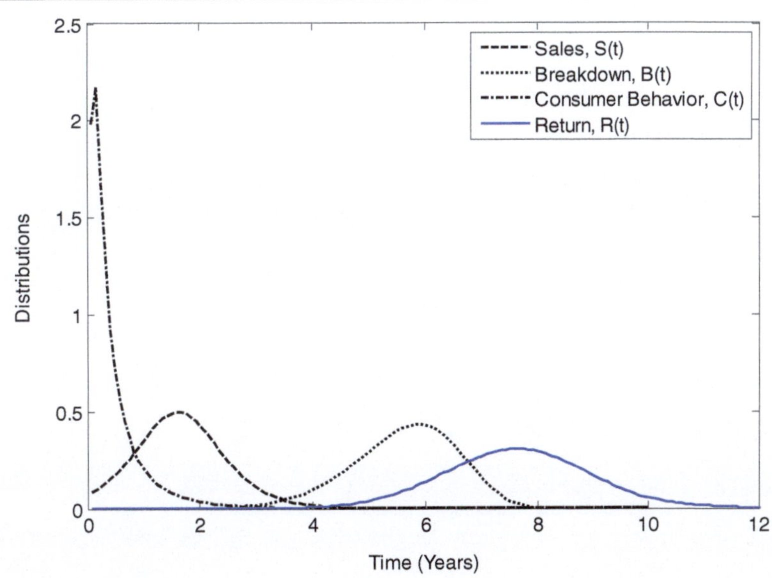

Figure 4 Return quantity forecasts.

$$\text{Return date} \quad \text{sale date} \quad \text{product life} \quad \text{return delay}$$

For aggregation, histogram is used to obtain the distribution of sale date, product life, return delay, and return date, as illustrated in Figure 5.

To compare the results from Monte Carlo simulation with analytical results, the difference between two curves is measured by *f*-divergence method. One of the *f*-divergence methods, KL-divergence value, is 0.028, and *Hellinger* distance is 0.035 which are all much less than 1. This indicates the results obtained by both methods are very close to each other.

Properties of predicted return function

Because of convolution and the general shapes of $S(t)$, $B(t)$, and $C(t)$, the convolution operation acts like a weighted moving average. This moving average can also be viewed as a low-pass finite impulse response filter. This means that only low frequency information will be preserved, and high frequency information will be eliminated by convolution. Here, frequency of a function roughly means how many times a function varies in a given time interval. By taking Laplace transform in the continuous case or Z-transform in the discrete case, it can be shown that distributions used by $S(t)$, $B(t)$, and $C(t)$ are essentially low-pass filters [28]. As a result, $R(t)$ is smoothed by each distribution, and any small noises of $S(t)$, $B(t)$, or $C(t)$ will be reduced. Figure 6 demonstrates a case where seasonality noise (a sine function), or high frequency function, is added to the sales function, $S(t)$. Sine function is chosen because it contains only a single frequency.

$$S_{noise}\ t \quad 1 \quad 0.3sin\ 8t \quad S\ t \qquad (17)$$

where t is in radian. Please note that the sine function has much higher frequency than the original function. Also note that after the sine function is added the mean, variance and the area under the curve of $S(t)$ is preserved. The resulting function $R(t)$ is exactly the same before and after adding this seasonality noise component (see Figure 6). Due

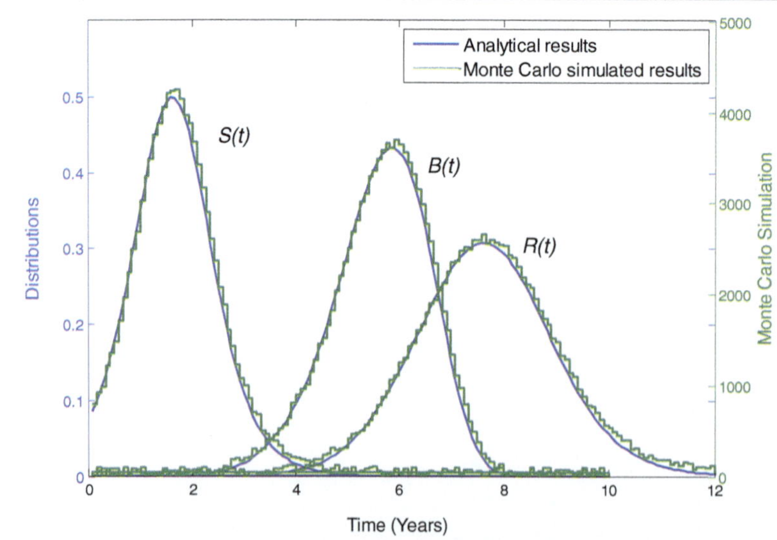

Figure 5 Analytical results (*smooth curves*) vs. Monte Carlo Simulation results (*step curves*).

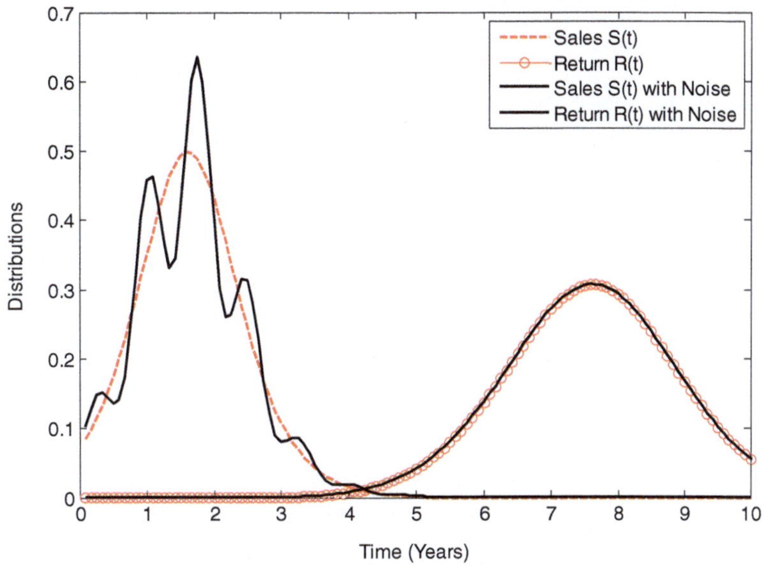

Figure 6 Adding high frequency noise to *S*(*t*) results in overlapping *R*(*t*)s.

to this smoothing effect, $R(t)$ is not sensitive to high-frequency changes in $S(t)$, $B(t)$, or $C(t)$.

Quality forecasting of returned products

Besides the quantity of the returned products, quality of them is another critical information that production planners would like to know beforehand for remanufacturing planning. In this study, the return product quality is defined as the remaining useful life (RUL) of the unbroken parts of a returned product. A typical EV battery pack usually consists of hundreds of battery cells. When the performance measurement of a battery pack, such as SOC or capacity, drops below certain threshold, only one or a few cells fails or degrades to unacceptable condition while the majority of the cells are still in good or reasonable quality. Because not all the cells degrade at the same rate, it is critical to estimate the RUL of good cells, so the remaining value can be assessed. RUL is essentially a conditional probability distribution which determines how long it takes to reach certain threshold given that the battery cells have already survived certain among of time (the time for the battery pack reaching the threshold). In order to express the quality information along with the quantity information, three-dimensional plots are used. The x-axis represents the battery return date, the y-axis is the expected quality indicated by remaining life, and z-axis is the quantity distribution.

Quality distribution, *Q*(*t*, *x*)

Remaining useful life (RUL) is defined as a conditional random variable $X_t = T - t$ when $T > t$, where t is the time where a part has survived so far and T is the time to failure. The conditional reliability function $R_t(t)$ contains all the information required for RUL. The reliability function is defined as

$$R_t\ x \qquad P\ T-t > x|T > t\ ;\ t,x,T\in\mathbb{R} \qquad 18$$

$$R_t\ x \qquad \frac{P\ T-t>x\ \cap P\ T>t}{P\ T>t} \qquad \frac{1-F\ T \quad \tau}{1-F\ T} \tag{19}$$

From Equation 19, the failure rate of RUL $Q(t, x)$ is then

$$f_t\ x \qquad -\frac{\partial}{\partial x}R_t\ x \qquad \frac{f\ T}{1-F\ t} \qquad h\ t \quad x\ R_t\ x \tag{20}$$

$$Q\ t,x \qquad f_t\ t,x \qquad f_t\ x|t\ f\ t \qquad h\ t \quad x\ R_t\ x\ f\ t \tag{21}$$

where $x = T - t$, and $f(t)$, $F(t)$, and $h(t)$ are time to failure probability density function (pdf), cumulative distribution function (cdf), and hazard rate functions defined in Equations 9 11. The conditional distribution becomes.

$$R_t\ x \qquad \frac{e^{-\ \frac{t\ x-a}{b}\ c}}{e^{-\ \frac{t-a}{b}\ c}} \qquad e^{-\ \frac{t\ x-a}{b}\ c -\ \frac{t-a}{b}\ c\ ]} \tag{22}$$

The joint probability becomes

$$Q\ t,x \qquad \frac{c}{b}\left(\frac{t\quad x-a}{b}\right)^{c-1}e^{-\left[\left(\frac{t\quad x-a}{b}\right)^{c}-\left(\frac{t-a}{b}\right)^{c}\right]}\ \frac{c}{b}\left(\frac{t-a}{b}\right)^{c-1}e^{-\left[\left(\frac{t-a}{b}\right)^{c}\right]}$$
$$\frac{c^2}{b^2}\left(\frac{t-a}{b}\right)^{c-1}\left(\frac{t\quad x-a}{b}\right)^{c-1}e^{-\left[\left(\frac{t\quad x-a}{b}\right)^{c}\right]} \tag{23}$$

For each breakdown time, t, there is a RUL distribution associated with it which gives the third dimension to the final plot.

Convolution

It is assumed that the quality is uniform when the product is out of the factory or at the time of purchase, so the third dimension of $S(t)$ is invariant. So we have $S(t, x) = S(t)$. Similarly, because consumer s behavior will not be affected by the RUL of returned battery, their return behavior would not be affected by return time either, i.e., $C(t, x) = C(t)$. Because of this invariance in both $S(t, x)$ and $C(t, x)$, the convolution is only performed in one dimension, and it is independent of the quality dimension. That means

$$f\ t,\tau * \ g\ t,\tau \qquad \int_{-\infty}^{\infty} f\ x,\tau\ g\ t-x,\tau\ dx \qquad \int_{-\infty}^{\infty} f\ t-x,\tau\ g\ t,\tau\ dx \tag{24}$$

The final result of $R(t, x)$ is the one-dimensional convolution of $S(t, x)$, $Q(t, x)$, and $C(t, x)$

$$R\ t,x \qquad S\ t,x * \ Q\ t,x * \ C\ t,x \tag{25}$$

Numerical examples

Sales distribution, S(t, x)

Continuing the numerical example from numerical examples and the sales function are essentially the same as Equation 17. The 3D shape of the function is shown in Figure 7.

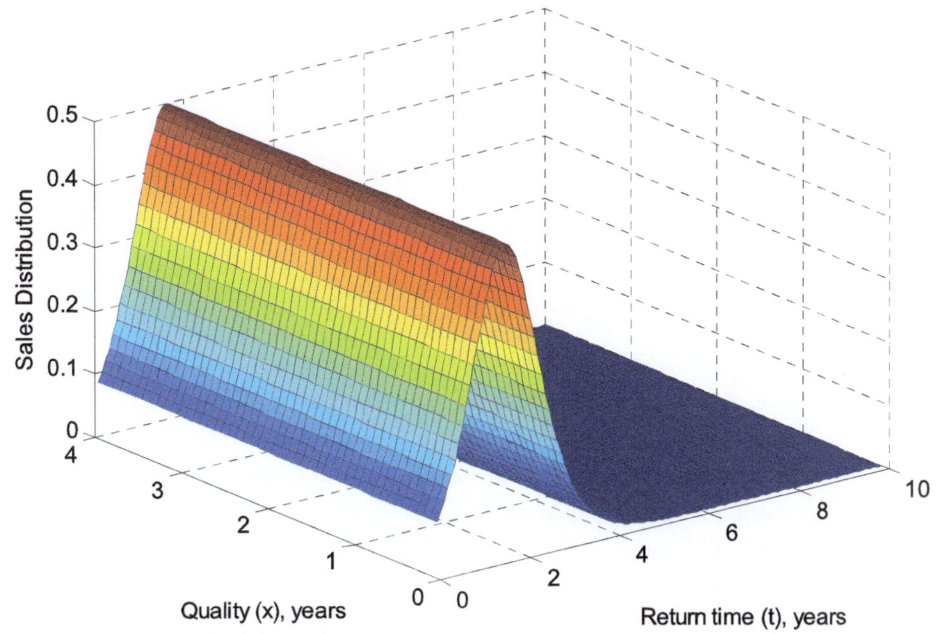

Figure 7 **Sales distribution,** ***S*(*t, x*).**

$$S(t,x) = 54.08\frac{e^{-2.08t}}{(1+25e^{-2.08t})^2} \qquad (26)$$

Quality distribution, Q(t, x)

This distribution is determined by two factors, the product breakdown time t and RUL x. The relationship is given by $Q(t, x)$ in Equation 23 for this example. It can be seen that when the shorter the product usage time t is, the longer the expected RUL x is, as in Figure 8. Because the mode of breakdown function $Q(t, x)$ along the t direction is around 7, the quality of returned product decreases dramatically towards 0.

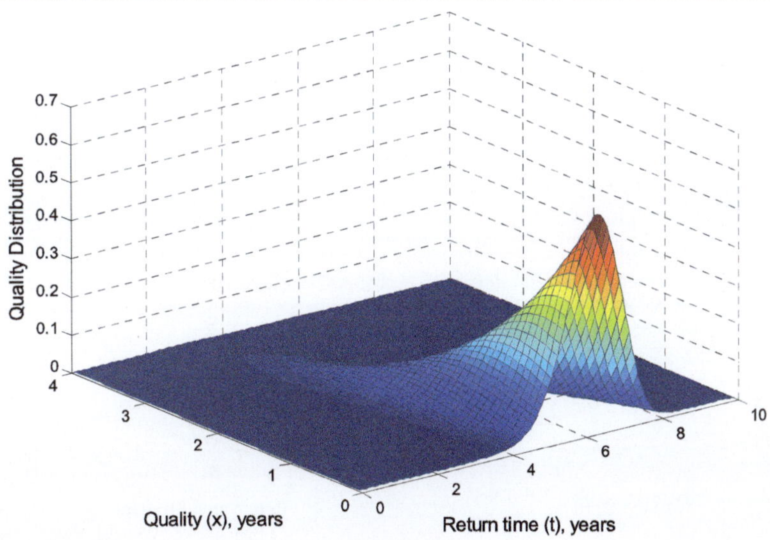

Figure 8 **Quality distribution,** ***Q*(*t, x*).**

$$Q \; t,x \quad 4\left(\frac{t \quad x-3.5}{2}\right)^3 \left(\frac{\tau-3.5}{2}\right)^3 e^{-\left[\frac{t \;\; \tau-3.5}{2} \;\; ^4\right]} \tag{27}$$

Customer return distribution, C(t, x)

Similar with the sales function, the customer return behavior function is also constant along the quality axis since return behavior would not alter the quality of the product itself. That is, $C(t, x) = C(t)$, which is shown in Figure 9.

Return distribution, R(t, x)

The final returned product quantity and quality plot is shown in Figure 10. The cross section of the plot along the time direction at $x = 0$ is the same as the $R(t)$ from Numerical examples.

From this plot, it can be shown that after return date (t) passes a certain time (9 years), or the sum of peak sales date (around 2 years) plus peak breakdown time (around 7 years), the returned products are largely in very low quality (x <1 year). This gives a rough time frame when a returned product is economical to be remanufactured. Some other end-of-life treatment, such as recycling or proper disposal, may be employed after certain time.

Verification with Monte Carlo simulation

For this illustrative example, one million samples are used for each of the random input generation. Generation of the sale dates, expected product life, and customer return behavior is the same as in Monte Carlo simulation verification. The RUL of returned products is generated by randomly drawing samples from $Q(t, x)$ where t of each sample is determined by the expected product life samples. Therefore, whenever a sample is generated, the $Q(t, x)$ distribution needs to be recalculated. As before, $Q(t, x)$ is not normalized, so it needs firstly be discretized then normalized for each sample.

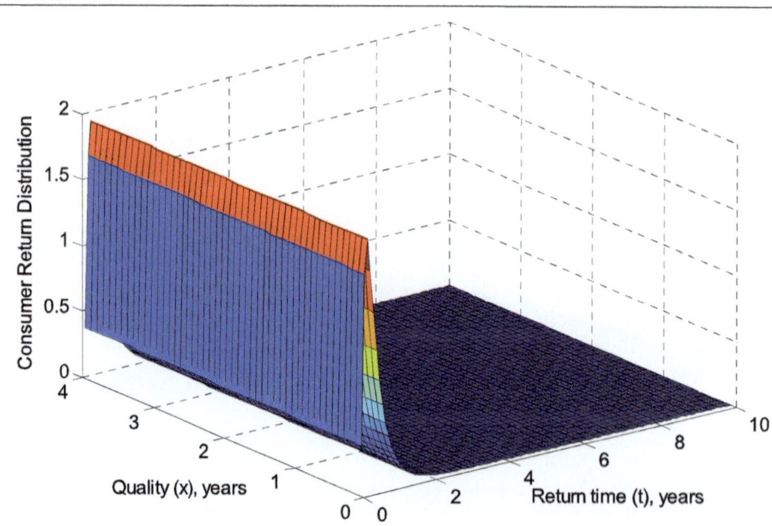

Figure 9 Customer return distribution, *C(t, x)*.

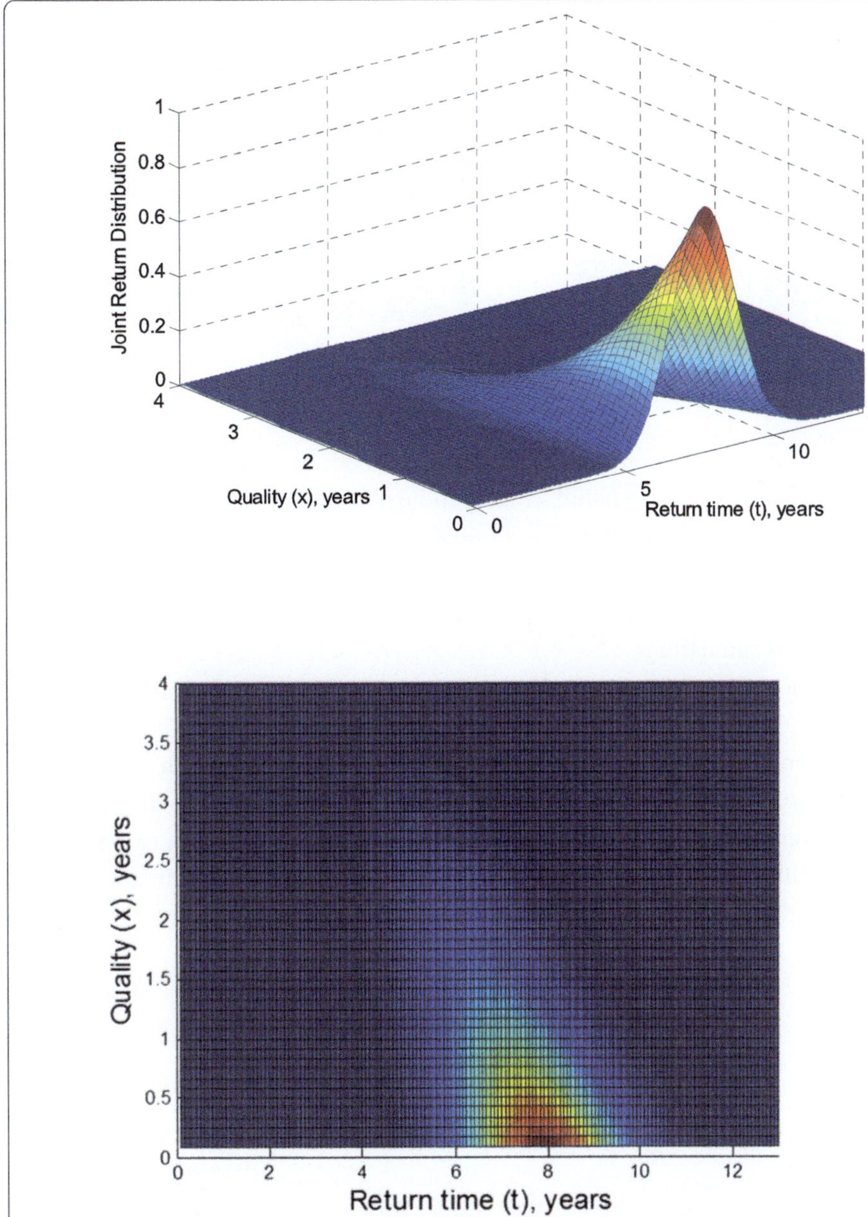

Figure 10 **Return quantity and quality distribution, *side view* and *top view*.**

Each sample has two components, the return date and return quality. Return date is calculated in the exact same way as in Monte Carlo simulation verification. The return quality is sampled directly from $Q(t, x)$ as shown in previous subsection. The distribution is obtained by using two-dimensional histogram or other scientific volume imaging techniques. The simulation results shown in Figure 11 agree well with $R(t, x)$ obtained using analytical convolution-based approach.

Conclusions and future work

This research provides a methodology to forecast long-term trend of both quantity and quality of product returns, or the supply of a remanufacturing system, by modeling three major influential factors (i.e., sales, life expectancy, and customer return behavior)

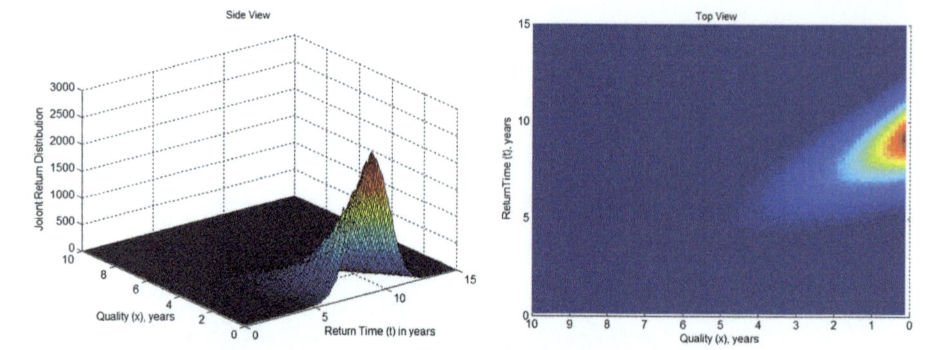

Figure 11 Monte Carlo for returned product quantity and quality.

and their combinatory formulation in a forecasting method. To meet the emerging needs of decision support in the remanufacturing industry, reliable forecasting of supply can help determine a reasonable estimate of used products attainable under a given set of conditions and further assist decision makings in remanufacturing operations. The effectiveness and accuracy of the forecasting model developed in this paper is verified and validated with simulations.

Generally, the typical goals of strategic forecasting are threefold: (1) estimate opportunity and outcome for future business actives, (2) find what influence and how to influence outcome, and (3) judge the potential risks associated with such business actives. The first two are covered by this research. The model developed in this paper only provides the most likely outcome based on a single set of assumptions, so potential risks are more difficult to discovery. Future works may include sensitivity analysis, scenario analysis, and different assumption management techniques.

Competing interests

The authors declare that they have no competing interests.

Authors contributions

XL carried out the construction of forecasting model and the numerical studies. XJ participated in the model refinement and validation. JN was responsible for defining the scope and objective of this study. All authors read and approved the final manuscript.

Authors information

Xavier Liang is a Ph.D. student at the University of Michigan. His research interest focuses on of modeling and analysis of supply and demand in remanufacturing systems and related operational strategies.
Xiaoning Jin received the B.S. degree in Industrial Engineering from Shanghai Jiaotong University, China in 2006 and the M.S. degree in Industrial and Operations Engineering from the University of Michigan in 2008. She is currently an assistant research scientist in the Industrial and Operations Engineering at the University of Michigan.
Her research interests include modeling and optimal control of production and maintenance systems, manufacturing/remanufacturing systems, control of production, and inventory systems with supply quality variation.
Jun Ni received the B.S. degree in Mechanical Engineering from Shanghai Jiaotong University, China in 1982 and the M.S. degree and Ph.D. degree in Mechanical Engineering from the University of Wisconsin-Madison in 1984 and 1987, respectively.
He is currently a Professor of Mechanical Engineering at the University of Michigan. His research interests include manufacturing process modeling, analysis and prediction; precision engineering and metrology, intelligent maintenance systems, monitoring and fault diagnosis, and remanufacturing systems.
He became a Presidential Faculty Fellow of the National Science Foundation in 2001 and a fellow of American Society of Mechanical Engineers in 2004. He is currently the Director of S. M. Wu Manufacturing Research Center at the University of Michigan, the Co-Director of Multi-Campus NSF Center for Intelligent Maintenance Systems, and the Deputy Director of US-China Clean Energy Research Consortium-Clean Vehicles Center.

Acknowledgements

This work was jointly supported by the US Department of Energy and General Motors Co. This work was also partially supported by the I/UCRC on Intelligent Maintenance Systems at the University of Michigan.

References

1. Lund, RT: Remanufacturing. Technol. Rev. **87**(2), 18 (1984)
2. Sutherland, JW, Adler, DP, Haapala, KR, Kumar, VA: Comparison of manufacturing and remanufacturing energy intensities with application to diesel engine production. CIRP. Annals Manuf. Technol. **57**, 5 8 (2008)
3. Guide, VDR: Production planning and control for remanufacturing: industry practice and research needs. J. Oper. Manag. **18**(4), 467 483 (2000)
4. Westkamper, E, Arndt, G: Life cycle management and assessment: approaches and visions towards sustainable manufacturing. CIRP. Annals- Manuf. Technol. **49**(2), 501 526 (2000)
5. Hauser, W, Lund, RT: The Remanufacturing Industry: Anatomy of a Giant. Department of Manufacturing Engineering. Boston University, Boston, USA (2003)
6. Srivastava, SK, Srivastava, RK: Managing product returns for reverse logistics. Int. J. Physical. Distri. Logist. Manage. **36**(7), 524 546 (2006)
7. Buxcey I: Remanufacturing in the Automotive Industry. [http://www.remanufacturing.org.uk/pdf/remanufacturing_in_automotive_industry_ian_buxcey_automotive_parts_rebuilders_associatio.pdf?-session=RemanSession:42F940711b84a138FAnHU1EE5043], Automotive Parts Remanufacturers Association 2003.
8. Hammond, R, Amezquita, T, Bras, B: Issues in the automotive parts remanufacturing industry: a discussion of results from surveys performed among remanufacturers. Eng. Des. Autom. **4**, 27 46 (1998)
9. Marx-G?mez, J, Rautenstrauch, C, N?rnberger, A, Kruse, R: Neuro-fuzzy approach to forecast returns of scrapped products to recycling and remanufacturing. Knowl.-Based Syst. **15**(1), 119 128 (2002)
10. Guide, VDR, Jayaraman, V, Srivastava, R: Production planning and control for remanufacturing: a state-of-the-art survey. Robot. Comput. Integr. Manuf. **15**, 221 230 (1999)
11. Ghoreishi, N, Mark, JJ, Nekouzadeh, A: A cost model for optimizing the take back phase of used product recovery. J. Remanuf. **1**(1), 1 15 (2011)
12. Jin, X, Hu, SJ, Ni, J, Xiao, G: Assembly strategies for product remanufacturing with variable quality returns. IEEE Trans. Autom. Sci. Eng. **10**(1), 76 85 (2013)
13. Jin, X, Ni, J, Koren, Y: Optimal control of reassembly with variable quality returns in a product remanufacturing system. CIRP. Annals-Manuf. Technol. **60**(1), 25 28 (2011)
14. Toktay, B: Forecasting product returns, in Business Aspects of Closed-Loop Supply Chains, pp. 203 219. Carnegie Mellon University Press, Pittsburgh, PA (2003)
15. Nahmias, S: Production and Operations Analysis. Irwin Professional Publishing, Glencoe, IL (1993)
16. Toktay, B, van der Laan, E, de Brito, MP: Managing Product Returns: The Role of Forecasting (22 2003 4,). *ERIM Report Series Reference* No. ERS-2003-023-LIS
17. Clottey, T, Benton, WC, Srivastava, R: Forecasting product returns for remanufacturing operations. Decis. Sci. **43**(4), 589 614 (2012)
18. Kelle, P, Silver, EA: Forecasting the returns of reusable containers. J. Oper. Manag. **8**, 17 35 (1989)
19. Goh, T, Varaprasad, N: A statistical methodology for the analysis of the life-cycle of reusable containers. IIE Trans. **18**, 42 47 (1986)
20. Toktay, B, Wein, LM, Zenios, SA: Inventory management of remanufacturable products. Manag. Sci. **46**, 1412 1426 (2000)
21. Hanafi, J, Sami, K, Hartmut, K: Generating fuzzy coloured petri net forecasting model to predict the return of products. Electronics & the Environment, Proceedings of the 2007 IEEE International Symposium on, pp. 245 250. IEEE, Orlando, FL (2007)
22. Urban, GL, Weinberg, BD, Hauser, JR: Premarket forecasting of really-new products. J. Mark. **60**(1), 47 60 (1996)
23. Mahajan, V, Muller, E, Bass, F: Diffusion of new products: empirical generalizations and managerial uses. Mark. Sci. **14**(3), 79 88 (1995)
24. Lilien, GL, Rangaswamy, A: Marketing engineering: computer-assisted marketing analysis and planning. Decision Pro. 253 261 (2004)
25. Rinne, H: The Weibull distribution: a handbook, 1st edn. Chapman and Hall/CRC, (2008)
26. Pinder III, JE, Wiener, JG, Smith, MH: The Weibull distribution: a new method of summarizing survivorship data. Ecology **59**, 175 179 (1978)
27. Nawaz, SM, Nazrul Islam, M: The Weibull distribution as a general model for forecasting technological change. Technol. Forecast Soc. Change **18.3**, 247 256 (1980)
28. Smith, SW: The scientist and engineer s guide to digital signal processing. California Technical Publishing **15**, 277 (1997)

15

A business process model of inspection in remanufacturing

Mark Errington* and Stephen J Childe

* Correspondence:
m.errington@ex.ac.uk
College of Engineering,
Mathematics and Physical Sciences,
University of Exeter, EX4 4QF,
Exeter, UK

Abstract

A crucial stage of the remanufacturing process is the inspection procedure. Surveys carried out in the automotive remanufacturing sector show that the industry is concerned about the need this causes for a large amount of specialist skills. Despite this, there has been little research into what is actually involved in the inspection process and its outcomes.

This paper presents case-based research that was carried out on the inspection procedures of both electronic and mechanical product remanufacturers. It presents generic inspection process diagrams, produced using case studies in UK companies engaged in remanufacturing activities. The models provide a greater understanding of the remanufacturing inspection procedures currently used. The models were tested with additional case studies.

The paper discusses the questions raised by the improved understanding of inspection processes in remanufacturing for operations managers and outlines some questions for future research.

Keywords: Remanufacture; Inspection; Case study

Background

Early results of this research were described in [1]. Due to global warming, the Climate Change Act [2] and the imminent peaking of world oil production (as described in [3] and [4]), closer attention is being paid to the impact of manufacturing and its energy consumption on the environment. There is now an increasing amount of legislation being developed by the European Union aimed at reducing energy use as well as our impact on the environment. As more and more countries join the European Union, more manufacturers will be required to comply with this legislation. This includes the so-called producer responsibility legislation including the WEEE directive [5], End of Life Vehicles Act [6], Packaging and Packaging Waste Directive [7] and the Batteries and Accumulators Directive [8]. This legislation requires the producers of certain products to be responsible for their disposal at end of life.

Producers are not only responsible for retrieving the used products from their end users but are also responsible for recycling them whilst meeting certain minimum recycling rates. With limited room remaining in landfill sites, material recycling is often preferable to land filling; however, most of the value added during manufacture and the energy embodied in the product is lost.

In order to take back post-consumer products, it is necessary to create a reverse supply chain. In cases when this is integrated into forward manufacture, in such a way that old products are used to produce new, at any level, this leads to a 'closed-loop' supply chain [9].

Many consumer products are still working at their end of life, and there are an increasing number of companies that collect these products and return them to market both in the UK and abroad. They provide a cheap alternative to buying new products and appear to have been very successful both in not-for-profit and for-profit organisations.

Other products reach their end of life because of an accumulation of rectifiable faults, the wearing of a small number of components within a product or sometimes because of their age or amount of use. It has been shown that there has been a long tradition of remanufacturing in the auto parts industry [10]. Manufacturers in other industries such as Hewlett Packard [11], Océ [12] and Xerox [13] have started to take advantage of this way of doing business and have started to run large remanufacturing operations. It is often stated that remanufacturing is 80% more energy efficient and 60% more cost efficient than traditional manufacturing. For this reason, a large amount of attention is being paid to these types of processes by old and emerging economies alike.

Many tools now exist which can be used to evaluate the environmental and financial performance of a given product against set criteria [14-17]. The aim of these tools is to identify areas of the design that could be changed in order to enable end of life strategies to be carried out more efficiently.

There has also been a large amount of research carried out on operations management within the remanufacturing process itself. Guide et al present an analysis of the performance of different static priority rules [18], while in an earlier paper, describe the application of Goldratt's drum-buffer-rope system in a remanufacturing environment [19,20]. Other authors look at the use of production planning approaches in remanufacturing environments [21-31].

Reverse logistics channels themselves have also attracted a large amount of interest. Jayaraman et al. present a deterministic integer programming model which can be used to determine the optimum location for remanufacturing facilities and collection centres [9]. A similar optimisation study has been carried out by Peng ZY and Zhong [32]. Blackburn et al. discuss the need for different reverse supply chains for different models [33]. They find that depreciation rates among high-value items mean that responsive supply chains are more appropriate in some cases.

There are also some studies that look at strategies for increasing the number of cores (end-of-life items) available for remanufacture. Ray et al. present a method for determining the correct buy-back pricing strategy for used products from consumers [34]. Xiaochen et al. present a method for calculating the optimum buy-back price under certain conditions [35].

Inspection and disposition decisions

'Inspection and disposition' is one of the five main parts of a remanufacturing business [36]. Despite this, research has tended to focus largely around product design for remanufacture rather than upon the operational questions related to how a specific product is actually inspected and remanufactured. Steinhilper [37] identifies that:

> "(A) step of great importance in remanufacturing is to assess the condition of the disassembled and cleaned parts as to their reusability or reconditionability."

He goes on to state that this is done in two parts: to define the objective criteria and accepted condition characteristics and to determine how this will be assessed.

Guide and Van Wassenhove [36] describe the inspection and disposition process as follows:

> "The testing, sorting and grading of returned products are labour-intensive and time-consuming tasks, but the process can be streamlined if a company subjects the returns to quality standards and uses sensors, bar codes and other technologies to automate tracking and testing."

There are many signs that the technology that is referred to in this text is beginning to be more widely used. Bosch have carried out tests with micro sensors to record data during the life of its power tool motors [38]. Their plan was to connect the tools to a data reader once they have been returned to them in order to decide whether the motors can be reused or not. At the time, similar technology was being developed as part of the Care Vision 2000 initiative to produce what they call a 'Green Port' [39].

Technology to make the inspection and disposition decision more streamlined is clearly useful in some cases. However, it is not applicable or relevant for many remanufacturing operations. These techniques are only useful to those companies who originally manufactured the product or have control over its design. Companies wishing to use such technology would have to invest with a view for a very long-term payback.

Inspection and disposition is carried out in all remanufacturing operations; however, little has been done to document current industry practices, to identify areas where savings could be made and to assess exactly which specific inspection tasks these technologies should be substituted for.

In contrast to new-product manufacture, where sampling methods are often used, remanufacturing always requires 100% inspection [40]. This is typically carried out on cores and disassembled parts as well as the finished products themselves. This is done to increase the second user's confidence in remanufactured products, and it is thought to explain why remanufactured products appear to have a better reliability than new products [40].

A comprehensive survey of manufacturers in the automotive industry was carried out in order to identify the problems the industry faces from the perspective of the remanufacturers themselves [10]. Twenty-nine percent of respondents stated that the factor which made inspection most difficult was the knowledge of the employee carrying out the work. A further 21% of respondents stated that identifying defects in cores was their main concern.

This suggests that there is a need for formal methods that can be used to simplify the processes involved in the inspection procedure. This would remove the dependence on such a high level of tacit knowledge held by the operators. As noted above, some attention has been focussed on the aspects of design that can facilitate remanufacture. In the survey, the authors note that design for remanufacturing is an issue with automotive remanufacturing [10]. 'Profit potential' and 'investment needed to repair' rank high as responses to how the decision to remanufacture a given product is made.

Research objectives

Researchers have established that inspection is one of the key stages in remanufacturing. The main aim of the inspection process is to determine the condition of the returned item (called the 'return' or the 'core') and select the most economically attractive re-use option [11]. Little has been done to look at the tasks involved in this process and how they can be carried out more efficiently. This paper describes work done to address this issue through case study research. It aims to establish how these processes are currently carried out as a basis for understanding and to provide a basis for analysis of potential improvements. Using the four purposes of research described by Voss et al. [41], the work is theory building since it attempts to develop new models or concepts that can help researchers and managers understand and deal with inspection. Using a broad range of cases involving different products, industries and business models but all centred on remanufacturing, the research was able to take a broad, general understanding of the issue which can be further tested and refined by future work.

The main objective of these case studies was to gain a detailed insight into the role and process of inspection operations carried out by remanufacturing firms. Despite the fact that research in this area is very immature, it is suspected that there is a huge quantity of technical expertise within the companies that in some cases have been carrying out these activities profitably for many years.

The work aimed to establish the objectives of inspection and testing at all stages of the remanufacturing process and to investigate the possibilities of gaining the same process outputs through different means.

Methodology

Case research is a very useful technique for carrying out work in immature research areas [42]. It has also been established that it can be used effectively where there is a large amount of expertise in the field that needs to be captured and formalised. Eisenhardt [43] shows how theory can be generated using case research and how it is a particularly useful method in new topic areas. Scientific rigour was ensured through following the framework for case research described by Voss et al. [41].

Discussion points were drawn from the literature and further developed using the findings from these cases. These included boundary questions to understand the characteristics of the case companies that were being studied. These included asking if the company was the original equipment manufacturers (OEM) of the products which it remanufactures and how diverse the mix of products remanufactured by each company was. The interviewees were asked to describe a remanufacturing process within their organisation and the role of inspection within it. For a full list of the questions used please see http://hdl.handle.net/10036/83380.

Cases were selected so that a broad range of industries were represented. These include OEMs and non-OEMs as well as for-profit and not-for-profit companies. Case studies were carried out with a number of companies engaged in remanufacturing in the UK. These operate different business models for a wide range of customers. Some carry out remanufacturing on a not-for-profit basis, and others remanufacture directly for OEMs on a large scale. The table gives some of the characteristics of the different companies that were included in the research.

It can be seen from Table 1 that the companies studied were from a wide range of industries with different products and different business objectives. The only not-for-profit organisation, CompCo, operates to give people in developing countries access to IT equipment. This adds an interesting characteristic since many of their staff work as volunteers, although this does not mean that labour is a cheap resource that may be used inefficiently. The remainder of the companies operate in a more conventional for-profit way. The prominence of automotive and defence sector remanufacturers in the study reflects the maturity of remanufacturing in these areas.

'Product type studied' describes the remanufactured component, module or product that was studied for this research. This is not necessarily the main product the company produces. Some of the cases remanufacture a large volume of products and operate in a highly competitive market while others remanufacture smaller numbers of items and are known for their expertise and competency.

Many of the cases studied were found not to describe their process as remanufacturing. For the purposes of this research, the process was considered to be remanufacturing if it met the definition developed by Ijomah and Childe [44]. This is a definition that is widely accepted in the UK remanufacturing industry. It defines a remanufacturing process as follows:

> "Remanufacturing is the only process where used products are brought at least to OEM performance specification from the customer's perspective and, at the same time, are given warranties that are equal to those of equivalent new products."

Companies were approached through the UK government-funded Centre for Remanufacturing and Reuse and through university contacts. The authors note that other definitions for remanufacturing exist, and a full discussion can be found in the

Table 1 Overview of case study companies

Company identifier	Customer type	Sector	Product type studied	OEM?	For profit?
DefCo	UK Ministry of Defence	Defence equipment	Mechanical/ electronic	OEM approved	For profit
CopyCo	Business	Copying/Printing equipment	Mechanical	OEM	For profit
PCCo	Business	IT equipment	Electronic	Non-OEM	For profit
GearCo	OEM	Automotive	Mechanical	OEM Approved	For profit
ClutchCo	Business	Automotive	Mechanical	Non-OEM	For profit
CoreCo	Remanufacturers	Automotive core broker	Mechanical/ Electronic	Non-OEM	For profit
TurbineCo	Business	Power generation	Mechanical	OEM	For profit
CompCo	NGOs	IT equipment	Electronic	Non-OEM	Not for profit
MilCo	**UK Ministry of Defence**	**Defence equipment**	**Mechanical**	**OEM**	**For profit**

PhD thesis Business Processes and Strategic Framework for Inspection in Remanufacturing which is available to download at: http://hdl.handle.net/10036/83380.

The main difference between this definition and others is its inclusion of the word 'warranty'. The authors of this definition felt this was vital as anyone claiming that their remanufactured product was as good as or better than the new but only willing to offer a shorter warranty may not be particularly confident in their performance claim.

Company visits were carried out in mid-2006, these included a meeting with the operations managers as well as the production staff followed by a tour of the facilities being used for remanufacturing. The purpose of these visits was to gain an overall understanding of the inspection processes carried out in each of the case companies.

Complete notes were compiled directly following each visit. These consisted of background information about the company and a description of the remanufacturing and inspection processes. These included flow charts of the processes that were studied. These notes were emailed directly to those who were interviewed, for checking. The purpose of this was to ensure their internal validity and correctness. Results from case studies were triangulated through visits to the factory floor and further interviews with the production staff. After each visit, the case notes were sent to the interviewees promptly to verify their accuracy.

An analysis of the data collected through the case study meetings is presented in the following section.

Results and analysis

In this section, we present four of the cases used in the research. Two are from the automotive industry, and two are from the computer industry. The cases show that there are similarities and differences within each industry and between the industries. It will be shown later how a general model can be structured to cover all the options.

GearCo

GearCo is a remanufacturer of automotive drive trains. They remanufacture engines, transmissions and gearboxes as an OEM approved remanufacturer.

Used items are stored when received, until the arrival of an order for that type of item. When an order arrives, an initial visual inspection of the core is made for obvious damage which would make the core uneconomical to repair, for example a cracked casing. The reason for failure of a returned core is not always obvious, and is not taken into consideration. The core is disassembled and cleaned using a three-stage cleaning process. A glass bead blast is used as well as high- and low-pressure chemical cleaning.

Some parts are always replaced, such as bearings or other wear items. These are removed and scrapped. Other parts are inspected to assess their potential for re-use. Experienced inspectors assess parts against criteria set in conjunction with the OEMs. This decision is made by taking measurements of the item and assessing the amount of wear it has experienced. This can be used to estimate the amount of life remaining in the part. For these products, there are no parts at all that are never replaced.

Some of the smaller parts are removed from the core and sent through a separate cleaning and inspection procedure. The main parts are placed in a tray together. An inspector uses a computer-based system to establish the part numbers of the replacement parts that will be required for the unit and uses the system to generate a stores order for them. It is at this point the failed parts are removed and scrapped.

All the replacement parts that are used are supplied by the OEMs or OEM-approved suppliers. Parts are never repaired. It would be possible for non-OEM parts to be used, but then the remanufactured gearbox would not be allowed to carry a full OEM warranty.

Once assembled, the unit is put through a testing procedure developed in conjunction with the OEM. The gearboxes are run both with and without load and analysed for noise and operation. Acceptable limits are set in conjunction with the OEM. One hundred percent of the units produced are inspected in this way.

Those units that fail this procedure are sent for rework by the operative who was responsible for assembling the unit.

ClutchCo

ClutchCo supplies 1,300 types of clutches used in cars in the UK and Europe. Half of these models, by part number, come from remanufacturing, and half are new imports, coming from China. The remanufactured clutches account for half the product lines but only 20% of sales volume. This strategy of using remanufacturing to supply lower volume items allows the company to hold less stock of slower moving parts at the same time as allowing it to have full market coverage. Thus, the imported clutches account for 80% of the total clutch sales, and these are cheaper. ClutchCo accepts that its remanufacturing area may now be running at a loss; however, it estimates this cost to be less than the cost of keeping the large stocks of slow moving parts that would be necessary if it was to import them from China.

As an independent remanufacturer, ClutchCo has had little cooperation from the OEMs of the products it remanufactures. This is despite the apparent legal obligation of the OEMs to cooperate. (The End of Life Vehicles Directive requires all information required for the correct and environmentally sound treatment of end-of life vehicles to be made available to authorised treatment facilities by vehicle manufacturers and component producers [6]. In order to establish their processes for remanufacturing, ClutchCo purchases OEM clutches and reverse-engineers them. They have found this to be a more effective method than obtaining information from the OEMs directly. None of the clutches that are remanufactured by ClutchCo are supplied to OEM manufacturers.

After the cores arrive at the plant they are sorted. They are identified by part type, against approved samples, and are inspected visually for any obvious damage that would make them unusable. The main fault inspectors are looking for is excessive wear to clutch diaphragms. Excessive wear would mean that the reassembled clutch would be unlikely to last a full second life. After inspection, the cores are formed into production batches and are stored in the warehouse.

When required, cores are collected from the warehouse and washed using a caustic solution. Sometimes the clutch diaphragm is tested before disassembly. The cover of the clutch is removed to facilitate access and all parts are cleaned in order to make them appear as new. Inspection of the clutch's component parts is carried out. Parts are worked upon or scrapped as necessary.

The clutch is then reassembled and tested for clamp load and clearance. Clutches that fail this test are reworked. Finally each clutch is ink-jet sprayed with its part number and batch code.

PCCo

The mission of PCCo is to extend the life of office equipment. They offer a wide range of services ranging from collection and auditing of unwanted equipment to the supply and installation of refurbished equipment. The majority of the work undertaken is to dispose of used IT equipment. Where possible, they do this through remanufacture and resale of the equipment.

Items that are deemed to be unmarketable are separated from rest of the load of incoming equipment and are sent directly to a recycler. At the time of study, this included 15′′ CRT monitors.

The remaining items are sent to the remanufacturing facility. On arrival, a batch number is allocated to the equipment, and a unique tracking number is given to each piece of equipment. Information regarding the type of equipment, manufacturer and model number is stored against the tracking number at this stage.

The batch is then sent to the workshop where it is processed. Equipment is connected to a network which automatically audits and tests it. The network stores data about the equipment, carries out data erasure in accordance with the UK government standards and creates a record of the wipe. Software is used to carry out a fault diagnosis of the equipment.

The majority of equipment is found to be working. The exact amount varies depending on its source, but the average is similar to the 80% figure quoted for Germany by Steinhilper [37]. A visual inspection is made, and the equipment is cleaned as necessary.

Information collected during the diagnostic 'bench' test is used in order to determine the potential for remanufacturing each item of the equipment. This is restricted by value, functional state and age. A triage principle is used to sort the equipment into three streams. These are 're-use', where the computer is reused without further treatment, 'recycle', where the computer is sent for materials recovery and 'process further', where the faults with the computer are not known but it may be possible to reuse after further operations. Diagnostic information, collected during the bench test, is then used to estimate the parts and labour cost of processing each item of equipment according to its individual requirements.

If the decision is made to process an item further, then the item is disassembled and an investigation is made into the cause of the equipment failure. Once the cause is identified, this module or part is replaced with a working one, and the equipment is reassembled An electrical safety test is carried out then each computer is packaged for sale and shipping.

CompCo

CompCo remanufactures between 1,400 and 1,700 computers per month. The vast majority of these are desktop computers, but they also process a smaller number of laptop computers. CompCo works on a not-for-profit basis and provides computers to educational establishments and NGOs in various locations throughout the developing world. The standard processing lead time is 2 weeks, and stock does not spend much time in the warehouse.

Computers are sorted at source as much as possible. Only high specification, working machines are accepted. Computers that are found not to work or to be the wrong specification are recycled with the cost, plus a premium charge for disposal, borne by the donor. This is meant to incentivise donors to supply the required machines, rather than use CompCo for general disposal.

Computers arrive at the warehouse in an unknown state. Information on the condition of the computers is sometimes given by the donor, but it usually turns out to be inaccurate. For this reason, the decision making about the suitability of a machine for remanufacture is made entirely on the basis of a visual inspection, the model of the computer and an estimation of its age. This is used to estimate the length of usable lifetime a computer will have once it is fully working. If it is decided that a computer is not suitable for remanufacturing it is sent for part recovery and/or materials recycling.

During the next stage of the process, the data wiping, the machine specifications are found using a diagnostic software. Based on this, it is decided if the computer is suitable for reuse as it is or if it requires upgrading or part replacement. If the equipment cannot be used, then it is sometimes cannibalised for spare parts.

No standard part replacements are made but computers are cleaned and tested for safety before they are shipped.

Generic inspection procedures in remanufacturing processes

It was found that the processes undertaken by the remaining cases were very similar to those described above. Electronic product inspection and remanufacture was carried out using the same steps at CompCo and PCCo whereas mechanical product inspection was carried out using the same steps at GearCo and ClutchCo. All of the processes were found to have certain stages in common.

The flow charts that were generated during each of the case studies were compared, and common stages of the process were extracted. Focussing particularly upon inspection processes, a name was selected for each stage of inspection that was common across the cases. This led to the creation of a model that could be described as a 'common ground' or 'consensus' model which applies across the cases. This could form the basis of a generic inspection procedure.

The model was shown graphically using the IDEF0 modelling standard. The IDEF0 standard was used in preference to other modelling techniques such as flow charts because it allows decomposition, is easy to understand due to the simple graphics that are used, is precise and can be used with data abstraction. This is a particularly useful feature in the creation of a generic process diagram such as the one described in this paper. Mechanism arrows were not included in the diagrams as these are often specific to each company. A company might choose to use a machine to carry out a given inspection whereas another might carry out the process using skilled labour. It is the activity or job that must be carried out that was of interest for this research rather than the exact way it was carried out in a specific company. For a full discussion of the benefits of the IDEF0 standard in process modelling, see [45] or [46].

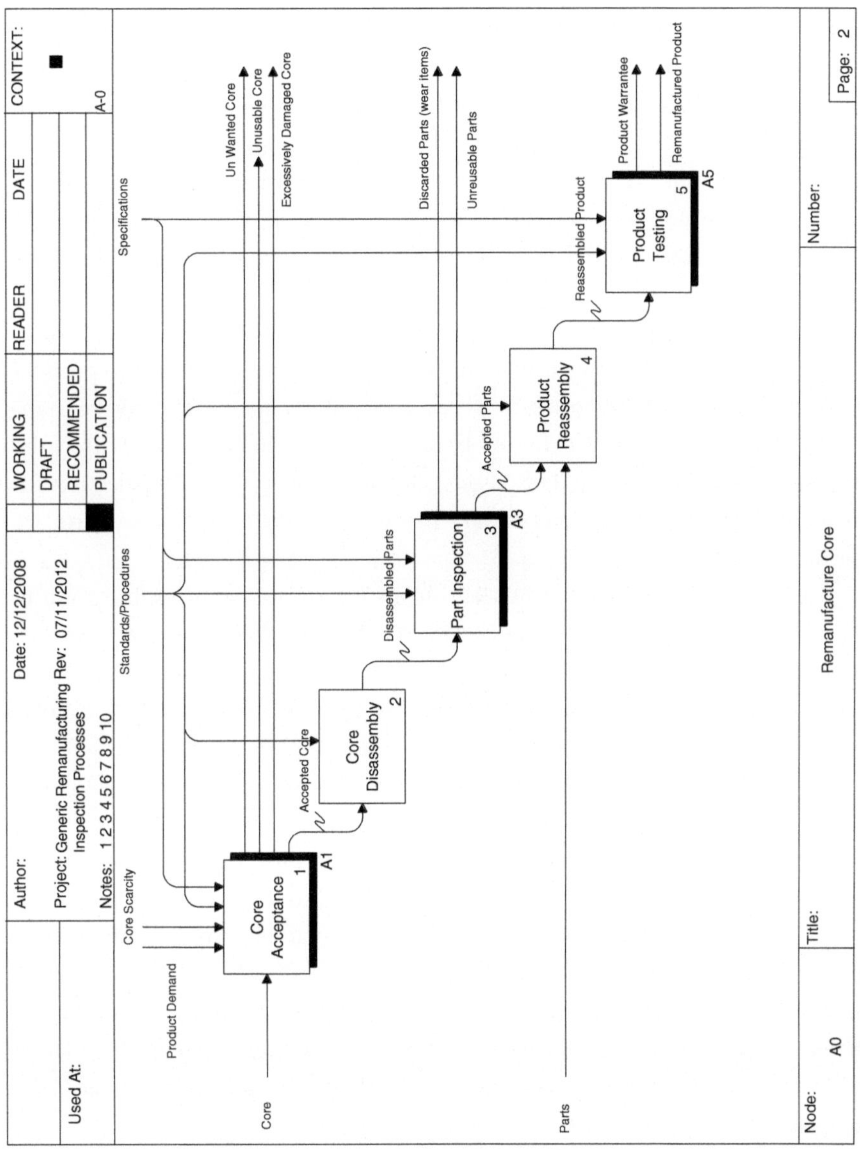

Figure 1 IDEF0 diagram of generic remanufacturing process.

Figure 1 shows the generic remanufacturing process that was observed in the cases.

It can be seen that three key areas of inspection were found. Each was carried out at a different stage in the remanufacturing processes. These were identified as follows:

- core acceptance;
- part inspection; and
- final product testing.

The different stages of inspection have different objectives. Due to these different objectives, the physical inspection processes themselves are different. The objective of *core* inspection and testing is to remove cores that will be uneconomical or impossible to remanufacture, accepting only those thought to be viable. This improves the reliability of the population of items that are produced and ensures that cores that are uneconomical to remanufacture do not enter the process. This process is often carried out by a triage-type process. This was referred to as 'scratch and sniff' by an expert in the automotive remanufacturing industry.

The second stage of inspection is carried out once the core has been disassembled. *Part* inspection and testing aims to remove non-reusable components from the product in order to increase its reliability once it is reassembled. These parts may be non-reusable because they have already failed or because they are likely to fail within the next working lifetime of the product. This is typically done using a visual inspection for wear, but in other processes, measurements are made using lasers, and some parts are tested for performance.

The *final product* inspection stage is carried out to ensure that the products are in full working order before they are shipped. Products that fail this stage are reworked before being retested and sold. This process is often similar to typical final product testing in traditional forward manufacturing. Products are operated and their performance assessed to ensure they are within acceptable limits.

All the inspection and testing procedures are carried out in order to increase confidence in the ability of a product to perform throughout its second life. Without these procedures, it would be difficult to ensure the quality of the products produced and to offer full 'as new' guarantees as required for remanufactured products. This could potentially lead to a lack of consumer confidence in remanufactured products and excessive guarantee costs to the remanufacturer.

The effect of the three stages of inspection was to remove from the process those items that were of less value either because of requiring more work or being more likely to fail in service. In some cases, these items were not discarded but were disassembled in order to reuse good components.

Core acceptance

Cores are usually inspected as soon as they arrive at the remanufacturing facility or at the first stage in the process once items are taken from storage for use. This is the process which Loomba proposes should be used more to reduce remanufacturing costs [47]. Figure 2 shows the details of how this inspection is carried out.

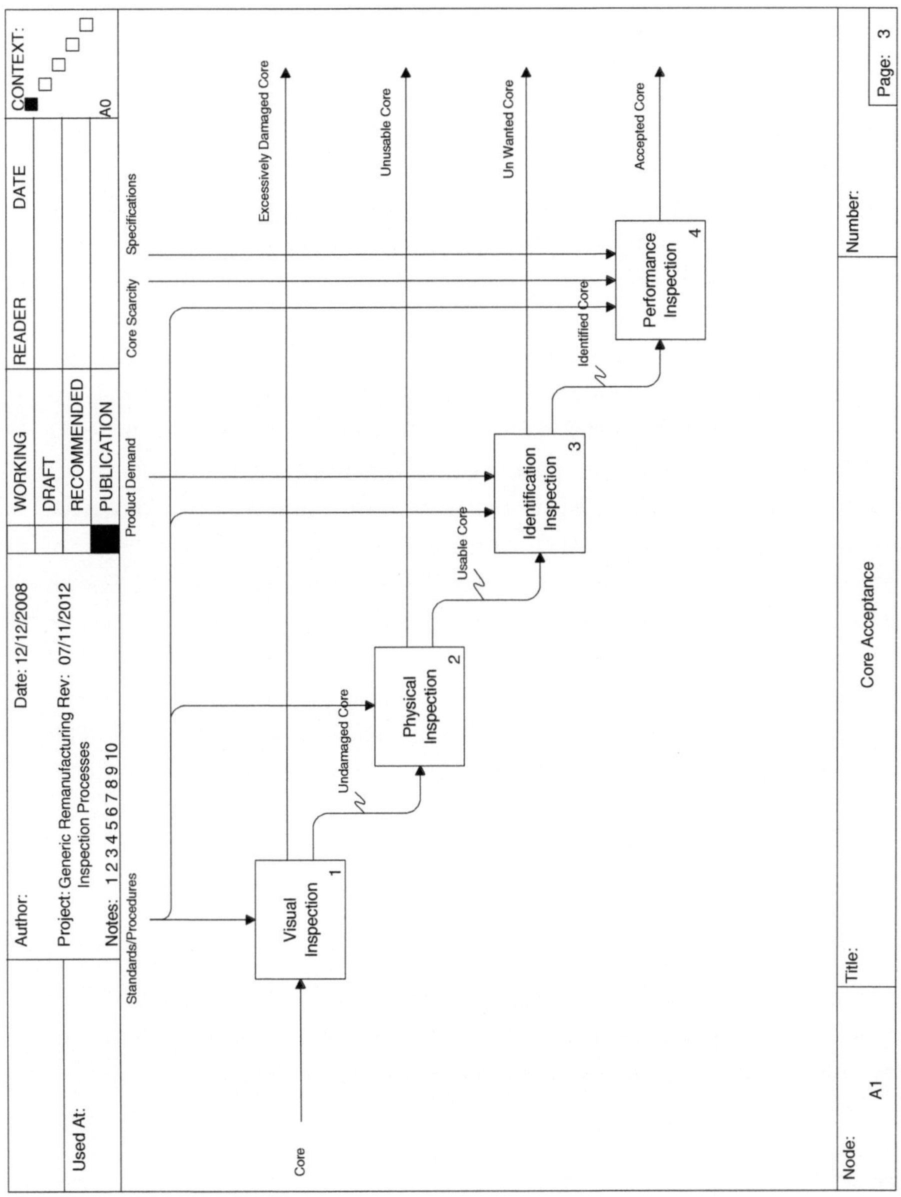

Figure 2 IDEF0 diagram of generic core inspection procedure.

The figure shows the four main tasks that were identified in the core acceptance procedure. The first is a visual inspection. This is carried out simply through looking at a core and deciding whether it has obvious major damage which would make it unusable. An example of products that would be rejected at this stage would be products that have been crushed during transit to the remanufacturing facility. All the cases used visual inspection with the aim of removing excessively damaged items from the process. However, the case of CompCo suggests visual inspection could be entrusted to suppliers given certain conditions.

Cores that have no obvious signs of major damage are subjected to the second test in the procedure. This has been described in Figure 2 as the physical inspection. Again this is done manually, but the actual process varies depending on the core being inspected. For automotive components, two main methods are commonly used. The first is to attempt to rotate any part of the product that should normally move and the second to smell electrical components to test for burn out. Cores that fail this test are sent for recycling and/or disposal.

The third stage of the process is to identify the part type and part number of the core. This is used to estimate the demand for and value of the product after it has been remanufactured. Cores for which there is no demand are disposed of at this stage. It may seem counter-intuitive that the identity of the part is only investigated at this stage of the process. One might assume that it would be sensible to identify a component before any analysis is made of its potential for remanufacture. The reason that was given for the identification belonging to this stage of the process was the relative difficulty of the first three stages of inspection. To carry out a visual and physical inspection, known by CoreCo as 'Scratch and sniff' is a very quick process whereas identifying a core through the use of part numbers and expertise can be relatively time consuming. This was stated by CoreCo as being particularly the case for steering columns which do not typically carry part numbers and must be identified by highly trained operatives. This identification operation allows inventory to be managed for future use.

The fourth and final task within the core acceptance procedure aims to assess the performance of the core. This is usually done using test rigs specifically designed for the procedure. The standards the product must meet are set by OEMs, industry bodies, international standards or the remanufacturing firm itself. The aim of this process is twofold: firstly, to establish whether the core is economical to process and secondly to establish whether the finished product is likely to conform to specifications once it has been remanufactured. This is another opportunity to remove unusable items from the process, occurring once simpler checks have been carried out.

Part inspection

Once cores have passed the core inspection and testing procedure, they are sent for disassembly and further inspection. After disassembly and cleaning, they are passed to the second set of inspections and tests. The diagram for this process is shown in Figure 3.

It can be seen from the diagram that some of the tasks carried out on the parts are similar to those carried out on the core. Firstly each part is identified. If, according to company procedures, it is a part that is always replaced, then it is discarded

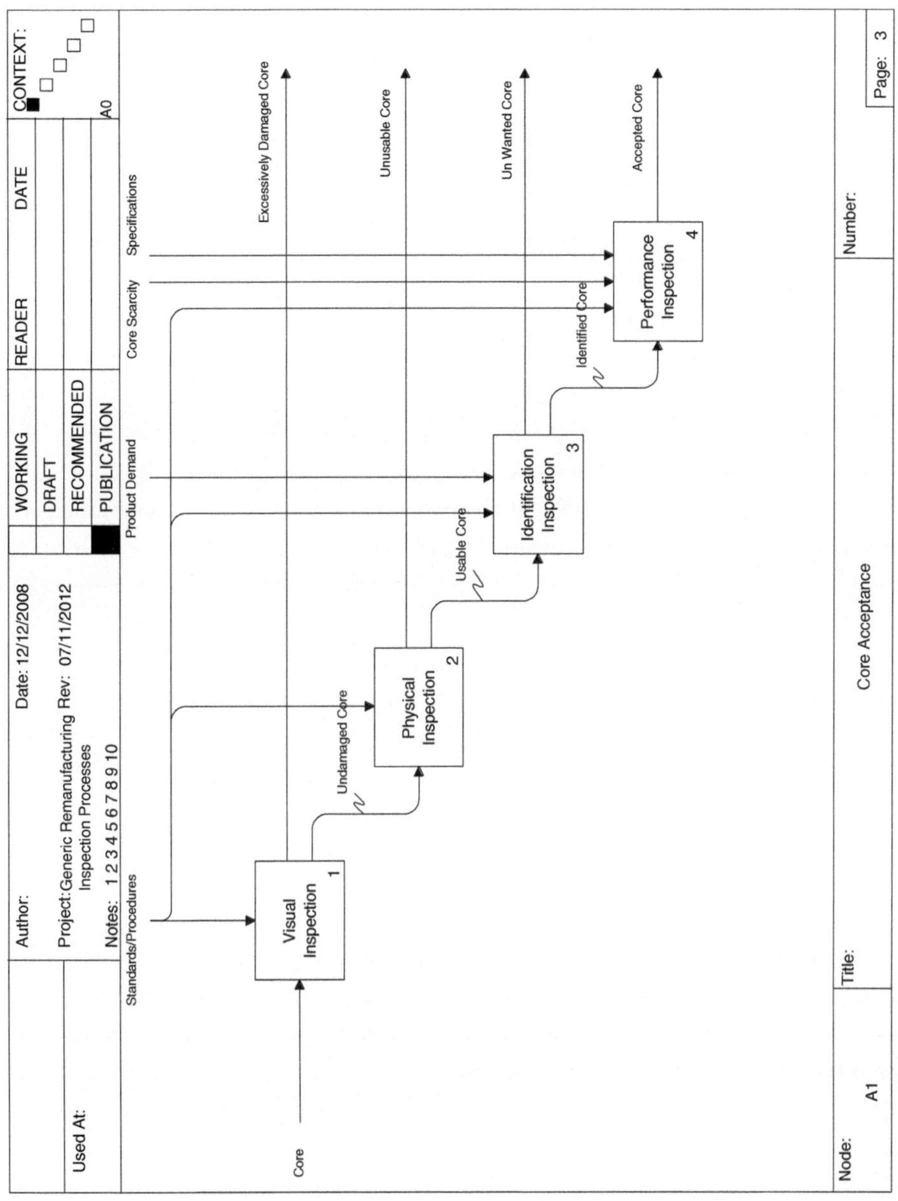

Figure 3 IDEF0 diagram of generic part inspection procedure.

immediately. These are usually wear components, such as bearings and bushes, and are often the reason for previous user discarding the product. When parts may be replaced, identification is clearly necessary.

Parts that could be reused are sent through a functional inspection process. Typical activities in this process include measurement and leakage testing. If they are found not to conform to required specifications, whether for functional or safety reasons, then they are either discarded or reconditioned. It was found that this decision was mainly made on the basis of cost and lead time.

During the case study interviews, it was stated that it is often quicker to recondition a part in house than it is to order a new one to be made. A similar comment is made by Depuy et al. [48]. This is especially the case when small numbers of parts which are not mass produced are concerned. In practice, this seems to make the reconditioning of parts a part of the inspection process, wherein the components are inspected until they are accepted or failed. This contrasts to manufacturing, where parts are normally processed until they are finished, with inspection as a final operation, and makes the remanufacturing process more similar to craft work or 'fitting'. After reconditioning, parts are inspected once more for performance in order to ensure they meet requirements.

It was observed at CompCo that no physical disassembly was ever performed, so physical parts were never assessed or replaced. In this case, the disassembly consisted only of removing data and reassembly only or installing of new software. Nevertheless, this case was regarded as remanufacturing (rather than re-use) since the confidence gained from the inspections and tests allowed the company to offer OEM level specification and warranty.

Once the quality of parts has been accepted, additional parts, either new or reconditioned from other sources, are used to replace the ones discarded in the previous inspection stage. The product is then reassembled. Once it is reassembled, it is passed to the third and final stage of the inspection procedure, the product testing stage.

Product testing

It can be seen that functional inspection is the key part of this inspection and testing process. This is the final stage at which the reliability of the final product can be estimated and/or ensured. The diagram for this process is shown in Figure 4. The process carried out and the standards applied during the functional inspection are different for each product. In the case of OEM-approved remanufacturers, these are developed with involvement of the OEM for the product concerned. This ensures satisfactory operation.

Products that fail the functional inspection are reworked and retested before they are sold. It would seem logical that products would be sent back to the disassembly process to be disassembled (where this involves a dedicated area or workplace). However, in practice, it was found to be more common for the product to be sent back to the person who reassembled it. GearCo stated that this was done because the reassembly operative would be likely to know what the cause of the fault would be. Thus, the corrective actions are again more closely linked to the inspections than in manufacturing.

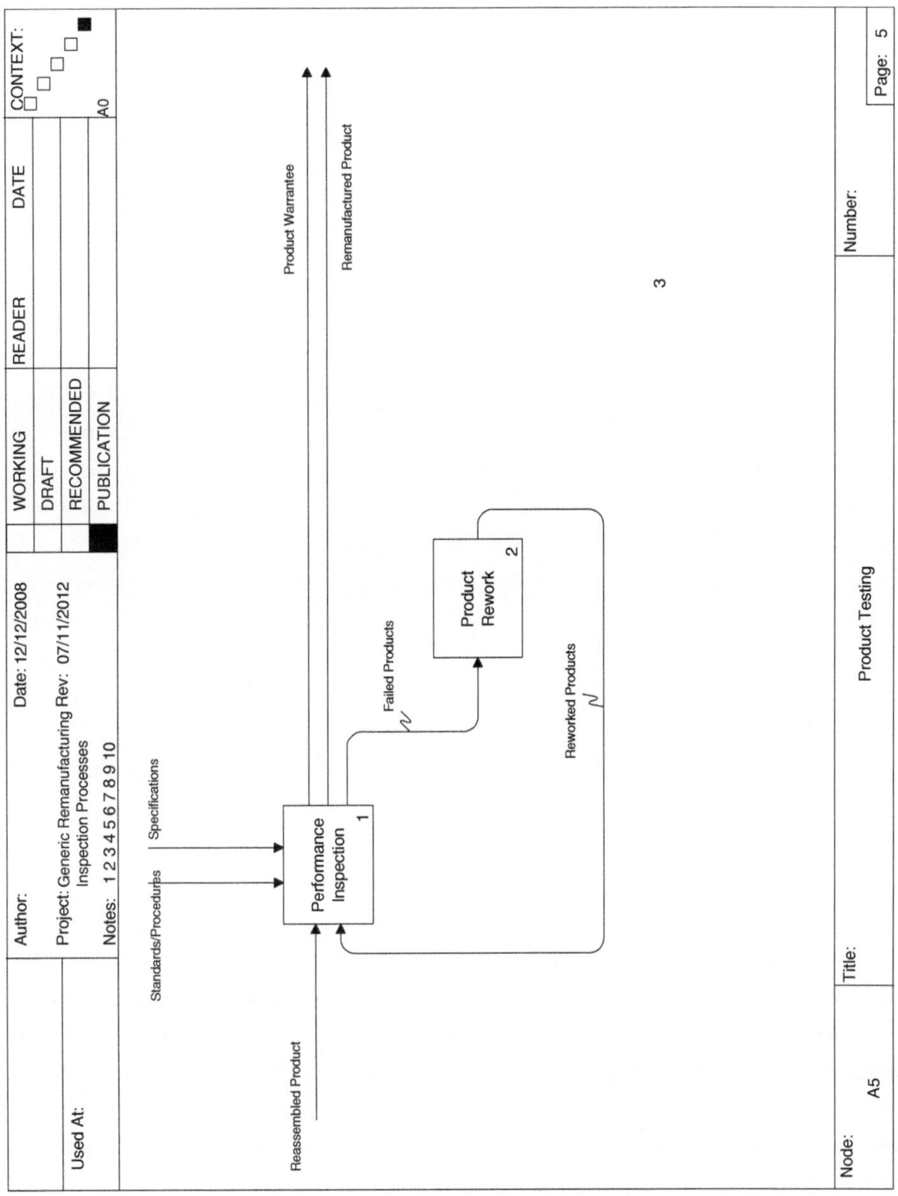

Figure 4 IDEF0 diagram of generic final product inspection procedure.

Discussion

There appear to be two motives for inspection: (1) to remove unreliable individuals from the population and (2) to show that enough work has been done to achieve the required level of performance.

By removing unreliable individuals, the confidence in the remanufactured products is raised. However, this does not mean that all products require all the possible inspection stages. The processing that units go through will be identical for all the units that are not obviously damaged or in a doubtful condition. It is simply not necessary to know the details of the state of a unit. By performing the simplest and cheapest inspection operations first, the parts that are removed by these operations do not undergo the more expensive or time-consuming activities. This kind of inspection could be termed inbound inspection and consists of identification and some kind of simple assessment of condition. Inbound inspection is not concerned with the units as individuals but as feedstock. This applies at the core level and at the part level.

By showing that enough work has been done, the second type of inspection becomes part of a cycle where products go through a loop of testing, re-work, re-assembly, testing until the performance is found to be adequate. At first glance, this looks like trying to 'inspect quality into a product' which Deming was against (Dodge quoted in Deming 1982 [49]). However, Deming was looking at the whole system of manufacturing, in which quality depends on product design and production system design. In this instance, it seems better to think of this cycle of inspection as adjustment or tuning of the product. For example, balancing is a normal requirement in the manufacture of rotating machines, and the activity of balancing continues until acceptable balance is achieved. The balancing cannot improve the quality of the components, but it provides correct performance of the product. This could be termed outbound inspection and is concerned with the units as individuals, providing specific attention to the condition of each one. This applies at the part level and at the final product.

It was found that, in most cases, inspection was carried out three times during the remanufacturing process. Although Brent and Steinhilper state that it is always necessary to carry out 100% inspection during remanufacturing processes [40], these cases show that this does not mean all inspection operations are done on all items. The key thing to consider is if the inspection will actually lead to any decisions being made. It is possible that without disassembling a core, sufficient information cannot be gathered to make a confident decision about the functionality of the part. In other cases, due to a shortage of cores, all items have to be disassembled anyway so there is no need to carry out the first inspection.

Some authors propose that information from green port systems or from the user directly can be used in place of some of the stages of inspection. These are particularly applicable to the core acceptance procedure but rely on accurate and reliable information being available. One way to ensure it is available is to lease rather than sell equipment to the customer. It has been widely reported that this has been a particularly successful method for Xerox Corporation.

In the case studies, no major differences between the way OEM and non-OEM remanufactures operated their inspection procedures were observed. OEM remanufactures are thought of as having fewer problems sourcing cores and information with regards to the design of the products. There is no doubt that these make the remanufacturing business easier to run; however, no impact was observed on the inspection procedures of the companies that were studied.

This section of the paper has presented a generic inspection procedure developed from case studies carried out with companies carrying out remanufacturing operations.

The following section will give details of how the models of the procedure that have been presented in this section were validated in order to gain confidence in their generalisability.

Model validation

Validation of the model was carried out using a group of six companies contacted through a consultancy firm specialising in remanufacturing. The consultancy firm in question is also responsible for running the UK government-funded Centre for Remanufacturing and Reuse. A visit was made to each of these remanufacturers during the late part of 2008, and a presentation was given describing the model, its origin and its potential uses. The practitioners were then asked to fill in a short questionnaire. Responses were recorded on paper. Interviewees were invited to comment on how they thought the models should be altered, where applicable, to show their processes more correctly.

Table 2 shows the companies that were studied for the validation stage of this research.

It can be seen that the cases were from a variety of industries which remanufacture a large variety of products. WasteCo is relatively new to remanufacturing, and at the time of writing, its processes were undergoing extensive development.

Interviewees were asked to respond to the statement *"The processes that have been described in the presentation closely match what is done in our process."* Each interviewee was asked to place their opinion on a six-point Likert scale. A score of 1 indicated that the interviewee strongly disagreed with the statement, and a score of 6 indicated that the interviewee strongly agreed. There was no neutral point. Table 3 shows the responses of the interviewees for each of the cases.

It can be seen that all the respondents agreed that the model closely matched what was done in their processes: 1 case strongly agreed, 5 agreed and 1 case slightly agreed. Overall, there was agreement among validation case interviewees to the view that the models matched what was done in their respective processes.

Conclusions

This work has identified the inspection operations used in nine case studies. The activities were grouped into a generic or common-ground model structured into three

Table 2 Case studies used for validation

Company identifier	Product	Electronic/Mechanical
CompressorCo	Compressors	Mechanical
VacCo	High vacuum pumps	Both
WasteCo	Televisions (also processes all types of WEEE)	Electronic
FlightCo	Aerospace flight control units	Mechanical
TonerCo	Toner cartridge	Mechanical
AltCo	Rotating automotive electrics	Mechanical

Table 3 Validation results

Company identifier	Role of interviewee	Remanufacturing volume units/ month	Close match to process (1 to 6 Likert)
VacCo	Global support manager remanufacturing	4,300	6 (Strongly agree)
WasteCo	Chairman	4,000	5 (Agree)
CompressorCo	**Marketing admin**	**200**	**5 (Agree)**
FlightCo	**Quality manager**	**2 to 3**	**4 (Slightly agree)**
TonerCo	**Production manager**	**700**	**5 (Agree)**
AltCo	**Managing director**	**10,000**	**5 (Agree)**

main stages: core acceptance, part inspection and final product testing. The model was subsequently assessed for its general applicability by representatives of a further six remanufacturing companies. The key contribution of this paper is a deeper understanding of the role inspection plays within remanufacturing processes. It can be used by practitioners to design a plan for remanufacturing their products, and it can be used by researchers as a framework for further research. It shows clearly where inspection technology may benefit the industry and how.

Another key finding was that in contrast to forward or traditional manufacturing, sample inspection is not usually used in remanufacturing processes. The state of the cores entering the system varies widely, each therefore needs to be inspected at the core level, part level or both to ensure that the components that are remanufacture are of suitable quality. One hundred percent inspection is potentially not necessary at the final product stage; however, this is almost always done. A possible reason for this is to increase confidence in remanufactured products.

The cases provide some insight into the use of inspection. Core acceptance procedures aim to remove unusable cores from the process, and the simplest inspections are used first so that most cores are rejected using the cheapest inspection procedure. Identification inspection is not generally the first activity in this stage since it can be time consuming. Functional inspection is generally left until last as it would be uneconomical to carry out this test on cores that could be rejected by cheaper tests. Core acceptance ensures, as far as possible, that only useful items proceed to the next stage, thus reducing the work required at subsequent stages, and removing problem items from the process, thereby improving the reliability of the population of items.

Part inspection is used to identify those parts that need replacing and to assess and recondition others. At this stage, there is a transition from accepting/rejecting items entering the process to specific assessment of individual items and the work required by each.

The final product inspection stage also sees products as individuals, the work done being tailored to achieving the correct performance from each unit. The activities of re-work and adjustment are only carried out on those units made from good cores, with new or reconditioned parts that have been re-assembled and from which fully acceptable performance can be expected.

The model that has been presented, although based on a limited set of cases, provides a structure that can be used to plan future remanufacturing operations. Some questions for the management of remanufacturing operations could be derived as follows: as operations managers might find these hard to answer, they may also be thought of as research questions.

Core acceptance

- What are the criteria for accepting a unit as a useable core (rather than as a source of parts or materials)?
- Which of these criteria can be most readily identified by the simplest inspection techniques? These should be carried out before the more complex, expensive or resource-intensive inspections.
- What are the relative advantages of using an item for remanufacture compared with the value of the parts it contains?

Part inspection

- Is dismantling necessary to confirm the acceptable condition of parts? Electronic components (such as in the IT cases) could be assessed by functional tests of the final product.
- What is the trade-off between the cost and energy saving of re-using a part compared to its expected reliability? If an item is part worn, its lifetime (and the lifetime of the product) can be assessed and related to the cost of a new part. Safety- and performance-critical parts may be replaced as a matter of policy without the need for inspection.
- What are the relative benefits and costs of re-conditioning parts in-house, subcontracting the reconditioning or making or purchasing new items?
- What are the inventory ramifications of storing large numbers of used parts?

Final product testing

- What degree of inspection and testing is required considering a product has been made up from known good parts?
- What is the relationship between assembly and adjustment for a particular product family? Should the adjustment form part of the same operation?

It is to be hoped that the models presented can be developed both by practitioners designing and managing remanufacturing operations and by researchers making further investigations in the area. Some research questions arising from this work are the following:

- Does the framework of core acceptance, part inspection and final product inspection apply across all types of remanufacturing operations?
- What activities may be added or deleted for particular technologies or market situations?
- What is the potential for incorporating remanufacturing activities into conventional manufacturing facilities?

The key contribution of this paper is a tool which can be used for further investigation into the area of inspection in remanufacturing processes within academia and consultancy. It also provides a framework to aid companies in the development of new remanufacturing processes.

Competing interests

The authors declare that they have no competing interests.

Authors' contributions

ME carried out the research and wrote the first and subsequent drafts of the paper. SC supervised the research work and case studies and reviewed, made suggestions and alterations to sections of the paper. Both authors read and approved the final manuscript.

Acknowledgements

The authors wish to thank the many industrial representatives who allowed their operations to be studied and who contributed to the research.

References

1. Errington, M, Childe, SJ: The Need for Inspection in Remanufacturing Operations. In: The Third World Conference on Production and Operations Management, Tokyo, Japan (2008)
2. Rooker, L, Benn, H: Climate Change Act 2008. HMSO, London (2008)
3. Richard, CD, Walter, Y: Encircling the peak of world oil production. Nat. Res. Res. **8**, 219–232 (1999)
4. Bentley, RW: Global oil & gas depletion: an overview. Energy Policy **30**, 189–205 (2002)
5. EU: Directive of The European Parliament and of The Council on Waste Electrical and Electronic Equipment (WEEE). European Parliament, Brussels (2003)
6. EU: Directive of The European Parliament and of The Council on End-of Life Vehicles. European Parliament, Brussels (2000)
7. EU: European Parliament and Council Directive on Packaging and Packaging Waste. European Parliament, Brussels (1994)
8. EU: Council Directive on Batteries and Accumulators Containing Certain Hazardous Substances. European Parliament, Brussels (1991)
9. Jayaraman, V, Guide, VDR, Srivastava, R: A closed-loop logistics model for remanufacturing. J. Oper. Res. Soc. **50**, 497–508 (1999)
10. Hammond, R, Amezquita, T, Bras, B: Issues in the automotive parts remanufacturing industry - a discussion of results from surveys performed among remanufacturers. (1998). Accessed 19 July 2013
11. Guide, VDR, Muyldermans, L, Van Wassenhove, LN: Hewlett-Packard company unlocks the value potential from time-sensitive returns. Interfaces **35**, 281–293 (2005)
12. van Nunen, JAEE, Zuidwijk, RA: E-enabled closed-loop supply chains. Calif. Manage. Rev. **46**, 40–54 (2004)
13. Maslennikova, I: Xerox's approach to sustainability. Interfaces **30**, 226–233 (2000)
14. Johnson, MR, Wang, MH: Planning product disassembly for material recovery opportunities. Int. J. Prod. Res. **33**, 3119 (1995)
15. Feldmann, K, Meedt, O, Trautner, S, Scheller, H, Hoffman, W: The "green design advisor": a tool for design for environment. J. Electron. Manuf. **9**, 17 (1999)
16. Rose, CM, Ishii, K: Product end-of-life strategy categorization design tool. J. Electron. Manuf. **9**, 41 (1999)
17. Bufardi, A, Gheorghe, R, Kiritsis, D, Xirouchakis, P: Multicriteria decision-aid approach for product end-of-life alternative selection. Int. J. Prod. Res. **42**, 3139–3157 (2004)
18. Guide, JV, Daniel, R, Souza, GC, van der Laan, E: Performance of static priority rules for shared facilities in a remanufacturing shop with disassembly and reassembly. Eur. J. Oper. Res. **164**, 341–353 (2005)
19. Guide, VDR: Scheduling using drum-buffer-rope in a remanufacturing environment. Int. J. Prod. Res. **34**, 1081 (1996)
20. Goldratt, EM: The goal: beating the competition. Hounslow, Creative Output (1986)
21. Chung, C-J, Wee, H-M: Short life-cycle deteriorating product remanufacturing in a green supply chain inventory control system. Int. J. Prod. Res. **129**, 195–203 (2011)
22. Richter, K: Pure and mixed strategies for the EOQ repair and waste disposal problem. OR Spektrum **19**, 123–129 (1997)
23. Kondoh, S, Salmi, T: Strategic decision making method for sharing resources among multiple manufacturing/ remanufacturing systems. J. Remanuf. **1**, 1–8 (2011)
24. Konstantaras, I, Skouri, K, Jaber, MY: Lot sizing for a recoverable product with inspection and sorting. Comput. Ind. Eng. **58**, 452–462 (2010)
25. Nenes, G, Panagiotidou, S, Dekker, R: Inventory control policies for inspection and remanufacturing of returns: a case study. Int. J. Prod. Econ. **125**, 300–312 (2010)
26. Uzsoy, R: Production planning for companies with product recovery and remanufacturing capability. In: Proceedings of the 1997 IEEE International Symposium on Electronics and the Environment, 1997 (ISEE-1997), San Francisco, 5-7 May 1997, pp. 285–290. IEEE, New York (1997)
27. Spengler, T, Schroter, M: Strategic management of spare parts in closed-loop supply chains– a system dynamics approach. Interfaces **33**, 7–17 (2003)

28. van der Laan, E, Salomon, M, Dekker, R: An investigation of lead-time effects in manufacturing/remanufacturing systems under simple PUSH and PULL control strategies. Eur. J. Oper. Res. **115**, 195–214 (1999)
29. Guide, JV, Daniel, R, Srivastava, R: An evaluation of order release strategies in a remanufacturing environment. Comput. Oper. Res. **24**, 37–47 (1997)
30. Guide, VDR, Jayaraman, V, Srivastava, R: The effect of lead time variation on the performance of disassembly release mechanisms. Comput. Indust. Eng. **36**, 759–779 (1999)
31. Guide, VDR, Srivastava, R: Inventory buffers in recoverable manufacturing. In: Proceedings of the 28th Annual Meeting of the Decision Sciences Institute Atlanta. Georgia USA **1–3**, 1405–1407 (1997)
32. Peng, ZY, Zhong, DY: Optimization model for closed-loop logistics network design in manufacturing and remanufacturing system. In: Zhong, DY (ed.) International Conference on Service Systems and Service Management, Chengdu, 9-11 June 2007, pp. 1–4
33. Blackburn, JD, Guide Jr, VDR, Souza, GC, Van Wassenhove, LN: Reverse supply chains for commercial returns. Calif. Manage. Rev. **46**, 6–22 (2004)
34. Ray, S, Boyaci, T, Aras, N: Optimal prices and trade-in rebates for durable, remanufacturable products. Manuf. Serv. Oper. Manag. **7**, 208–228 (2005)
35. Xiaochen, S, Yancong, Z, Yuling, N, Guangwei, S, AGS: Optimal Control for a Remanufacturing Reverse Logistics System under Buy-Back Policy. In: Yancong, Z ISDA '06 Sixth International Conference on Intelligent Systems Design and Applications 2006, vol. 1, pp. 1197–1202. IEEE, Washington (2006)
36. Guide, VDR, Van Wassenhove, LN: The reverse supply chain. Harv. Bus. Rev. **80**, 25–26 (2002)
37. Steinhilper, R: Remanufacturing: The Ultimate Form of Recycling. Fraunhofer IRB, Stuttgart (1998)
38. Goldberg, LH: Green Electronics / Green Bottom Line: Environmentally Responsible Engineering. Woburn, Butterworth-Heineman (2000)
39. Nagel, C, Meyer, P: Caught between ecology and economy: end-of-life aspects of environmentally conscious manufacturing. Comput. Ind. Eng. **36**, 781–792 (1999)
40. Brent, AC, Steinhilper, R: Opportunities for remanufactured electronic products from developing countries: hypotheses to characterise the perspectives of a global remanufacturing industry. In: Steinhilper, R (ed.) IEEE AFRICON Conference in Africa, Gaborone, vol. 892, pp. 891–896. IEEE, Piscataway (2004)
41. Voss, C, Tsikriktsis, N, Frohlich, M: Case research in operations management. Int. J. Oper. Prod. Manag. **22**, 195–219 (2002)
42. Meredith, J: Building operations management theory through case and field research. J. Oper. Manag. **16**, 441–454 (1998)
43. Eisenhardt, KM: Building theories from case study research. Acad. Manage. Rev. **14**, 532 (1989)
44. Ijomah, WL, Childe, SJ: A model of the operations concerned in remanufacture. Int. J. Prod. Res. **45**, 5857–5880 (2007)
45. Aguilar-Savén, RSRS: Business process modelling: review and framework. Int. J. Prod. Res. **90**, 129–149 (2004)
46. Ang, CL, Luo, M, Gay, RKL: Knowledge-based approach to the generation of IDEF0 models. Comput. Integr. Manuf. Syst. **8**, 279–290 (1995)
47. Loomba, APS, Nakashima, K: Enhancing value in reverse supply chains by sorting before product recovery. Prod. Plan. Control **23**, 205–215 (2011)
48. Depuy, DW, Usher, JS, Walker, RL, Taylor, GD: Production planning for remanufactured products. Prod. Plan. Control **18**, 573–583 (2007)
49. Deming, WE: Out of the Crisis. MIT, Cambridge (1982)

Permissions

All chapters in this book were first published in JR, by Springer; hereby published with permission under the Creative Commons Attribution License or equivalent. Every chapter published in this book has been scrutinized by our experts. Their significance has been extensively debated. The topics covered herein carry significant findings which will fuel the growth of the discipline. They may even be implemented as practical applications or may be referred to as a beginning point for another development.

The contributors of this book come from diverse backgrounds, making this book a truly international effort. This book will bring forth new frontiers with its revolutionizing research information and detailed analysis of the nascent developments around the world.

We would like to thank all the contributing authors for lending their expertise to make the book truly unique. They have played a crucial role in the development of this book. Without their invaluable contributions this book wouldn't have been possible. They have made vital efforts to compile up to date information on the varied aspects of this subject to make this book a valuable addition to the collection of many professionals and students.

This book was conceptualized with the vision of imparting up-to-date information and advanced data in this field. To ensure the same, a matchless editorial board was set up. Every individual on the board went through rigorous rounds of assessment to prove their worth. After which they invested a large part of their time researching and compiling the most relevant data for our readers.

The editorial board has been involved in producing this book since its inception. They have spent rigorous hours researching and exploring the diverse topics which have resulted in the successful publishing of this book. They have passed on their knowledge of decades through this book. To expedite this challenging task, the publisher supported the team at every step. A small team of assistant editors was also appointed to further simplify the editing procedure and attain best results for the readers.

Apart from the editorial board, the designing team has also invested a significant amount of their time in understanding the subject and creating the most relevant covers. They scrutinized every image to scout for the most suitable representation of the subject and create an appropriate cover for the book.

The publishing team has been an ardent support to the editorial, designing and production team. Their endless efforts to recruit the best for this project, has resulted in the accomplishment of this book. They are a veteran in the field of academics and their pool of knowledge is as vast as their experience in printing. Their expertise and guidance has proved useful at every step. Their uncompromising quality standards have made this book an exceptional effort. Their encouragement from time to time has been an inspiration for everyone.

The publisher and the editorial board hope that this book will prove to be a valuable piece of knowledge for researchers, students, practitioners and scholars across the globe.

List of Contributors

Shinichi Fukushige
Graduate School of Engineering, Osaka University, Osaka, Japan

Kazuhiro Yamamoto
Graduate School of Engineering, Osaka University, Osaka, Japan

Yasushi Umeda
Graduate School of Engineering, Osaka University, Osaka, Japan

Erik Sundin
Division of Manufacturing Engineering, Department of Management and Engineering, Linköping University, Linköping SE-58183, Sweden

Otto Dunbäck
Division of Manufacturing Engineering, Department of Management and Engineering, Linköping University, Linköping SE-58183, Sweden

Gokula Vijaykumar Annamalai Vasantha
Design, Manufacture and Engineering Management, University of Strathclyde, Glasgow G1 1XQ, UK

Hitoshi Komoto
Advanced Manufacturing Research Institute, National Institute of Advanced Industrial Science and Technology, Tsukuba, Ibaraki 305-8564, Japan

Romana Hussain
School of Engineering, Cranfield University, Cranfield, Bedfordshire MK43 0AL, UK

Rajkumar Roy
Manufacturing and Materials Department, Cranfield University, Cranfield, Bedfordshire MK43 0AL, UK

Tetsuo Tomiyama
Manufacturing and Materials Department, Cranfield University, Cranfield, Bedfordshire MK43 0AL, UK

Steve Evans
Institute for Manufacturing, University of Cambridge, Cambridge CB3 0FS, UK

Ashutosh Tiwari
Manufacturing and Materials Department, Cranfield University, Cranfield, Bedfordshire MK43 0AL, UK

Stewart Williams
Welding Engineering and Laser Processing Centre, Cranfield University, Cranfield, Bedfordshire MK43 0AL, UK

Ana Paula Barquet
São Carlos School of Engineering, University of São Paulo (USP), Av. Trabalhador São-carlense, 400, São Carlos, Sao Paulo 13566-590, Brazil

R Henrique
São Carlos School of Engineering, University of São Paulo (USP), Av. Trabalhador São-carlense, 400, São Carlos, Sao Paulo 13566-590, Brazil

Fernando A Forcellini
Industrial Engineering Department, Federal University of Santa Catarina(UFSC), Campus Reitor João David Ferreira Lima, Florianópolis Santa Catarina88040-970, Brazil

Joanna Daaboul
Department of Mechanical Systems Engineering, Université de Technologie de Compiègne, UMR CNRS 7337 Roberval, rue du Dr Schweitzer, CS60319, 60203

Julien Le Duigou
Department of Mechanical Systems Engineering, Université de Technologie de Compiègne, UMR CNRS 7337 Roberval, rue du Dr Schweitzer, CS60319, 60203

Diana Penciuc
Department of Mechanical Systems Engineering, Université de Technologie de Compiègne, UMR CNRS 7337 Roberval, rue du Dr Schweitzer, CS60319, 60203

Benoît Eynard
Department of Mechanical Systems Engineering, Université de Technologie de Compiègne, UMR CNRS 7337 Roberval, rue du Dr Schweitzer, CS60319, 60203

Nurul Hanna Ismail
Université Grenoble Alpes, G-SCOP, F-38000 Grenoble, France
CNRS, G-SCOP, F-38000 Grenoble, France

Guillaume Mandil
Université Grenoble Alpes, G-SCOP, F-38000 Grenoble, France
CNRS, G-SCOP, F-38000 Grenoble, France

Peggy Zwolinski
Université Grenoble Alpes, G-SCOP, F-38000 Grenoble, France
CNRS, G-SCOP, F-38000 Grenoble, France

Xiaojun Wang
Department of Management, University of Bristol, Bristol BS8 1TN, UK

Hing Kai Chan
Norwich Business School, University of East Anglia, Norwich NR4 7TJ, UK

Junjun Wang
School of Mechanical Engineering,Shanghai Jiao Tong University,Shanghai 200240, People's Republicof China

Ming Chen
School of Mechanical Engineering,Shanghai Jiao Tong University,Shanghai 200240, People's Republicof China

Leila Hamzaoui-Essoussi
Telfer School of Management, University of Ottawa, 55 Laurier E, Ottawa K1N 6 N5, Ontario, Canada.

Jonathan D Linton
Power Corporation for the Management of Technological Enterprises, University of Ottawa, 55 Laurier E, Ottawa K1N 6 N5, Ontario, Canada

Sebastian Lind
Division of Manufacturing Engineering, Department of Management and Engineering, Linköping University, Linköping
SE-58183, Sweden

David Olsson
Division of Manufacturing Engineering, Department of Management and Engineering, Linköping University, Linköping
SE-58183, Sweden

Erik Sundin
Division of Manufacturing Engineering, Department of Management and Engineering, Linköping University, Linköping
SE-58183, Sweden

Yun Arifatul Fatimah
Sustainable Engineering Group, Curtin University, Perth, Australia

Wahidul Biswas
Sustainable Engineering Group, Curtin University, Perth, Australia

Ilyas Mazhar
Department of Mechanical Engineering, Curtin University, Perth, Australia

Mohammad Nazrul Isla
Department of Mechanical Engineering, Curtin University, Perth, Australia

Monsuru O Ramoni
Department of Industrial Engineering, Texas Tech University, Lubbock, TX 79409, USA

Hong-Chao Zhang
Department of Industrial Engineering, Texas Tech University, Lubbock, TX 79409, USA

David Opresnik
Department of Management, Economics and Industrial Engineering, Politecnico di Milano, Piazza Leonardo Da Vinci 32, 20133 Milano, Italy

Marco Taisch
Department of Management, Economics and Industrial Engineering, Politecnico di Milano, Piazza Leonardo Da Vinci 32, 20133 Milano, Italy